215
Advances in Polymer Science

Editorial Board:
A. Abe · A.-C. Albertsson · R. Duncan · K. Dušek · W. H. de Jeu
H.-H. Kausch · S. Kobayashi · K.-S. Lee · L. Leibler · T. E. Long
I. Manners · M. Möller · O. Nuyken · E. M. Terentjev
B. Voit · G. Wegner · U. Wiesner

Advances in Polymer Science
Recently Published and Forthcoming Volumes

Self-Assembled Nanomaterials II
Nanotubes
Volume Editor: Shimizu, T.
Vol. 220, 2008

Self-Assembled Nanomaterials I
Nanofibers
Volume Editor: Shimizu, T.
Vol. 219, 2008

Interfacial Processes and Molecular Aggregation of Surfactants
Volume Editor: Narayanan, R.
Vol. 218, 2008

New Frontiers in Polymer Synthesis
Volume Editor: Kobayashi, S.
Vol. 217, 2008

Polymers for Fuel Cells II
Volume Editor: Scherer, G. G.
Vol. 216, 2008

Polymers for Fuel Cells I
Volume Editor: Scherer, G. G.
Vol. 215, 2008

Photoresponsive Polymers II
Volume Editors: Marder, S. R., Lee, K.-S.
Vol. 214, 2008

Photoresponsive Polymers I
Volume Editors: Marder, S. R., Lee, K.-S.
Vol. 213, 2008

Polyfluorenes
Volume Editors: Scherf, U., Neher, D.
Vol. 212, 2008

Chromatography for Sustainable Polymeric Materials
Renewable, Degradable and Recyclable
Volume Editors: Albertsson, A.-C., Hakkarainen, M.
Vol. 211, 2008

Wax Crystal Control · Nanocomposites Stimuli-Responsive Polymers
Vol. 210, 2008

Functional Materials and Biomaterials
Vol. 209, 2007

Phase-Separated Interpenetrating Polymer Networks
Authors: Lipatov, Y. S., Alekseeva, T.
Vol. 208, 2007

Hydrogen Bonded Polymers
Volume Editor: Binder, W.
Vol. 207, 2007

Oligomers · Polymer Composites Molecular Imprinting
Vol. 206, 2007

Polysaccharides II
Volume Editor: Klemm, D.
Vol. 205, 2006

Neodymium Based Ziegler Catalysts – Fundamental Chemistry
Volume Editor: Nuyken, O.
Vol. 204, 2006

Polymers for Regenerative Medicine
Volume Editor: Werner, C.
Vol. 203, 2006

Peptide Hybrid Polymers
Volume Editors: Klok, H.-A., Schlaad, H.
Vol. 202, 2006

Fuel Cells I

Volume Editor: Günther G. Scherer

With contributions by

S. Alkan Gürsel · M. Eikerling · L. Gubler · B. Gupta · S. Holdcroft
D. J. Jones · A. A. Kornyshev · T. J. Peckham · J. Rozière
G. G. Scherer · A. Siu · E. Spohr · A. Watakabe
Y. Yang · M. Yoshitake

 Springer

The series *Advances in Polymer Science* presents critical reviews of the present and future trends in polymer and biopolymer science including chemistry, physical chemistry, physics and material science. It is adressed to all scientists at universities and in industry who wish to keep abreast of advances in the topics covered.

As a rule, contributions are specially commissioned. The editors and publishers will, however, always be pleased to receive suggestions and supplementary information. Papers are accepted for *Advances in Polymer Science* in English.

In references *Advances in Polymer Science* is abbreviated *Adv Polym Sci* and is cited as a journal.

Springer WWW home page: springer.com
Visit the APS content at springerlink.com

ISBN 978-3-540-69755-8   e-ISBN 978-3-540-69757-2
DOI 10.1007/978-3-540-69757-2

Advances in Polymer Science ISSN 0065-3195

Library of Congress Control Number: 2008933499

© 2008 Springer-Verlag Berlin Heidelberg

This work is subject to copyright. All rights are reserved, whether the whole or part of the material is concerned, specifically the rights of translation, reprinting, reuse of illustrations, recitation, broadcasting, reproduction on microfilm or in any other way, and storage in data banks. Duplication of this publication or parts thereof is permitted only under the provisions of the German Copyright Law of September 9, 1965, in its current version, and permission for use must always be obtained from Springer. Violations are liable to prosecution under the German Copyright Law.

The use of general descriptive names, registered names, trademarks, etc. in this publication does not imply, even in the absence of a specific statement, that such names are exempt from the relevant protective laws and regulations and therefore free for general use.

Cover design: WMXDesign GmbH, Heidelberg
Typesetting and Production: le-tex publishing services oHG, Leipzig

Printed on acid-free paper

9 8 7 6 5 4 3 2 1 0

springer.com

Volume Editor

Dr. Günther G. Scherer
Paul Scherrer Institut
Labor für Elektrochemie
5232 Villigen, Switzerland
guenther.scherer@psi.ch

Editorial Board

Prof. Akihiro Abe
Department of Industrial Chemistry
Tokyo Institute of Polytechnics
1583 Iiyama, Atsugi-shi 243-02, Japan
aabe@chem.t-kougei.ac.jp

Prof. A.-C. Albertsson
Department of Polymer Technology
The Royal Institute of Technology
10044 Stockholm, Sweden
aila@polymer.kth.se

Prof. Ruth Duncan
Welsh School of Pharmacy
Cardiff University
Redwood Building
King Edward VII Avenue
Cardiff CF 10 3XF, UK
DuncanR@cf.ac.uk

Prof. Karel Dušek
Institute of Macromolecular Chemistry,
Czech
Academy of Sciences of the Czech Republic
Heyrovský Sq. 2
16206 Prague 6, Czech Republic
dusek@imc.cas.cz

Prof. Dr. Wim H. de Jeu
Polymer Science and Engineering
University of Massachusetts
120 Governors Drive
Amherst MA 01003, USA
dejeu@mail.pse.umass.edu

Prof. Hans-Henning Kausch
Ecole Polytechnique Fédérale de Lausanne
Science de Base
Station 6
1015 Lausanne, Switzerland
kausch.cully@bluewin.ch

Prof. Shiro Kobayashi
R & D Center for Bio-based Materials
Kyoto Institute of Technology
Matsugasaki, Sakyo-ku
Kyoto 606-8585, Japan
kobayash@kit.ac.jp

Prof. Kwang-Sup Lee
Department of Advanced Materials
Hannam University
561-6 Jeonmin-Dong
Yuseong-Gu 305-811
Daejeon, South Korea
kslee@hnu.kr

Prof. L. Leibler
Matière Molle et Chimie
Ecole Supérieure de Physique
et Chimie Industrielles (ESPCI)
10 rue Vauquelin
75231 Paris Cedex 05, France
ludwik.leibler@espci.fr

Prof. Timothy E. Long
Department of Chemistry
and Research Institute
Virginia Tech
2110 Hahn Hall (0344)
Blacksburg, VA 24061, USA
telong@vt.edu

Prof. Ian Manners
School of Chemistry
University of Bristol
Cantock's Close
BS8 1TS Bristol, UK
ian.manners@bristol.ac.uk

Prof. Martin Möller
Deutsches Wollforschungsinstitut
an der RWTH Aachen e.V.
Pauwelsstraße 8
52056 Aachen, Germany
moeller@dwi.rwth-aachen.de

Prof. Oskar Nuyken
Lehrstuhl für Makromolekulare Stoffe
TU München
Lichtenbergstr. 4
85747 Garching, Germany
oskar.nuyken@ch.tum.de

Prof. E. M. Terentjev
Cavendish Laboratory
Madingley Road
Cambridge CB 3 OHE, UK
emt1000@cam.ac.uk

Prof. Brigitte Voit
Institut für Polymerforschung Dresden
Hohe Straße 6
01069 Dresden, Germany
voit@ipfdd.de

Prof. Gerhard Wegner
Max-Planck-Institut
für Polymerforschung
Ackermannweg 10
55128 Mainz, Germany
wegner@mpip-mainz.mpg.de

Prof. Ulrich Wiesner
Materials Science & Engineering
Cornell University
329 Bard Hall
Ithaca, NY 14853, USA
ubw1@cornell.edu

Advances in Polymer Science
Also Available Electronically

For all customers who have a standing order to Advances in Polymer Science, we offer the electronic version via SpringerLink free of charge. Please contact your librarian who can receive a password or free access to the full articles by registering at:

springerlink.com

If you do not have a subscription, you can still view the tables of contents of the volumes and the abstract of each article by going to the SpringerLink Homepage, clicking on "Browse by Online Libraries", then "Chemical Sciences", and finally choose Advances in Polymer Science.

You will find information about the

- Editorial Board
- Aims and Scope
- Instructions for Authors
- Sample Contribution

at springer.com using the search function.

Color figures are published in full color within the electronic version on SpringerLink.

Preface

The concept to utilize an ion-conducting polymer membrane as a solid polymer electrolyte offers several advantages regarding the design and operation of an electrochemical cell, as outlined in Volume 215, Chapter 1 (L. Gubler, G. G. Scherer). Essentially, the solvent and/or transport medium, e.g., H_2O, for the mobile ionic species, e.g., H^+ for a cation exchange membrane, is taken up by and confined into the nano-dimensional morphology of the ion-containing domains of the polymer. As a consequence, a phase separation into a hydrophilic ion-containing solvent phase and a hydrophobic polymer backbone phase establishes. Because of the narrow solid electrolyte gap in these cells, low ohmic losses reducing the overall cell voltage can be achieved, even at high current densities.

This concept was applied to fuel cell technology at a very early stage; however, performance and reliability of the cells were low due to the dissatisfying membrane properties at that time. The development of perfluoro sulfonate and carboxylate-type membranes, in particular for the chlor-alkali process, directly fostered the further development of proton-conducting membranes and, as a consequence, also the progress in this type of fuel cell technology (polymer electrolyte fuel cell, PEFC).

Within the past 20 years, tremendous progress has been achieved in PEFC technology, in particular since the automotive industry has joined forces to further develop this energy conversion technology with its advantages in efficiency and environmental friendliness. This development has brought about a much deeper understanding of the various functions of the polymer electrolyte in the cell, particularly under duty cycle conditions of automotive applications. As a further and utmost prerequisite, the cost issue came to every one's attention.

Many national and international research programs have recently initiated work on proton-conducting polymer membranes for fuel cell applications. The contributions in these two volumes aim to summarize some major efforts, without claiming to be exhaustive.

Hence, M. Eikerling, A. A. Kornyshev, and E. Spohr start out in Volume 215, Chapter 2 with a general description of proton-conduction in polymer membranes, elucidating the influence of water and charge-bearing species in the polymer environment. Y. Yang, A. Siu, T. J. Peckham, and S. Holdcroft give an overview on implications of design approaches for synthesis of fuel cell membranes in Volume 215, Chapter 3. Some recent progress in the most prominent class of these membranes, the perfluoro sulfonic acid-type membranes, is described in Volume 215, Chapter 4 from an industrial perspective by M. Yoshitake and A. Watakabe. The development of the radiation grafting process to yield fuel cell membranes is covered in Volume 215, Chapter 5 by S. Alkan Gürsel, L. Gubler, B. Gupta, and G. G. Scherer, based on their long-term experience working in this area. The requirement for operating cell temperatures above 100 °C has led to the approach of composite membranes, combining the advantageous properties of inorganic and polymeric proton conductors (D. J. Jones, J. Rozière, in Volume 215, Chapter 6) to control the water content at these temperatures. On the basis of the promising properties of polymeric aromatic engineering materials and their modification to proton-conducting membranes, G. Maier and J. Meier-Haack review the state-of-art in sulfonated aromatic polymers in Volume 216, Chapter 1. High-temperature applications are also in the focus of the next two contributions. Polymer blends with phosphoric acid allow operating temperatures well above 100 °C, with advantages in water management and electrocatalysis (CO-tolerance), as pointed out in the contribution by J. Mader, L. Xiao, T. J. Schmidt, and B. C. Benicewicz in Volume 216, Chapter 2. A similar approach was followed, introducing the phosphonic acid group directly onto the polymer chain, by A. L. Rusanov, P. V. Kostoglodov, M. J. M. Abadie, V. Y. Voytekunas, and D. Y. Likhachev in Volume 216, Chapter 3. Two new classes of polymers and their properties are addressed in the last two Chapters 4 and 5 in Volume 216. R. Wycisk and P. N. Pintauro describe their view on polyphosphazene-based membranes for fuel cell applications, while C. Marestin, G. Gebel, O. Diat, and R. Mercier report on their and others' work on polyimide-based membranes.

As documented in and expressed by these various contributions, the topic "Polymers for Fuel Cells" is a vast one and concerns numerous synthetic and physico-chemical aspects, derived from the particular application as a solid polymer electrolyte. In this collection of contributions, we have emphasized work which has already led to tests of these polymers in the real fuel cell environment. There exist other synthetic routes for proton-conducting membrane preparation, which are not discussed in this edition. Furthermore, certain polymers are utilized as fuel-cell structure materials, e.g., as gaskets or additives (binder, surface coating) to bipolar plate materials. These aspects are not covered here.

In summary, we endeavored to bring together contributions from several expert groups who have worked in this area for many years to summarize the current state-of-the-art. There still exist many challenges down the road to bring at least some of these developments to commercial fuel cell technology. For an ultimate success, a comprehensive polymer *materials* approach has to be adopted to rationalize all the various aspects of this highly interdisciplinary task.

The editor wishes to thank all the authors for their contribution and the Paul Scherrer Institut for its support of membrane work over many years.

Villigen, May 2008   Günther G. Scherer

Contents

A Proton-Conducting Polymer Membrane as Solid Electrolyte –
Function and Required Properties
L. Gubler · G. G. Scherer 1

Proton-Conducting Polymer Electrolyte Membranes:
Water and Structure in Charge
M. Eikerling · A. A. Kornyshev · E. Spohr 15

Structural and Morphological Features
of Acid-Bearing Polymers for PEM Fuel Cells
Y. Yang · A. Siu · T. J. Peckham · S. Holdcroft 55

Perfluorinated Ionic Polymers for PEFCs
(Including Supported PFSA)
M. Yoshitake · A. Watakabe 127

Radiation Grafted Membranes
S. Alkan Gürsel · L. Gubler · B. Gupta · G. G. Scherer 157

Advances in the Development of Inorganic–Organic Membranes
for Fuel Cell Applications
D. J. Jones · J. Rozière 219

Subject Index 265

Contents of Volume 216

Fuel Cells II

Volume Editor: Scherer, G. G.
ISBN: 978-3-540-69763-3

Sulfonated Aromatic Polymers for Fuel Cell Membranes
G. Maier · J. Meier-Haack

Polybenzimidazole/Acid Complexes as High-Temperature Membranes
J. Mader · L. Xiao · T. J. Schmidt · B. C. Benicewicz

Proton-Conducting Polymers and Membranes
Carrying Phosphonic Acid Groups
A. L. Rusanov · P. V. Kostoglodov · M. J. M. Abadie
V. Y. Voytekunas · D. Y. Likhachev

Polyphosphazene Membranes for Fuel Cells
R. Wycisk · P. N. Pintauro

Sulfonated Polyimides
C. Marestin · G. Gebel · O. Diat · R. Mercier

A Proton-Conducting Polymer Membrane as Solid Electrolyte – Function and Required Properties

Lorenz Gubler (✉) · Günther G. Scherer

Electrochemistry Laboratory, Paul Scherrer Institut, 5232 Villigen, Switzerland
lorenz.gubler@psi.ch

1	Introduction	1
2	The Polymer Electrolyte Fuel Cell	7
3	Required Membrane Properties for Fuel Cell Application	8
3.1	Interfacial Characteristics	9
3.2	Bulk Characteristics	11
3.2.1	Ion Exchange Capacity – Water Sorption – Conductivity	11
3.2.2	Water Management	12
3.2.3	Durability	12
3.2.4	Cost	13
4	Conclusion	13
	References	14

Abstract Fuel cells are considered as a major energy conversion technology of the future, due to certain inherent advantages of electrochemical conversion processes as compared to thermal combustion processes. Polymer electrolyte fuel cells (PEFCs), operating with hydrogen and air or oxygen at temperatures of around 100 °C, utilize a proton-conducting polymer membrane as solid electrolyte. In this configuration, the proton-conducting polymer membrane has to fulfill several functions: (i) the electrolyte function for surface and bulk ion conduction and (ii) the separator function for gas (reactant) separation. Furthermore, the membrane is part of the gasket system, requiring certain specific mechanical properties. This ensemble of required specifications asks for a comprehensive approach in membrane development for this application. In this short introductory chapter, we summarize some of the general aspects of membrane development for polymer electrolyte fuel cells.

Keywords Polymer electrolyte fuel cell · Interfacial properties · Bulk properties · Proton exchange membrane · Membrane electrode assembly

1 Introduction

Interest in fuel cells and their applications has grown tremendously over the past two decades, primarily due to energy conversion and environmental concerns [1].

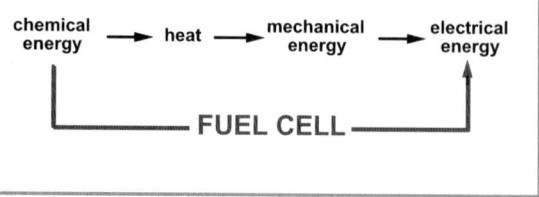

Fig. 1 Scheme of electrochemical energy conversion in a fuel cell

The idea of a gaseous voltaic battery or fuel cell dates back to Grove, who in 1839 described the first hydrogen/oxygen fuel cell consisting of platinized platinum electrodes immersed in sulfuric acid [2]. Generally speaking, a fuel cell is an *electrochemical device that continuously converts the chemical energy of a fuel (and oxidant) directly into electrical energy*, as shown schematically in Fig. 1. Heat is generated as a byproduct. The fuel cell process has the major advantage of not being Carnot-limited, thus allowing a theoretical efficiency higher than that of a heat engine. Fuels can include, for example, H_2, N_2H_4, NH_3, CH_3OH, coal gas, or hydrocarbons. In the case of pure hydrogen, a fuel cell acts as a local zero emission converter. Normally, the oxidant for terrestrial applications is air, and in some cases pure oxygen is utilized.

The basic design of a fuel cell, an ionically conducting *electrolyte and separator layer* sandwiched between two electronically conducting *gas diffusion electrodes* (the fuel anode and the oxidant cathode, respectively), is shown schematically in Fig. 2 for a polymer electrolyte fuel cell with an acidic electrolyte and hydrogen and oxygen as the corresponding reactants. Typically, under open circuit conditions, H_2/air fuel cells exhibit a cell voltage of

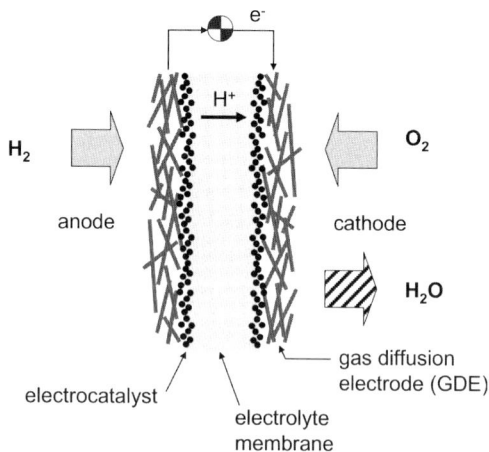

Fig. 2 Operating principle of a H_2/O_2 fuel cell with acidic electrolyte membrane. Protons are transported from anode to cathode, where water is formed

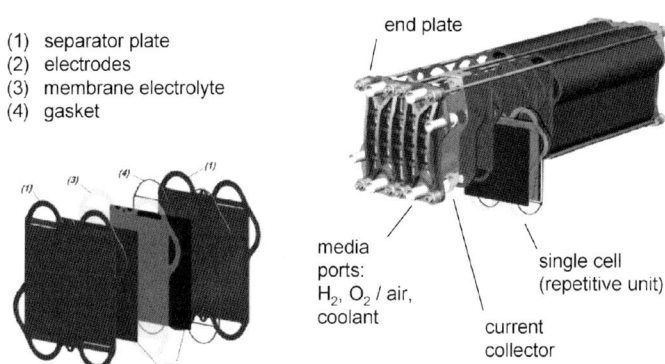

(1) separator plate
(2) electrodes
(3) membrane electrolyte
(4) gasket

Fig. 3 Components of the polymer electrolyte fuel cell (PEFC): membrane electrode assembly (MEA) on the *left*, including separator plates and gasket. A schematic of a PEFC stack is shown on the *right*, comprising a number of single cells in series

around 1 V and less than 1 V under current flow. Higher voltages, necessary for any application, are obtained by stacking individual cells into a bipolar arrangement (Fig. 3). Within the repetitive unit of the cell, the membrane also functions as sealing material, in combination with gaskets introduced in the periphery of the active area (Fig. 4). This plate-and-frame design is typical for filter press type cells.

Historically, fuel cells are classified by the nature of the electrolyte and/or by the temperature of operation. Thus, one separates fuel cells into alkaline or acidic, or low temperature (up to 100 °C), medium temperature (up to 200 °C), and high temperature (up to 1000 °C) fuel cells. Currently, interest focuses on the fuel cell families depicted schematically in Fig. 5. In general terms, the nature of the oxidant as well as the type of fuel set restrictions on

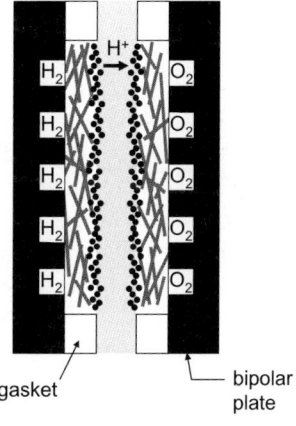

Fig. 4 Repeating unit in a fuel cell stack with bipolar arrangement

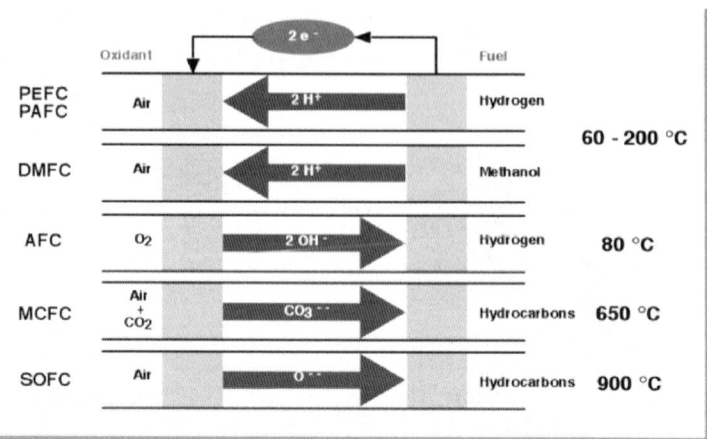

Fig. 5 Important fuel cell families considered today

the operating conditions of the various fuel cell types [1]. Polymer electrolyte fuel cells (PEFCs), phosphoric acid fuel cells (PAFCs), and direct methanol fuel cells (DMFCs or sometimes referred to as DM-PEFCs) are acidic fuel cells that can utilize air as oxidant. Alkaline fuel cells (AFCs) must operate on pure oxygen so as to avoid carbonisation of the alkaline electrolyte. Most of the PEFCs, as well as DMFCs and AFCs operate at temperatures up to 100 °C, while some PEFCs utilize so-called high temperature polymer membranes, which allow operation beyond 100 °C. PAFCs, with working temperatures of around 200 °C, tolerate fuels with CO levels in the range of several hundred ppm, in contrast to PEFCs with platinum-based anodes, which require high purity hydrogen. Operation of PEFCs with CO-containing hydrogen produced from a reforming process, as well as DMFCs fed with gaseous or liquid methanol, require anode catalysts with high CO-tolerance.

Molten carbonate fuel cells (MCFCs) and solid oxide fuel cells (SOFCs) are high temperature fuel cells, which allow the utilization of fuels other than hydrogen, e.g., methane.

Grove early discovered one of the major obstacles in designing an efficient fuel cell: It is necessary to have a high interfacial area of the three-phase boundary between gas, electrolyte (ionic conductor), and electrode (electronic conductor) (Fig. 6). He expressed this prerequisite in his second publication by stating that a fuel cell *needs a notable surface of action* [3]. Even today, *the design and preparation of three-phase boundaries of high interfacial area is one of the major challenges for fuel cell research*. In solid electrolyte fuel cells, i.e., the SOFC and PEFC, the three-phase boundary is designed as an interpenetrating network of electronic and ionic conductor material, with finite porosity to allow the access of reactant. With regard to this three-phase boundary, an aqueous electrolyte has an advantage in comparison to a ceramic one, e.g., in SOFCs, in that the reactant gases are soluble in the aqueous medium.

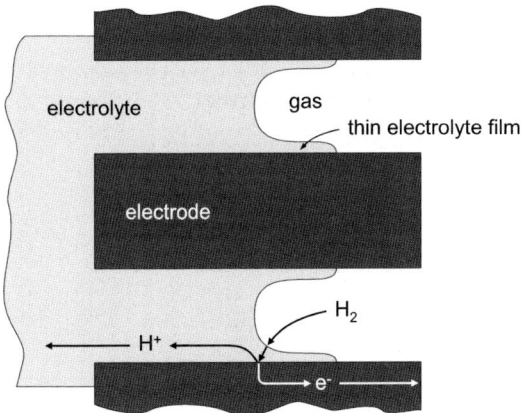

Fig. 6 Scheme of a three-phase boundary with an aqueous electrolyte

This *notable surface of action* is achieved by making use of a porous gas diffusion electrode (GDE) that fulfills two essential prerequisites: A high electrochemically active surface area and a possible mass flow perpendicular to the electrode/electrolyte plane. As an illustration, a cross-section of a commercial GDE (typically 400 μm thick), originally developed for PAFCs and today also widely utilized in PEFCs, is shown in Fig. 7 (Kuhn, Paul Scherrer Institut, unpublished results). A carbon cloth or carbon paper serves as support (middle) for the active layer (ca. 50 μm, right side) and the wet-proofing gas diffusion layer (left side). The active layer is composed of a mixture of the electrocatalyst, either platinum black or highly dispersed platinum particles (2 to 5 nm) deposited onto carbon black (Pt/C electrodes), and the proton conducting ionomer for combined electronic–ionic conductivity. A precious metal catalyst is required for high activity and corrosion resistance due to contact with the acidic electrolyte. The interphase is tailored by selective impregnation of the three layers with PTFE, which acts as a hydrophobizing as well as a bonding agent between the carbon particles.

The design of an optimum interface is strongly dependent on the pore structure of the active layer of the gas diffusion electrode. According to Fig. 6, the liquid electrolyte has to penetrate into the pores and to wet the pores, so that a thin layer of electrolyte covers the pore wall (low contact angle). This electrolyte film should be as thin as possible to allow a short diffusion path for the reactant gases to exist. A high solubility of the reactant gases in the electrolyte film is also favourable.

Furthermore, the pores of the support and the wet-proof layer have to allow mass flow of reactants (fuel and oxidant) *to*, and products (liquid water, at temperatures below 100 °C) *from* the wetted pore of the active layer, where the electrochemical reactions take place. This requires a balance of hydrophilic (carbon surface) and hydrophobic pores (PTFE), achieved by

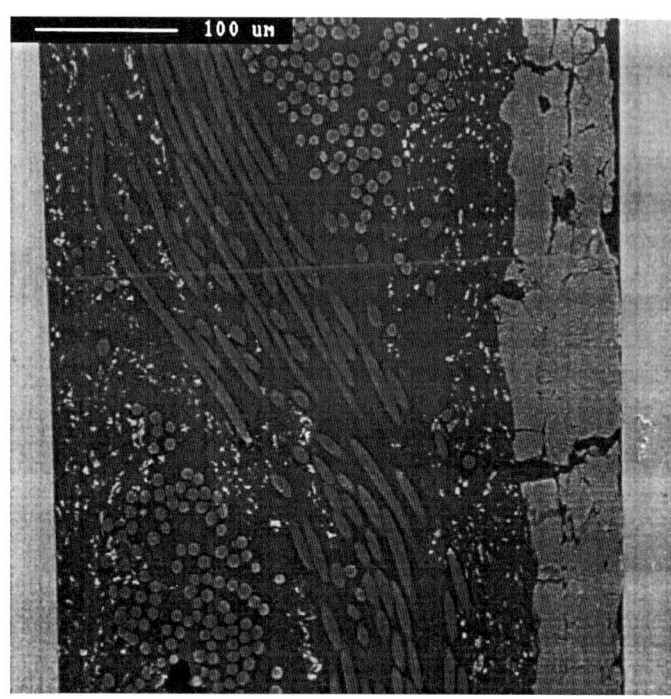

Fig. 7 Gas diffusion electrode (E-Tek). Scanning electron micrograph of a cross section. *Left side*: carbon cloth; *right side*: active layer with platinum on carbon particles

selective impregnation of the different layers with PTFE-particles and subsequent processing (heat treatment, rolling, etc.). The optimum interfacial design for anode and cathode may be different, although today in many cases the same electrode type is used for both electrodes.

The optimization of interfaces for low temperature fuel cells has always depended on the availability of specific materials to control the porosity and wetting behaviour of the respective pores by the electrolyte. For example, one of the breakthroughs in the development of AFCs was the work of Bacon at a time when the chemically stable hydrophobizing agent, PTFE, was not yet available [4]. Bacon managed interface control with a dual-layer, dual-porosity electrode, made out of Ni powders of different grain size. The later availability of PTFE opened up new possibilities for improved GDE designs.

This introductory chapter provides a brief outline and history of the PEFC technology, and important requirements and aspects in the development of polymer membranes for fuel cells. To obtain materials that meet specific requirements, the relationship of composition/structure and the properties has to be established. For the preparation of the membrane electrode assembly, it is important to understand the interfacial properties between membrane and electrodes.

2
The Polymer Electrolyte Fuel Cell

The idea of using a proton-conducting ion exchange membrane as _solid polymer electrolyte_ (SPE) in a fuel cell was first demonstrated by Grubb [5] (Fig. 2). Later on, SPE became a tradename for the technology (fuel cells, electrolyzers, etc.) developed by General Electric. Due to the harsh environment prevailing in these cells, the lifetime of PEFCs was limited by the stability of the mainly hydrocarbon-based membranes available at that time [6]. A real breakthrough was the development of perfluorinated cation exchange membranes (Nafion, DuPont, USA) by Grot, which extended the lifetime to several thousands of hours at operation temperatures below 100 °C [7, 8].

The required properties of solid polymer electrolyte membranes may be divided into interfacial and bulk properties [9]. As described above, the interfacial characteristics of these membrane materials are important for the optimum formation of the three-phase boundary. Hence, flow properties, gas solubility, wetting of carbon supported catalyst surfaces by the polymer, etc. are of paramount importance. The bulk properties concern proton conductivity, gas separation, and mechanical properties. This whole ensemble of properties has to be considered and balanced in the development of novel proton-exchange membranes for fuel cell application.

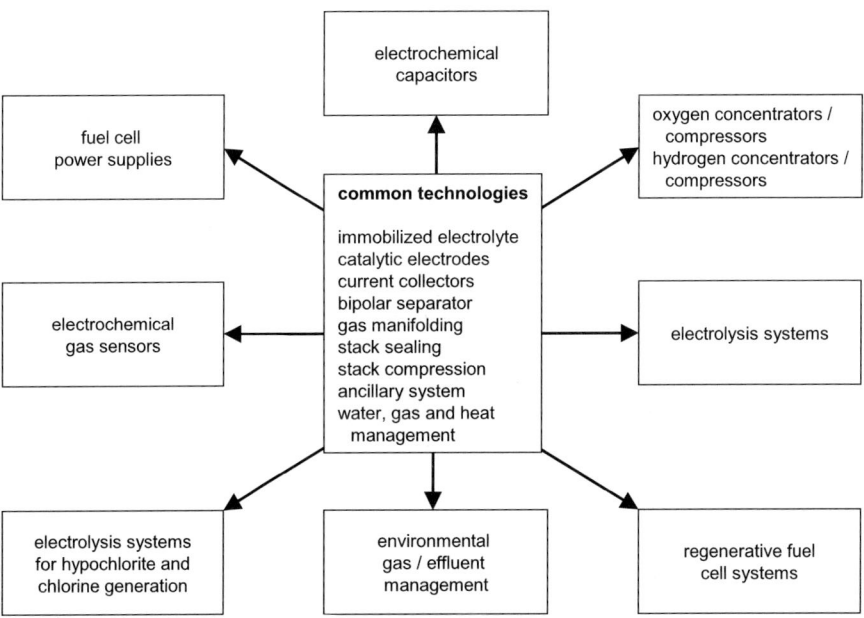

Fig. 8 Solid Polymer electrolyte Technologies (LaConti AB, Giner Inc., USA, personal communication)

Interestingly, the concept of a solid polymer electrolyte can be applied to a variety of electrochemical cells, as depicted in Fig. 8 (LaConti AB, Giner Inc., USA, personal communication). This range of opportunities emphasizes the importance of membrane research in specific applications, as well as the significance of membrane research in general.

3
Required Membrane Properties for Fuel Cell Application

In the PEFC, the membrane, together with the electrodes, forms the basic electrochemical unit, the membrane electrode assembly (MEA). The first and foremost function of the electrolyte membrane is the transport of protons from anode to cathode. On one hand, the electrodes host the electrochemical reactions within the catalyst layer and provide electronic conductivity, and, on the other hand, they provide pathways for reactant supply to the catalyst and removal of products from the catalyst. The components of the MEA need to be chemically stable for several thousands of hours in the fuel cell under the prevailing operating and transient conditions. PEFC electrodes are wet-proofed fibrous carbon sheet materials of a few 100 μm thickness. The functionality of the proton exchange membrane (PEM) extends to requirements of mechanical stability to also ensure effective separation of anode and

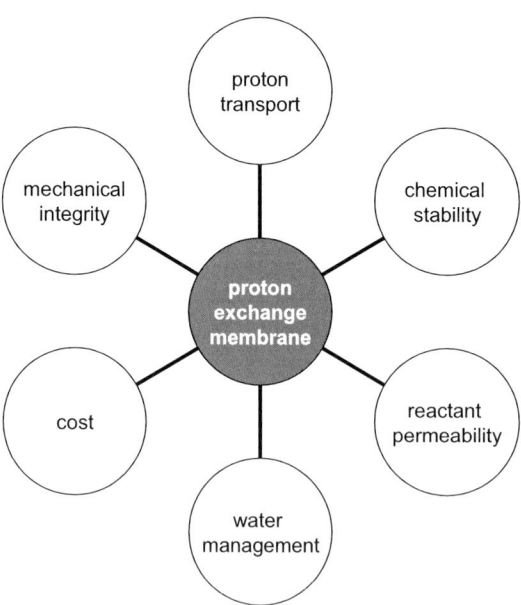

Fig. 9 Requirements for fuel cell membranes

cathode under aggravated conditions, such as operation on reactant gases below the water vapour saturation point, fuel cell start-up/shut-down, and transient load.

Generally, compared to the total number of articles on the subject, few fuel cell tests using membranes other than the commercially available perfluorinated ones have been reported in the literature. Frequently, characterization is restricted to the membrane, and not extended to include fabrication of membrane electrode assemblies (MEAs) and fuel cell testing. Therefore, important insights relating to electrochemical performance, membrane-electrode interface properties, membrane integrity, and lifetime are missed out. Significant necessary requirements for fuel cell membranes are sketched in Fig. 9 and will be discussed in the following paragraphs.

3.1
Interfacial Characteristics

First modelling results of a solid polymer electrolyte interface suggested that incorporation of solid polymer electrolyte particles into the active layer, thus forming a spatially extended inter*phase* rather than an planar inter*face*, extending the electrode–electrolyte interface into the third dimension, should improve the electrochemical polarisation behaviour [10]. In the past, this was realized to some extent by pressing the GDEs onto the surface of the hydrated membrane at a temperature above its glass transition, thereby bringing about the penetration of electrocatalyst particles into the membrane surface. A solubilized form of the perfluorinated polymer membranes [11] offered the possibility to impregnate the active layer of GDEs and thereby increase the electrochemically available platinum surface, as shown schematically in Fig. 10 [12, 13]. Platinum particles dispersed on carbon, which have previously not been in contact with the solid electrolyte, become electro-

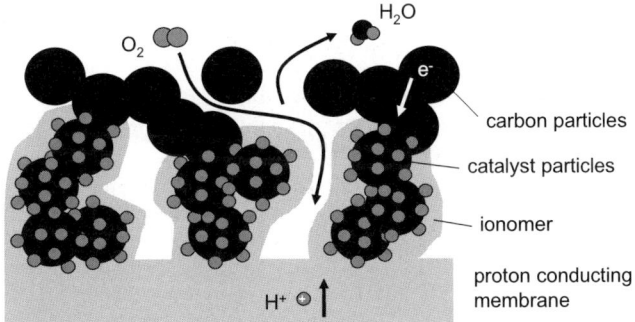

Fig. 10 Catalyst layer of the PEFC, with co-existing electronic, ionic, and gas phase (porosity). The cathode reaction, the reduction of oxygen and formation of water, is shown for illustration

chemically active due to the continuous thin film of solid electrolyte now covering their surface. Combining both methods results in fuel cell polarisation curves much superior to those obtained with non-treated electrodes. This approach is still used to test new membranes. For the preparation of electrodes also containing the novel ionomer in the catalyst layer, the material has to be available in a solubilized form.

Modelling work has addressed the optimization of the ionomer content in the active layer, finding a balance between the limited transport of reactant at high ionomer content and poor ionic conductivity at low ionomer content [14].

A more recent approach in preparing electrodes of low platinum loading (~ 0.1 mg/cm^2) involves casting, doctor blading, or screen printing of thin films, typically with a thickness of a few μm, from a suspension or paste ("ink") of electrocatalyst particles in solubilized membrane material onto an inert support, and, subsequently, hot-pressing the dried film onto the membrane surface [15] or directly coating the ink onto the membrane surfaces. Another promising approach has been taken by the company 3M [16]. Whiskers of a polymer (perylene red) with a thickness of around 70 nm and a length of a few microns are thereby grown onto a substrate and subsequently sputter-coated with the nobel metal catalyst. Thus, a continuous and corrugated catalyst surface with a high roughness factor is obtained. The whisker layer is transfer-coated onto the electrolyte membrane. A conventional gas diffusion layer can be attached next to the (thin) active layer. In this catalyst layer design, high surface area carbon support is not required, rendering carbon corrosion problems obsolete. Also, impregnation with soluble ionomer does not appear to be necessary, and the entire catalyst area, being a continous Pt or Pt-alloy surface, is electrochemically active.

In catalyst layers of conventional design, comprising ionomer-impregnated catalyst particles, the gases have to permeate through the thin impregnated solid electrolyte layer to react at the platinum surface (Fig. 5). Permeation of oxygen in water swollen perfluorinated membranes has been studied at the interface of a platinum micro-electrode in contact with membranes of different equivalent weights (EW)[1] [17]. Chronoamperometry facilitated a separation of the oxygen permeability into its solubility and diffusivity components. Solubility is favoured by a higher EW, i.e., a higher content of the perfluorinated backbone phase, while diffusivity is favoured by a higher water content (swelling caused by ionic content) of the membrane. These results have to be further explored with respect to impregnation of GDEs, particularly for the cathode, where reduction of oxygen occurs at rather high overpotential. More recently, other types of membranes were also characterized using this method [18].

---

[1] The equivalent weight (EW) is the dry ionomer weight per sulfonic acid site, unit g(polymer)/mol(H$^+$).

3.2
Bulk Characteristics

3.2.1
Ion Exchange Capacity – Water Sorption – Conductivity

The ion exchange capacity (IEC) of a polymer is defined as the dry mass equivalent to one mol of acid groups, unit g/mol(H^+), which is the reciprocal value of the equivalent weight (EW). The IEC or EW are quantities often used to characterize a proton-conducting polymer material or to compare different materials to each other. One aspect worth noting is that this mass-mased IEC may be misleading, due to the fact that a different polymer backbone chemistry leads to different masses in relation to the acid functionality. Therefore, a volume based IEC (unit mol/cm^3) may be more meaningful to quote, but that does not seem to be customary. Furthermore, the IEC, being measured as a bulk quantity, gives no indication about the distribution of the exchange sites across the membrane thickness, which is, evidently, of paramount importance for the protons to be transported all the way from anode to cathode.

As the acidic groups need to dissociate for the proton to become mobile, one can expect that the water content of the membrane will also have a strong influence on conductivity. Proton transport occurs either via hopping of protons from one water molecule to the next (*Grotthus* mechanism) or via the net transport of H_3O^+ or other aggregates of water and H^+ [19]. Evidently, as the number of ion exchange sites increases, so will the hydrophilicity of the material, resulting in an increase of the water uptake.

In addition to the conductivity in water swollen state, the conductivity of the fuel cell membrane under non-saturated water vapour conditions is of importance, as partial drying of the membrane and electrodes may occur during cell operation. Also, fuel cell operation with partially humidified or even dry reactant gases is highly desirable to minimize system complexity.

The requirement of water within the polymer structure as a proton transport medium limits the operating temperature of such membranes to below 100 °C at moderate pressure. Alternative membrane concepts using anhydrous proton conduction, and hence operation at temperatures above 100 °C, are under development. Among the approaches, phosphoric acid doped polybenzimidazole (PBI) appears to be among the most promising. Here, protons are transported via a phosphoric acid network [20]. The concept of a water free proton conducting polymer material has been studied intensively with sulfonic acid, phosphonic acid, and imdazole compounds [21]. The key properties to be optimized in this concept are the anhydrous conductivity and electrochemical compatibility with available electrocatalysts to yield high exchange currents.

3.2.2
Water Management

In the characterization of fuel cell membranes, there are a number of important materials and component properties that have to be assessed in order to determine the applicability and operability in the fuel cell environment. Since proton mobility within the polymer structure is a strong function of the water content, the water uptake and transport properties of the membrane are of paramount importance, determining the water profile through the thickness of the membrane as well as in-plane. Water transport mechanisms in the polymer include diffusion due to a gradient in water content, hydraulic permeation as a consequence of a pressure gradient between anode and cathode, and electroosmotic drag, i.e., water flux coupled to proton transport [22]. Recently, in-plane neutron imaging of liquid water in an operating fuel cell has provided insight into the water distribution across the membrane electrode assembly dependent on the operating conditions, i.e., relative humidities of the feed gases [23].

3.2.3
Durability

Maintaining the chemical and mechanical integrity of the membrane over the anticipated lifetime is a key requirement, and it deserves attention already in the early stage of membrane development. Membrane aging is associated with the loss of one or both of the electrolyte and separator functionalities. The loss of ion exchange groups leads to gradual decrease in membrane conductivity, whereas the loss of the mechanical integrity, by pinhole formation or rupture, is perceived as a more "dramatic" event, because it represents catastrophic MEA failure.

PEFC membranes undergo chemical degradation through polymer chain scission, loss of functional groups or constituents (blocks, side chains, blend component), caused by HO and HOO radicals, which are formed in situ through interaction of H_2 and O_2 with the noble metal catalyst on both anode and cathode side [24]. In this context, the permeability of the membrane material for H_2 and O_2 is of importance. Higher rates of gas crossover will lead to higher amounts of radical species formation and, thus, higher rates of membrane degradation. In addition, hydrolysis may be the cause of chain deterioration for some polymer types with respective susceptible functional units.

The mechanical properties of the membrane are equally important, although in the scientific literature this aspect is largely underrepresented. In addition to tensile strength and elongation at break values, dimensional stability upon swelling, resistance to crack formation, and propagation also have to be considered. Creep of the polymer is likely to occur because the water swollen membrane is plasticized and the membrane is under a constant compaction force in the cell [25]. This may lead to membrane thinning and, eventually, puncturing and pinhole formation. An effect especially pertaining to swelling of the polymer upon water sorption is a fatigue-type phenomenon, where the membrane electrode assembly is subjected to dry-wet cycles, leading to periodic stress build up-relaxation in the membrane and, ultimately, to crack formation. This has been observed to be a membrane failure mode [26].

3.2.4
Cost

Last but not least, cost of the material is an important factor in membrane commercialization. The most widely used class of membrane materials today for the PEFC are of the perfluorinated sulfonic acid (PFSA) type, e.g., Nafion® (Dupont, USA), Flemion® (Asahi Glass, Japan), Aciplex® (Asahi Kasei, Japan), and derivatives thereof, such as the GORE-SELECT® membranes (W.L. Gore, USA). However, they have the disadvantage of being inherently expensive due to the complex fluorine chemistry involved in their fabrication. Although the projected membrane cost scales dramatically with increasing production volume, the development of intrinsically more cost-effective alternative membrane materials that are partially fluorinated or even fluorine-free is of high interest.

4
Conclusion

The development of membranes for fuel cells is a highly complex task. The primary functionalities, (i) transport of protons and (ii) separation of reactants and electrons, have to be provided and sustained for the required operating time. Optimization of the composition and structure of the material to maximize conductivity and mechanical robustness involves careful balancing of synthesis and process parameters. The ultimate membrane qualification test is the fuel cell experiment. It is evident that the membrane is not a stand-alone component, but is combined with the electrodes in the membrane electrode assembly (MEA). Interfacial properties, influence on anode and cathode electrocatalysis, and water management are the key aspects to be considered and optimized in this ensemble.

Successful membrane development and demonstration in fuel cells of technical relevance involves collaborations of experts from various fields, including polymer chemists, physicists, electrochemists, and process engineers, in order to allow development of successful polymer materials.

References

1. Vielstich W, Gasteiger HA, Lamm A (eds) (2003) Handbook of Fuel Cells – Fundamentals, Technology and Application. John Wiley & Sons, Chichester
2. Grove WR (1839) Philos Mag Ser 3 14:127
3. Grove WR (1842) Philos Mag Ser 3 21:417
4. Bacon FT (1954) BEAMA J 6:122
5. Grubb WT (1959) US Patent 2 913 511
6. Fickett AP (1970) Proc Symp Battery Separators, Columbus Section of the Electrochemical Society, p 354
7. Grot WR (1972) Chem Ing Technol 44:167
8. Grot WR (1975) Chem Ing Technol 47:617
9. Scherer GG (1990) Ber Bunsenges Phys Chem 94:1008
10. Chernyshov SF (1982) J Res Inst Catal Hokkaido Univ 30:179
11. Grot WR (1982) European Patent 0 066 369
12. Srinivasan S, Ticianelli EA, Derouin CR, Redondo A (1988) J Power Source 22:359
13. Raistrick ID (1989) US Patent 4 876 115
14. Xie Z, Navessin T, Shi Z, Chow R, Wang Q, Song D, Andreaus B, Eikerling M, Liu Z (2005) J Electrochem Soc 152:A1171
15. Wilson M, Gottesfeld S (1992) J Appl Electrochem 22:1
16. Debe M (2003) Novel Catalysts, Catalyst Supports and Catalysts Coated Membrane Methods. In: Vielstich W, Gasteiger HA, Lamm A (eds) Handbook of Fuel Cells – Fundamentals, Technology and Application, Part 3. John Wiley & Sons, Chichester, p 576
17. Büchi FN, Wakizoe M, Srinivasan S (1996) J Electrochem Soc 143:927
18. Xie Z, Holdcroft S (2004) J Electroanal Chem 568C:247
19. Kreuer KD (1992) Conduction mechanisms in materials with volatile molecules. In: Colomban P (ed) Proton Conductors. Cambridge University Press, Cambridge, p 474
20. Schmidt TJ, Baurmeister J (2006) ECS Trans 3 1:861
21. Scharfenberger G, Meyer WH, Wegner G, Schuster M, Kreuer KD, Maier J (2006) Fuel Cells 6:237
22. Kreuer KD, Paddison SJ, Spohr E, Schuster M (2004) Chem Rev 104:4637
23. Boillat P, Kramer D, Seyfang BC, Frei G, Lehmann E, Scherer GG, Wokaun A, Ichikawa Y, Tasaki Y, Shinohara K (2008) Electrochem Commun 10:546
24. Mittal VO, Kunz HR, Fenton JM (2006) Electrochem Soc Trans 3:507
25. Makharia R, Kocha SS, Yu PT, Gittleman C, Miller D, Lewis C, Wagner RT, Gasteiger HA (2005) Meeting Abstracts 208th Meeting of The Electrochemical Society, Los Angeles, USA, Oct 16–21, Abstract #1165
26. Gasteiger HA (2005) Durability of Polymer Electrolyte Membrane Fuel Cell Materials. Int Conf Solid State Ionics (SSI-15), Baden-Baden, Germany, July 17–22, Oral Contribution #72

Proton-Conducting Polymer Electrolyte Membranes: Water and Structure in Charge

Michael Eikerling[1,2] (✉) · Alexei A. Kornyshev[3] (✉) · Eckhard Spohr[4,5] (✉)

[1] Department of Chemistry, Simon Fraser University,
Burnaby, British Colombia V5A 1S6, Canada
meikerl@sfu.ca

[2] Institute for Fuel Cell Innovation, National Research Council,
Vancouver, British Colombia V6T 1W5, Canada

[3] Department of Chemistry, Faculty of Physical Sciences, Imperial College London,
London SW7 2AZ, UK
a.kornyshev@imperial.ac.uk

[4] Lehrstuhl für Theoretische Chemie, Universität Duisburg-Essen, S05 V06 E15,
Universitätsstr. 5, 45141 Essen, Germany
eckhard.spohr@uni-due.de

[5] Institut für Energieforschung 3 – Brennstoffzellen, Forschungszentrum Jülich,
52425 Jülich, Germany

1	Fuel Cell Membranes as Matrices for Aqueous Proton Transfer	16
2	State-of-the-Art PEMs and Recent Membrane Development	18
3	Membranes: Structural Complexity at Different Scales	19
4	The Role of the Theory: From Molecular Structure to Operation	24
4.1	Proton Transport in Water and Aqueous Networks	26
4.2	Models of Proton Transport at Mesoscopic Scale	30
5	Proton Transport Near the Polymer–Water Interface	31
6	From Meso-to-Macroscale: Effective Transport Properties	36
7	Macroscopic Network Models	41
8	Membrane Operation in the Cell	43
9	Concluding Remarks	47
References		50

Abstract This article reviews the structure and properties of aqueous-based proton-conducting membranes for fuel cell applications. We will discuss (1) structure of and phase separation in the membranes, (2) principles of proton mobility in hydrogen-bonded aqueous networks, (3) proton conductance through the percolating aqueous phase inside the membrane, and (4) the effects determining membrane performance in a fuel cell.

Keywords Atomistic simulations · Fuel cells · Ionomer membranes · Network models and theory · Pore morphology and proton conductance · Proton transfer

1
Fuel Cell Membranes as Matrices for Aqueous Proton Transfer

The best proton-conducting membranes for fuel cell applications are currently phase-separated polymer electrolyte membranes (PEMs). They utilize the anomalous proton mobility of aqueous-networks inside the membrane [1–6]. Proton transfer in water has long been regarded as a sequential concerted mechanism along the hydrogen-bond network, the original idea of which dates 200 years back to the legendary Baron von Grotthuss [7]. Today, it is widely accepted that fast proton mobility in water is triggered by hydrogen-bond coordination fluctuations that destabilize localized proton states and, thus, facilitate proton relocation between water molecules [8].

As we will see below, this briefly sketched mechanism of proton transfer in liquid water could be highly relevant for important types of PEMs under realistic operating conditions in the fuel cell. In many cases, in particular when it comes to advanced membrane design, it is, however, not enough to consider the membrane as an "inert container for water pathways". Indeed, the dependence of membrane conductivity on the global or local water content is determined not only by the connectivity of the aqueous percolation network but also by the varying ability of each pathway to conduct protons.

A major incentive of this article will be to stress the complicating traits of the membrane environment on effective proton transport and fuel cell performance. The polymer affects distribution and structure of water and dynamics of protons and water molecules at multiple scales. In order to describe the conductivity of the membrane, one needs to take into account explicit polymer-water interactions at molecular level, interfacial phenomena at polymer–water interfaces at mesoscopic scale and the statistical geometry and topology of randomly distributed aqueous and polymeric domains at macroscopic scale.

Appropriate membrane materials are usually composed of perfluorinated (hydrophobic) polymer backbones with sidechains containing acid groups, in most cases sulfonic acid ($- SO_3H$) end groups. The most well-known example of this class is Nafion® by DuPont de Nemours. Recently, polyether and polyketo polymers with statistically sulfonated phenylene groups such as sPEK, sPEEK and sPEEKK or polymers on the basis of benzimidazole (sulfonated PBI) have been applied as well. When exposed to water, the sulfonic acid groups in all these materials dissociate and release protons as charge carriers into the aqueous sub-phase. At the same time, the remaining SO_3^- groups become hydrated and form a neutralizing environment at the polymer–water interface in which protons and water molecules move.

In order to sustain high fuel-cell power densities, PEMs primarily require high proton conductivity (>0.1 S cm^{-1}) and impermeability to gases supplied on anode and cathode sides. Membranes should be chemically stable and mechanically robust. They have to survive in a chemically aggressive environment over the lifetime of envisaged applications, e.g. several 1000s of operating hours for automobile applications. The high acidity that confers the good proton conductivity could adversely affect the chemical stability of PEMs and catalyst layers (CLs), causing problems of degradation. High levels of hydration, required by present PEMs, limit the temperature of operation, since water evaporates above 90 °C, even under pressurization (which is anyway undesirable as it costs energy and, thereby, reduces system efficiency). In fuel cell vehicles, PEMs should easily adapt to widely varying operation conditions, e.g. start-up at −40 °C. On the other hand, fuel cell operation at temperatures above 150 °C would significantly enhance electrode kinetics and decisively increase CO tolerance of catalysts as well as improve heat transfer (thus, facilitating cooling in, e.g., automotive applications).

The seemingly unavoidable coupling between good proton and high water mobilities is a negative factor, as the resulting electro-osmotic flux shuffles water molecules from the anode to the cathode. Although back diffusion or hydraulic permeation partly balances this flux, the remnant water flux depletes water at the anode side. Drying the membrane material there drastically diminishes its conductive abilities, since the local conductivity of presently available PEMs significantly decreases with decreasing local water content. On the other hand, water traffic towards the cathode could cause excessive membrane swelling and flooding of the catalyst layer and porous transport layer on the cathode side. These effects could lead to dramatic voltage losses and cause limiting current behavior [9–16]. Overall, this poses an issue of water management in the membrane, catalyst layers and the whole cell [17]. Moreover, the dimensional instabilities due to excessive swelling and deswelling (or even shrinking) of different membrane parts impair the durability of membrane electrode assemblies [18].

Proton transfer (PT) is of fundamental importance to a plethora of processes in biology, chemistry and physics, and is as ubiquitous as electron transfer (ET) [19]. In biological organisms, coupled ET and PT processes are the key to understanding the energy metabolism of the cell. In the different context of electrochemical energy conversion in fuel cells, coupled ET and PT processes in electrodes control the conversion of chemically stored energy into electrical energy. The separation of electron and proton fluxes on macroscopic scales makes the free energy released available to external appliances. These aspects of electrochemical performance oblige us, generally, to give special attention to the proton conductance not only in the bulk of the membrane material, on which this article is focused, but also to the membrane performance in the electrode. As the microscopic mechanisms of proton transport are believed here to be essentially the same as in the bulk membrane (except for possibly a contribution of surface diffusion on the catalyst surface) the problem here is rather in the "network" performance of the randomly scattered ionomer domains and the interactions of ionomer with other components in the random composite electrodes [20].

2
State-of-the-Art PEMs and Recent Membrane Development

Presently, the most widely used and tested polymer electrolyte membranes are Nafion® (DuPont), Dow (Dow Chemicals), Flemion® (Asahi Glass), Aciplex® or various modifications thereof, which differ in chemical structure, ion exchange capacity and thickness. These perfluorinated sulfonated polymers all have Teflon-like backbones. Pendant sidechains contain charged hydrophilic segments or endgroups (usually SO_3^-). Spontaneous aggregation of polymer backbones, occurring to some degree even in the dry state, results in phase-separation at nm-scale. The hydrophobic polymer phase forms the stable membrane matrix, while a hydrophilic phase consisting of solvent, mobile protons and fixed acid groups forms the pathways for proton conduction. This phase separation leads to highly heterogeneous morphologies and complex charge distributions in the hydrated state.

Proton conductivity in PEMs is usually a strong function of the degree of hydration. The maximum conductivity in Nafion® (~ 0.1 S cm^{-1}) is attained with water contents $\lambda \sim 15$ ($H_2O : SO_3^-$ ratio) under typical operation conditions. Conductivity decreases monotonically towards lower water content. It usually exhibits a quasi-percolation transition for $\lambda < 5$ [21, 22].

Current membrane development focuses on perfluorinated ionomers, hydrocarbon and aromatic polymers and acid-base polymer complexes. Good recent reviews on membrane synthesis and experimental characterization can be found in this volume and for example in [23–29].

Straightforward strategies for improvement of the membrane operation in fuel cells involve increasing the ion exchange capacity (e.g. by increasing the volume density of sidechains or by increasing the number of acid groups per sidechain) in order to enhance the density of protons, water uptake and proton mobility. Because of the higher density of hydrophilic groups, such membranes often show increased solubility in water thereby decreasing mechanical stability during operation. Reducing the membrane thickness to ~ 20–50 μm improves membrane conductance and fuel cell power density. Fluorine-free and therefore cheaper PEMs can be synthesized using aromatic hydrocarbon polymers, e.g. polyether ether ketone and polystyrene. Increased membrane stability (thermal, mechanical and electrochemical) can be achieved by impregnating Nafion into inorganic matrices of clays, silica or phosphotungstic acid or into porous Teflon. Such hybrid membranes exhibit reduced swelling, improved water retention and improved aptitude for operation at $T > 120\,°C$. Higher power densities could thus be achieved. Other directions of research pursue the development of water-free systems, by utilizing acid-base complexes [24] or simply by replacing water in Nafion by other proton conducting groups such as imidazole, which can be immobilized covalently to the polymer backbone [30].

3
Membranes: Structural Complexity at Different Scales

The structure of PEMs, in particular their phase-separated morphology at nm-scale, has been studied with a number of experimental techniques, including small- and wide-angle X-ray and neutron scattering, infrared and Raman spectra, time-dependent FTIR, NMR, electron microscopy, positron annihilation spectroscopy, scanning probe microscopy, and scanning electrochemical microscopy (SECM) (for a review of this literature see [31]). Structural studies of PEMs have mainly focused on Nafion. A thorough recent review on this particular membrane is provided in [32].

Based in large part on small-angle X-ray scattering techniques (SAXS) and later supplemented by small-angle neutron scattering (SANS), the first morphological models of membrane structure emerged during the early 1980s. The group at DuPont concluded from SAXS studies that the morphology of hydrated ionomers is best described by a model of inverted spherical micelles of nanoscopic dimension, confined by anionic head groups of the sidechains [33, 34]. In a dry state, the diameter of these micelles is ~ 2 nm and they are disconnected from each other. With water-filling, the micelle diameter grows up to ~ 4 nm. In order to form pathways for proton and water transport, it was inferred without any experimental evidence that aqueous necks should form between spherical micelles at intermediate water contents. After formation of a critical number of these necks, an uninterrupted pathway through the water sub-phase emerges in the network of micelles and necks, and the membrane percolates. During water uptake, the cluster network topology continuously reorganizes by swelling and merging of individual clusters and by formation of additional necks [22, 31, 33, 34].

Crystallinity in Nafion was probed by detailed wide-angle X-ray diffraction (WAXD) experiments [35]. A small degree of crystallinity (<12%) was found in sulfonated 1100 EW Nafion.

The cluster network model, described above, which is mostly referred to as the Gierke model, motivated numerous studies of the phase-segregated morphology of Nafion, utilizing SAXS and SANS techniques. However, those techniques have difficulties in determining more than just the position of the Bragg or *ionomer* peak characterizing the short-range periodical motif in the arrangement of the micelles. Thus, only information about the distance between the centers of the micelles is available, but not on the size of the micelle. For the latter information, a Guinier analysis is needed, which is difficult due to the closeness of this part of the spectrum to the ionomer peak. The Porod part of the spectrum (q^{-4} power law dependence of scattering intensity on wave vector) can aid the determination of the overall surface of the micelles and the average micelle diameter. Additional information about the size distribution of micelles can also be obtained from the capillary pressure isotherms [36] and infrared spectroscopy [37], whereas electron microscopy [38] discerns micellar shapes. For the necks there is no method to detect them directly.

The weaknesses of the cluster-network model become apparent when it is employed to describe the structural evolution of membranes over a wide range of humidification conditions, from the highly diluted polymer solution (water volume fraction $w \to 1$) to the dry membrane state ($w \to 0$). An earlier conceptual model for this morphological reorganization, suggested by G. Gebel [39], had to involve a rather vague conjecture about the structural inversion from a colloidal dispersion of rod-like polymer aggregates in solution to a cluster network of water-filled ionic domains (or inverted micelles) embedded into a polymer matrix at $w < 0.5$. These considerations evolved later into a new structural model of Nafion-type membranes, which gives a continuous transition between dry membrane state and highly diluted polymer solution [40]. Gebel's group utilized SAXS, SANS and USAXS (Ultra Small-Angle X-ray Scattering) techniques to study characteristic dimensions of hydrated Nafion membranes from 1 nm to 1000 nm. The USAXS upturn in scattering curves was related to large-scale inhomogeneities in electron density ($\sim$100 nm). The so-called matrix-knee corresponds to the supralamellar distance in the crystalline domains. The main feature in the small-angle scattering curves is the ionomer peak, corresponding to the first maximum of the structure factor. It is attributed to local ordering between ionic clusters, widely studied as a function of water content, ion exchange capacity, and temperature. The scattering intensity in the region between USAXS upturn and ionomer peak follows a q^{-1} power law. This is interpreted as scattering from rodlike particles with diameter $\sim$6 nm and length $\geq$100 nm. At large scattering angles the intensity follows a Porod power law ($I(q) \propto q^{-4}$).

The emerging structural model consists of elongated polymeric aggregates with cylindrical or ribbonlike shape, lined on the surface by an array of acidic dissociated sidechains and surrounded by solvent and mobile counterions. The model suggests that polymer rods and their aggregation should be considered explicitly as the structure forming elements. This view opens new intriguing perspectives for predictive theoretical modeling of the formation of random heterogeneous membrane structures, based on the consistent treatment of interactions between polymer components in the presence of the solvent.

Scattering and diffraction techniques are useful tools for structural analysis. However, the transformation of the measured data into structural models is often intricate and ambiguous. Direct visualization of domain sizes and random morphology of ionomer membranes can be obtained from microscopy studies. In the latter case, the obtained information is impaired by the need to consider thin films (~100 nm thick films in TEM investigations, cast from solution or microtomed) or by the limitation to surface studies (AFM investigations). Holdcroft and coworkers used TEM pictures to compare the morphologies of hydrocarbon-based graft copolymer membranes with random copolymer membranes with the same ion content [41, 42]. For the hydrated grafted membranes, a continuous phase-separated network of water-filled channels with diameters of 5–10 nm was observed. The random copolymer membranes exhibit a weaker tendency toward microphase separation with water being dispersed more randomly within the polymer domains. The water uptake in the grafted membranes is limited to smaller values, since the hydrophobic nature of channel walls impedes the penetration of water into the polymer domains. Nevertheless, their ionic conductivity is an order of magnitude larger than that of the random copolymer membranes, probably due to increased connectivity of the aqueous domains.

Spatial organization of distinct membrane domains determines the state of water, fundamental interactions in the polymer/water/ion system, vibration modes of fixed sulfonate groups and mobilities of water molecules and protons. Dynamic properties of the membrane can be probed at microscopic scale with spectroscopic techniques (FTIR, NMR) [32]. FTIR measurements provide information about sidechain motions [43–45]. Differential scanning calorimetry (DSC) and NMR revealed different types of water environments, corresponding to non-freezable tightly bound water, freezing weakly bound water and free water molecules [46, 47]. NMR can be used to probe mobilities of protons and water molecules at nm resolution [48]. In fully hydrated membranes these molecular mobilities were found to be in a similar range as values obtained for bulk water. Macroscopic proton transport can be studied with electrochemical impedance spectroscopy (EIS). Cappadonia et al. [49, 50] used EIS to study variations of proton conductivity with water content and operating temperature for Nafion 117. Arrhenius representation of conductivity data revealed activation energies between 0.36 eV at the lowest degree of hydration and 0.11 eV at the highest degree of hydration. For a given water content, the transition between low and high activation energies was observed at a transition temperature ~260 K for well-hydrated membranes. This was interpreted as a freezing point suppression due to the confinement of water in small membrane pores.

In spite of tremendous experimental activities aiming at unraveling structure and dynamics in fuel cell membranes, the quest for a universally accepted model continues. Primordially, the ambiguity of structural models stems from the random morphology of the membrane. Indeed, due to this complexity it is not surprising that the most notable improvement in membrane "design" in recent years has been the reduction of membrane thickness from ~175 μm to ~25 μm! On the contrary, it is highly surprising in view of this prohibitive structural complexity, that only very few experimental studies exist which utilize well-defined "model" membranes or membrane substructures with controlled molecular architectures. A few notable exceptions can be found in [41, 42, 51, 52] where grafted block copolymers and random copolymers are compared, and in [26, 27], where the role of the protogenic group is explored.

At low water content the quasi-crystalline motif in the distribution of inverted micelles is strong; the size of the micelle is smaller than the persistence length of the bundles of backbone-chains forming the membrane skeleton. A legitimate question arises then: how can one build a quasi-crystalline structure of inverted micelles (aqueous droplets supported by hydrated sidechains), if the sidechains are attached to the backbones? In an attempt to answer this question, a more detailed morphological model of Nafion-type ionomers was suggested [31]: a quasi-crystalline arrangement of units cells as depicted in Fig. 1.

This model managed to rationalize the observed correlation between the macroscopic swelling of the membrane and the distance between micelles. Channels that could be built through the windows in the cages will be very narrow. Upon membrane swelling the cages expand by strings sliding along each other. At the beginning, the size of the droplets grows but each droplet is still encapsulated within each cage, disconnected from other droplets. With further water uptake, the droplet ejects water into the windows building small cathenoids adjoining the neighboring droplets. The analysis shows, that this

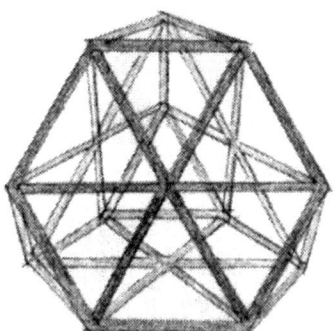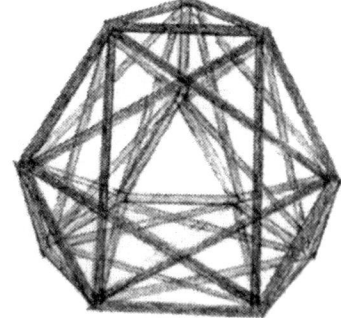

Fig. 1 Possible types of cages where the strings represent the backbones or their bundles, that can provide a short-range-ordered system of fourfold coordinated inverted micelles composed of hydrated sidechains (not shown) pointing from the strings towards the interior of the cages [31]. These micelles keep water droplets (not shown) with protons of dissociation encapsulated inside the cage. The channels, when they form, bridge water droplets, through the windows in the cages

will take place as a first-order transition. At the transition, the system may slightly shrink as some water from the droplets will be taken to build channels. With further water uptake, the system will swell again, and both the droplets and cathenoids will continuously increase in size. A theory of this phenomenon is in progress. The fact that in this configuration the emerging channels are very narrow, could help to explain the data by Cappadonia et al. of much higher activation free energy at low water content, and essentially the sharp increase of activation free energy only for $\lambda < 5$.

Indeed, if the channels evolve at the beginning as *extremely narrow* units (less than half nm radius of the narrowest part of the cathenoid) and remain narrow even in the "mature state", it is clear why the activation energy of the proton mobility, entirely controlled by the necks, will depend, dramatically, on water content, as experimentally observed (see Fig. 2). The bottleneck of conductance would be the proton transfer through the aqueous necks, or the fluctuational formation of the neck itself (in the spirit of a conjecture made by Vishnyakov and Neimark [53]). Furthermore, the more flexible the sidechains are, the higher the proton mobility, since fluctuations of the chains will support the necks, reducing their surface tension, in addition to possible sidechain-fluctuation promoted proton transport.

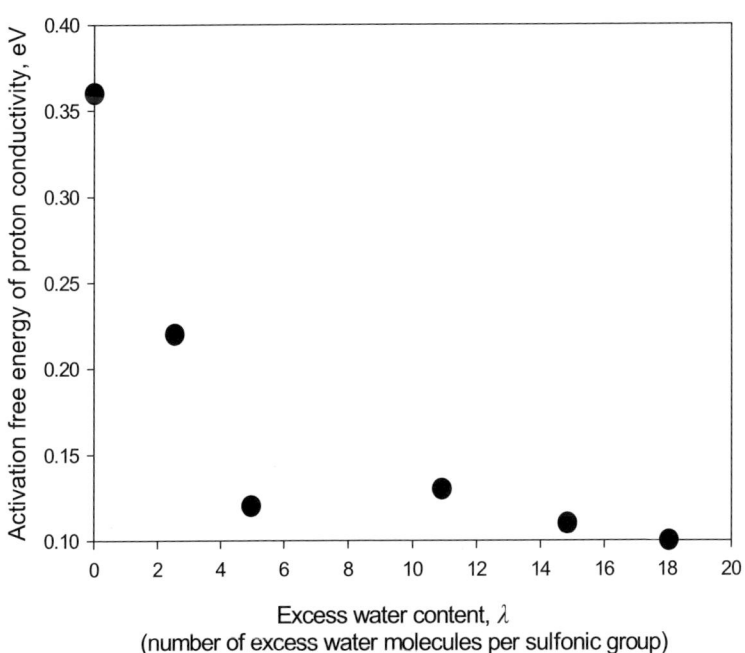

Fig. 2 Replotted high temperature data of Cappadonia et al. [49, 50] on activation free energy of similarly prepared Nafion 117 samples with different water content

4
The Role of the Theory: From Molecular Structure to Operation

The anchor point of any of the structural models, discussed so far, is a very simple, possibly oversimplified picture. It suggests that PEMs consist of two well-separated phases, viz. the hydrophobic part of the polymer that forms elongated polymeric aggregates, and the solvent phase hosting protons of dissociation. Unfortunately, the interfacial region complicates this picture, since sidechains that are fixed to the backbones protrude into the solvent-containing domains. It can thus be expected that sidechain–sidechain and sidechain–water interactions strongly affect the hydrogen-bonded network of the aqueous phase and the mobilities of nearby protons and water molecules. This complicated interfacial region thereby largely determines differences in performance of membranes with different chemical architectures. Indeed, the picture of a "polyelectrolyte brush" could be more insightful than the picture of a well-separated domain structure in order to rationalize such differences [54].

Feasible theoretical approaches, aimed at establishing a relation between the chemical architecture and membrane performance, have to combine several concerted steps. In the first major step, one needs to understand, how synthesis procedures and principal chemical structure at the primary structural level of solvated ionomers control the self-organization of structural units and their effective properties at the mesoscopic scale. Solvent–solvent, solvent–polymer and polymer–polymer interactions and correlations determine resulting conformations of the polymer phase and charge distributions in the surrounding solvent phase [54–56]. Genuine parameters of the polymer are chemical architecture of monomers (hydrophobic/-philic properties), length and separation of sidechains, density of acid groups on sidechains, ion exchange capacity, ionomer concentration (water content of the membrane), degree of dissociation. Field theories [56–58], scaling arguments [54, 55] and numerical simulations [53, 59–64] can be employed here. By minimizing the free energy of the system using these approaches, metastable conformations of the polymer-solvent system can be found. The majority of heterogeneous structures encountered in such studies exhibit nanophase segregation between polymer and aqueous phases, in qualitative agreement with experimental results reported above.

Structural descriptors at the secondary level (mesoscale) are topology and domain size of polymeric aggregates (persistence lengths and radius), effective length and density of charged polymer sidechains on the surface, properties of the solution phase (percolation thresholds and critical exponents, water structure, proton distribution, proton mobility and water transport parameters). Moreover, n-point correlation functions could be defined that statistically describe the structure, containing information about surface areas of interfaces, orientations, sizes, shapes and spatial distributions of the phase domains and their connectivity [65]. These properties could be partially determined experimentally and utilized for the reproduction of the membrane structure in computer models. In order to accomplish the transition from chemical architectures at microscopic scale to the formation of structural conformations at mesoscopic and macroscopic scales, relevant concepts of polymer physics [54, 55, 66] have to be fused with those of the physics of random heterogeneous media [65].

Upon the transition from primary polymer architectures to secondary structural units at the mesoscopic scale interactions of solvent–solvent, solvent–polymer, polymer–polymer types are renormalized into effective interactions between sidechains and aqueous domains, as indicated in Fig. 3. These interactions control proton distribution and mobilities as well as the coupling between proton and water transport. At the mesoscopic level of the theory, the hydrophobic polymer phase formed by the backbones can already be considered as an inert, structureless matrix.

The two-step approach, outlined in Fig. 3, leads first from the definition of primary structural parameters at molecular scale to the formation of structural elements at the mesoscale and in a subsequent step to effective properties at the macroscopic scale. Overall, it could establish predictive relations between membrane architecture and performance. The function of existing highly heterogeneous membranes with multiple length scales contributing to structure formation and dynamics of protons could be explained. Equally importantly, though, such a theoretical foundation, could expedite the routes towards the design of cheap, highly performing membranes, with a major focus on membranes that could operate stably at minimal hydration and elevated temperatures (>120 °C).

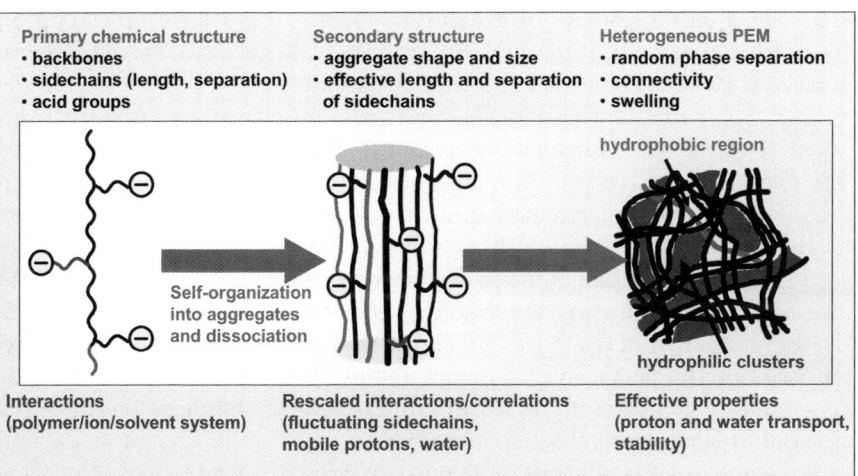

Fig. 3 Structural evolution of PEMs from primary chemical architecture to aggregate formation at mesoscale to random heterogeneous medium at macroscale

The difficulties in implementing such a coherent approach stem from the fact that one has to deal with complex interactions in a heterogeneous many particle system and that proton transfer in water is a genuinely complicated many-particle phenomenon itself. In the beginning of everything lies the fundamental phenomenon of PT in aqueous networks.

4.1
Proton Transport in Water and Aqueous Networks

Water was "born" to conduct protons (see Special Issue: "Is life possible without water?" [67]). The conductance of distilled water is miserable due to a negligible concentration of free protons (10^{-7} mol/liter), but the proton mobility in water is approximately five times higher than the mobility of an alkali cation (e.g. Na^+), an object of similar size as the hydronium (H_3O^+) ion [68]. So, "donated" protons can run fast through the aqueous phase. Excess protons result from dissociation of acidic molecules or molecular groups, e.g. in solutions of strong acids, hydrated polymer-electrolytes, or proteins. In acidic solutions both the protons and counter-anions are mobile. In polymer-electrolyte membranes and in proteins only protons are mobile in the connected aqueous phase while the counter anions are mostly a part of an immobile skeleton.

Experimental studies of temperature-dependent proton mobility have a long and dramatic history. In a modern sense they date back to the works of Johnston [69] and Noyes [70, 71], followed much later by the studies of the pressure dependence by Eucken [72, 73], Gierer and Wirtz [74], Gierer [75], and Franck, Hartmann and Hensel [76]. Reference [77] gives a comprehensive overview of aqueous proton conductivity and the early experimental data, based on the concept of the *excess mobility*, responsible for the difference of the observed proton mobility from the one provided by the classical hydrodynamic motion of the hydronium ion.

The excess mobility-vs.-temperature curve was found to exhibit a maximum at elevated temperatures near 150 °C, achievable at elevated pressure. The magnitude of the proton mobility in pure water was not addressed in those studies, although attempts to determine it were made by Kohlrausch at the end of the 19th century [78]. Focus was instead on the conductance of strong acids such as HCl in the limit of infinite dilution. The difference of the measured conductance and the limiting conductance of a salt of a cation with size similar to that of H_3O^+ was attributed to *excess* proton mobility, based on the assumption that the hydrodynamic radius of both ions is similar. The excess mobility was taken to represent non-classical proton hops on top of the classical hydrodynamic motion of the H_3O^+.

Proton conductivity in bulk aqueous solution can be contrasted with proton conductance in water-saturated polymer electrolyte membranes, such as perfluorinated sulfonic acids [49, 50, 79], where protons are the only charge carrier species. Their volume density can be determined through the polymer equivalent weight (the number of SO_3^- groups per gram of dry polymer), density of the polymer and the measured water uptake. Proton conductance can therefore be measured directly, determining the average values of single-proton mobility.

While such results are of obvious interest for membrane science and technology, they do not apply straightforwardly to proton mobility in pure water. However, for high water uptake the proton mobility in a membrane can approach that of bulk water, although it can never be higher than in pure water, unless a synergetic Davydov solitonic mechanism of collective proton transport [80] in proton-conducting "wires" exists, which is the subject of many speculations in solid-state protonics [81]; but the latter type of mechanism is unlikely to exist in disordered polymer media. To explain this fact one may suggest that protons have only limited ability to move along the surfaces of the water-filled channels [82].

The precise proton distribution is, however, not known, as well as the precise structure of the pore space. Mean field theory of proton distribution in "sample" pores [68] suggests a smooth radial distribution with a finite proton population in the center of the pore, where they would move very much like in bulk water. Simulation results show strong pinning of protons to immobile negative charges on the pore walls if they are considered point-like and static (see also Fig. 7 below). The localization of protons is, however, much weaker for distributed and fluctuating counter charges or models of real sidechains [83]. Membrane data have, further, been obtained at ambient pressures in a temperature range from 310 down to 170 K. The data of Cappadonia et al. [49, 50] reveal two conductance regimes, with a change in activation energy between 225 and 260 K. Straight Arrhenius plots were obtained above 0 °C. The apparent activation energy decreases significantly with water uptake, reaching 0.1 eV in water-saturated membranes. This is the value usually attributed to proton mobility in bulk water. Mean field theory explains this readily, but other explanations may be needed if protons are localized more strongly near pore surfaces [84]. All in all it could, therefore, be misleading to assess the proton mobility in water on the basis of membrane conductance.

There are other caveats with the simplest notion of excess proton mobility, and the comparison with membrane proton conductance at water saturation. Subtracting the limiting conductance of Na^+ ions, which move only by the classical mechanism, essentially cancels the classical conductance of H_3O^+. However, the cancelation cannot be complete, because H_3O^+ exhibits classical motion only for *part of the time*. Nevertheless, the difference can be used as a measure of non-classical contributions to proton conductance. It is more precise the smaller the classical, hydrodynamic contribution to the proton mobility.

Another observation relates to the contrasting temperature variation of the excess proton mobility in water and proton mobility in saturated polymer electrolyte membranes. The former variation is strong and non-monotonic in the high-temperature region. The latter is not only monotonic but also Arrhenius-like, at least above the freezing point of bulk water. At first glance, a plausible assumption would be that classical H_3O^+ diffusion compensates for the decrease of the non-classical *excess* mobility at high temperatures. This is, however, inconsistent with the assumption that the classical H_3O^+ contribution is small.

Fast proton mobility in water attracted theoretical attention early, beginning with the works of von Grotthuss [7], at a time when the existence of the proton was not known, the chemical formula of water not settled, the notion of molecules was new, and little was known about the electricity laws. Modern landmarks were set by Bernal and Fowler [85], Eigen and de Mayer [86], Conway et al. [87], and Zundel and Metzger [88]. This was followed by more detailed molecular mechanisms, and analytical and computational models, see [68, 77, 79, 84, 89–93] and a conceptual essay by Agmon [1] which stimulated a new round of activities in this area.

The fundamental importance of proton transfer (PT) in biology [94–99], and in the development of polymer electrolyte fuel cells for vehicle and portable applications [100, 101], continues to press for a deeper understanding of PT mechanisms in hydrogen-bonded systems.

Recent computer simulations have highlighted the nature of the elementary act of PT in water [8, 92, 93, 102]. They have, particularly, provided new evidence for the crucial effect of the dynamics of solvation water molecules of the PT clusters. In one suggested mechanism PT is initiated by the breaking of a hydrogen bond between the acceptor water molecule and a water molecule in its solvation shell. Hydrogen bond breaking ushers the acceptor molecule to a favorable configuration for accepting the proton, while the donor molecule forms a new hydrogen bond with a water molecule in *its* solvation shell. Regardless of details, together with classical H_3O^+ ion motion, this mechanism gives rise to *translocation of hydronium*.

Computer simulations [8] and infrared spectroscopy [91] have shown, however, that the H_3O^+ ion is not the only and possibly not even the most stable excess proton state. The proton spends similar amounts of time in the Zundel state $H_5O_2^+$ as in the Eigen state $H_9O_4^+ = (H_3O)^+(H_2O)_3$, which is the hydrated H_3O^+ ion. In the Zundel state, the excess proton is delocalized between two water molecules, in the Eigen state the proton charge is largely centered on one oxygen atom and the central H_3O^+ ion is strongly hydrated by three water molecules, forming one strong hydrogen bond to each H_3O^+ proton. Rapid fluctuative interconversions between these two limiting states takes place. In fact, "an unambiguous distinction between $H_5O_2^+$ and $H_9O_4^+$ can no longer be achieved" [8]. The equilibrium O–O distance in the Zundel complex is significantly shorter ($r_{OO} = 2.45$ Å) than the average O–O distance in bulk water (2.85 Å). As the proton potential energy profile depends crucially on the O–O distance, it reduces to a single or a shallow double well potential at the short O–O distance of the Zundel complex. In that way, the Zundel complex can be regarded as a transient structure or relay group between two different hydronium ion states, according to the scheme

$$H_2O...(H_3O^+)...H_2O...H_2O \rightarrow$$
$$H_2O...(H_2O...H^+...H_2O)...H_2O \rightarrow$$
$$H_2O...H_2O...(H_3O^+)...H_2O.$$

Because of the rapid conversion between the two complexes, an alternate view is also possible: A hydronium ion is a transient intermediate structure between two distinct Zundel ions according to the scheme

$$(H_2O...H^+...H_2O)...H_2O \rightarrow H_2O...(H_3O^+)...H_2O \rightarrow$$
$$H_2O...(H_2O...H^+...H_2O).$$

Since the relative abundance of Eigen vs. Zundel complexes is comparable, both views are equivalent and involve the displacement of two protons. This PT mechanism is different from H_3O^+ translocation. The rate-determining step in this *structural diffusion* mechanism is either the transformation of a given $H_5O_2^+$ cluster into an adjacent $H_5O_2^+$ cluster, or, alternatively, the transformation of a $H_9O_4^+$ (or H_3O^+) cluster to an adjacent $H_9O_4^+$ (or H_3O^+), and the activation energy is determined by the concomitant reorganization of the solvent around these ions.

Computer simulations of excess proton conductivity in water have reached a powerful level [8, 92, 93, 102]. Importantly, simulations extend to *quantum-mechanical* proton dynamic features, so that proton motion can be coupled to details of the molecular environmental dynamics. A recent feature article explored an analytical theory in order to rationalize these complex processes that involve interconversion of proton-bearing clusters and proton transfers [103]. With a simple two-state empirical valence bond model (see below for details), which "implements" in a classical way the above-mentioned idea of two limiting protonated structures, namely the $H_5O_2^+$ and the H_3O^+ cluster, it was indeed observed that the two alternative sequences are equivalent with similar life times for both clusters, and that conversions between the two clusters are purely fluctuative.

Do quantum effects dominate proton transfer, as one might expect due to relatively small proton mass? Recent computer simulations and quantum chemistry calculations of small proton-bearing clusters show that they don't. The act of proton transfer is preceded by the fluctuational preparation (reduction) of the barrier for tunneling, since protons do not like to tunnel over distances larger than 1 Å. The proton transfer adiabatically follows the fluctuations of the slower nuclear degrees of freedom of the environment, but it occurs in a transition configuration characterized by a small barrier along the proton coordinate. Eventually, the motion along the proton coordinate proceeds almost barrierless. The only quantum effect present here is the resonance splitting of the adiabatic potential energy surfaces caused by hydron tunneling in the double well potential spanned by reaction coordinate. These features are seen in the kinetic isotope effect, as discussed in detail in Sects. 7.3 and 7.4 of [103].

How will this process be modified when water is confined by the membrane interior? This can be parameterized [107] or simulated [83, 84, 109] for a given picture of the membrane interior particularly its structural units that comprise the bottleneck for the proton transport. To be able to do this one must have a clear vision of the structure of the membrane interior, which remains the most difficult part of the whole story. In addition, the underlying description needs to be sensitive to the chemical architecture of the polymer.

4.2
Models of Proton Transport at Mesoscopic Scale

There are different ways to depict membrane operation based on proton transport in it. The oversimplified scenario is to consider the polymer as an inert porous container for the water domains, which form the active phase for proton transport. In this scenario, proton transport is primarily treated as a phenomenon in bulk water [1, 8, 90], perturbed to some degree by the presence of the charged pore walls, whose influence becomes increasingly important the narrower are the aqueous channels. At the molecular scale, transport of excess protons in liquid water is extensively studied. Expanding on this view of molecular mechanisms, straightforward geometric approaches, familiar from the theory of rigid porous media or composites [104, 105], could be applied to relate the water distribution in membranes to its macroscopic transport properties. Relevant correlations between pore size distributions, pore space connectivity, pore space evolution upon water uptake and proton conductivities in PEMs were studied in [22, 107]. Random network models and simpler models of the porous structure were employed.

At the other extreme, proton transport could be regarded primarily as an interfacial phenomenon at the water–polymer interface, with lots of ramifications due to the complex random structure of this interface and complex interactions in the polymer/water system. These effects become increasingly important with decreasing water content. A number of model calculations focus primarily on effects of electrostatic surface charges on proton transport in the hydrogen-bonded water network. Electrostatic contributions to activation Gibbs free energies could be calculated using mean field Poisson–Boltzmann theories [84, 107], or concepts of statistical mechanics [6, 108].

At the molecular level, refined approaches could involve explicit correlations between sidechain fluctuations and proton mobility. More recently, molecular dynamics simulations, utilizing force-field parameterizations [60, 62–64, 83, 84, 109] or ab initio calculations based on a quantum mechanical description, were employed in order to study such correlations and examine direct proton exchange between water of hydration and surface groups or sidechains [110–114].

In real membranes, transport properties of protons and water molecules can be obtained by an effective interpolation between the bulk and the interfacial mechanisms, as explored in [82]. Subsequently, we will consider theoretical approaches along this general scheme, increasing in complexity and covering scales from molecular to macroscopic.

5
Proton Transport Near the Polymer–Water Interface

The confinement of water in nanometer-size pores between the walls of the polymer material affects the structure of water. The observed freezing-point suppression [49, 50] and reduced dielectric constant of water [115–117], are macroscopic manifestations of this effect. The interfacial area between polymer and water provides a complex environment for proton transport. The complications for the theoretical description are caused by the flexibility of the sidechains, their random distributions and their partial penetration into the bulk of water-filled pores. The charged polymer sidechains contribute elastic ("entropic") and electrostatic terms to the free energy [54, 55]. Distribution and mobilities of protons depend strongly on the resulting sidechain–water interactions [54, 56, 118].

In simplest approximation, models of interfacial phenomena in membrane pores factor in static continuous charge distributions on the polymer walls. Continuum dielectric approaches have been utilized to calculate electrostatic effects of charged pore walls on proton distribution and proton mobility. In [107], mean-field Poisson–Boltzmann theory was applied for the model system of a slab-like pore. Fixed anionic charges (SO_3^--groups) were represented as a square lattice array of point charges on the opposite surfaces. The surface charge density distribution determines the depths of attractive potential wells for protons along the array of sulfonic groups. The proton density steeply decreases from locations near the charged surface towards the center of the pore. In this model, strong Coulomb barriers for proton mobility arise in the vicinity of the surface. Electrostatic contributions to the Gibbs free energy of activation of the elementary act of PT in the Pekar–Marcus equation [119],

$$G_a = \frac{(E_r + \Delta G)^2}{4E_r}. \tag{1}$$

with reorganization Gibbs free energy, E_r, and work terms (reaction Gibbs free energy), ΔG, of the elementary act.

In the language of reaction dynamics, E_r is the work needed to move the whole system from the state that corresponds to the initial state of the proton to the state that corresponds to the final state of the proton, but staying on the potential energy surface of the initial state. These potential energy surfaces are multi-dimensional. All modes, including local translational, librational, 'hydrogen-bond-deformational', as well as collective inertial polarization modes contribute to these potential energy surfaces, but with one restriction. Their characteristic frequencies should be smaller than $4kT/h$ [106]. Only then they contribute to the Franck–Condon activation barrier, all faster modes either contribute to tunneling mode pre-factors or slightly affect the effective mass of the proton.

ΔG is the free energy difference between initial and final equilibrium states of the proton; it is zero in the bulk liquid. Near the surface ΔG is dominated by the Coulomb energy profile and therefore it is approximately equal to the difference of the electrostatic potentials ψ at the proton positions before and after the transfer, i.e. $\Delta G \approx e \left(\psi_{\text{final}} - \psi_{\text{initial}} \right)$. This difference depends strongly on the distance of the proton from the surface and thus causes a strong position dependence of proton mobility. Values of G_a were found in the range of 0.5 eV. This value decreases, however, to the bulk water value of the activation energy for proton-surface separations exceeding $\sim$3 Å (the thickness of one monolayer of water). Moreover, the electrostatic activation energy is a strong function of the separation between surface charges. In [107], mean separations of surface charges in the range of 7–20 Å were considered.

In a later work [84] several shortcomings of this approach were addressed. With a more realistic atomic description of the sulfonate groups (charge smearing via a form factor) and its motion (via a Debye–Waller factor), and by considering the finite size of the proton complex (Zundel or Eigen cation), the strong dependence of the activation energy of the proton mobility on the size of the pores was not recovered. Additional MD simulations with a dynamic all-atom model of the sulfonate group, which also include correlations between protons by the very nature of the MD simulation technique, support this finding. The large activation energies observable in dry membranes thus appear to be less related to the properties of a single pore but more to polymer dynamics.

As corroborated above the effect of arrangement and fluctuations of charged surface groups on proton transport in membranes is the more pronounced the smaller the water content in the membrane is. Squeezing the pore leaves no room for bulk-like transfer. Upon deswelling of pores, sidechain separations are likely to decrease [33]. The proton concentration, surface charge density and corresponding electrostatic interactions will thus increase, and explicit sidechain-sidechain correlations will become more pronounced. The process will lead to with an array of micelles loosely connected by ultrathin aqueous necks, as described in [31].

Other experimental data for membranes with varying IEC suggest that mean sidechain separations in the dry state should not drop below $\sim$7 Å, since otherwise random ionomers could not form a stable phase-separated structure upon water uptake. At allowed separations, direct proton transfer between stiffly fixed SO_3^- groups is still not possible. This situation could change, however, when random configurations of sidechains, sidechain flexibility, and fine structure and smearing of charges of sulfonate groups are taken into consideration [84].

Under realistic conditions in lowly hydrated PEMs, sidechain fluctuations could trigger concerted proton transport mechanisms. A trifluoromethane sulfonic acid monohydrate (TAM) solid was explored in [112] as a basic model system for studying mechanisms of proton transfer, that are relevant to a regime of minimal water content and high concentration of anionic groups in PEMs. The regular structure of the crystal [120] provides a proper basis for performing controlled ab initio molecular dynamics. The unit cell consists of 48 atoms ($4\times[CF_3SO_3^-H_3O^+]$). The crystal melts into a "room temperature" ionic liquid at 309 K and ordinary pressures. A protocol of geometry optimizations and MD production runs was performed. The experimental crystal structure was found to be stable during a simulation time of >40 ps. All four acid protons in the unit cell reside on H_3O^+ moieties, each stabilized by three hydrogen-bonds with neighboring sulfonate oxygens, similar to the localized proton state in $H_9O_4^+$ complexes in bulk water.

In seeking activated states relevant to proton conduction, a proton-hole defect was introduced intermittently and, thereafter, restored. This protocol led to the formation of a metastable intermediate state with two delocalized protons, as depicted in Fig. 4. One proton is delocalized between the two oxygen atoms in a Zundel-ion ($H_5O_2^+$). Two neighboring sulfonate groups re-arrange in order to accommodate a second proton between them. Overall, the reorganization occurs within $\sim$2 ps of the creation of the defect. It was observed

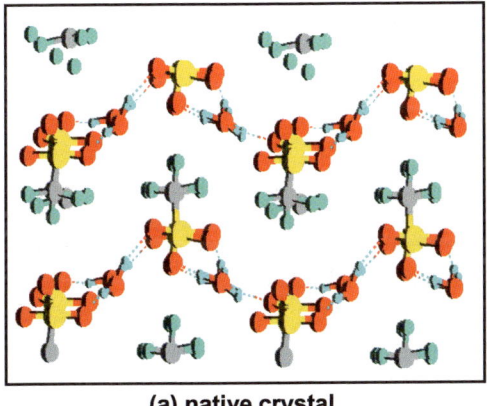
(a) native crystal

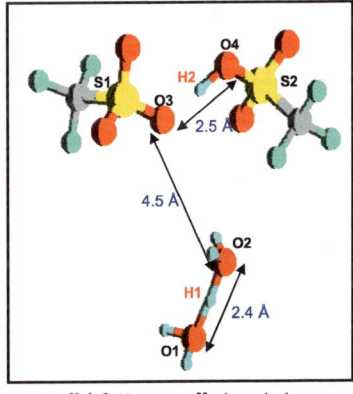
(b) Intermediate state

Fig. 4 Ab initio molecular dynamics simulation of a triflic acid monohydrate crystal. The intermediate state (*right*) with two delocalized protons is $\sim$0.3 eV higher in energy than the ordered conformation of the native crystal (*left*) [112]

that the Zundel-ion drifts between the sulfonate sites. It, thereby could act as a relay group that mediates effective proton transfer in the system.

The free energy of formation of the defect state was calculated to be ~ 0.25–0.3 eV [112]. This value agrees reasonably with activation energies of proton conductivity in the TAM solid determined from experimental data [121]. It is moreover in the range of activation energies of proton conductivity in Nafion 117 under conditions of minimal humidity, ~ 0.36 eV, reported in [50]. The sequence of formation and destabilization of such defects could, thus, establish a mechanism of long-range proton transport along densely packed arrays of anions.

Notably, the outlined defect mechanism of proton transfer resembles the protocol of proton transfer in water, viz. transformation between localized and delocalized proton states (water: $H_9O_4^+ \leftrightarrow H_5O_2^+$, TAM solid: stable crystal configuration $\leftrightarrow$ metastable intermediate state), triggered by hydrogen-bond fluctuations and molecular rotations. Sequences of these transformations, hydrogen-bond rotations and proton transfers generate the net proton motion.

The unconstrained dynamics of anionic groups in the TAM solid enables the substantial local reorganizations by which the intermediate state could form. This indicates that an appropriate flexibility of anionic sidechains could be vital for high proton mobility in PEM under conditions of minimal hydration and high anion density. This was also demonstrated independently by MD simulations on the basis of an empirical valence bond model [83] (see below).

The merit of the approach [112] is that it applies high-level ab initio molecular dynamics simulations to an insightful and experimentally feasible model system. Such studies could help to establish basic mechanisms of proton transport in PEMs, involving strong correlations between protons and charged polymer groups in the interfacial layer. There are however several limitations to this approach. Because of the disparity of the fs-scale of AIMD simulations and the time interval between consecutive proton transfer events ($\gg 1$ ps) a meaningful statistical sampling of those events cannot be obtained. Moreover, the considered model system is rather stiff, in the sense that chemical composition and water content of the crystal are fixed. Only minimal hydration of the system could be considered.

Currently, work is being performed on a model system that consists of a 2D regular hexagonal array of acidic surface groups as depicted in Fig. 5 [122]. For this system, chemical composition, lengths and separations of sidechains could be varied. Optimization studies of such layers based on DFT calculations (using Vienna Ab initio Simulations Package, VASP [123–127]) provide the energy of formation as a function of the packing density and chemical structure of the acid-functionalized sidechains. Systematic studies reveal a wealth of structural conformations and transitions between them under conditions of minimal hydration, as shown in Fig. 5. Upon increasing the separation of fixed sidechain endpoints (surface group separations d_{CC}), the system proceeds through a series of transitions from fully dissociated min-

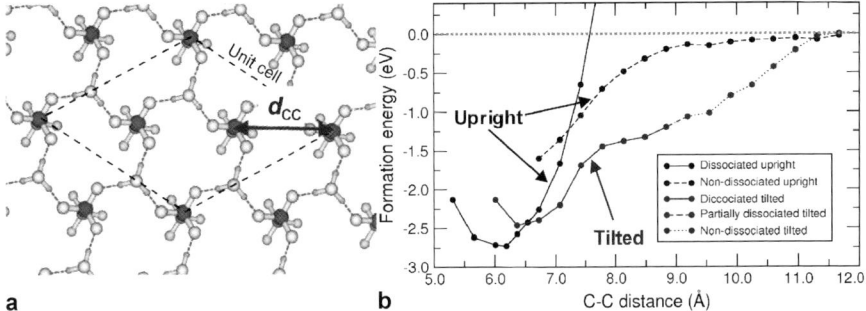

Fig. 5 Regular hexagonal array of surface groups with fixed endpoints (unit cell: $3\times[F_3C-SO_3H+H_2O]$). On the *left* the most stable conformation of the minimally hydrated array is shown at a high packing density of surface groups ($d_{CC}\approx 7$ Å). The figure on the *right* shows the formation energy per unit cell. It reveals the transition from the stable upright conformation with long-range interfacial correlations to a clustered conformation with tilting of surface groups towards each other in order to maximize the number of strong hydrogen bonds [122]

imum energy conformations for $d_{CC} < 7.7$ Å (Fig. 5a) via partially dissociated conformations to fully non-dissociated structures for $d_{CC} > 8.6$ Å. In dissociated conformations the H_3O^+ are fully immersed in the interfacial layer.

At high density, $d_{CC} < 6.7$ Å, ionized SGs and H_3O^+ ions form an ordered "upright" conformation with full dissociation of all acid groups. At $d_{CC} > 6.7$ Å, we observed cluster-like "tilted" conformations. The conformational transition at $d_{CC} \simeq 6.7$ Å, apparent in Fig. 5b, is accompanied by a sharp transition from weak (<0.1 eV) to strong binding (>0.6 eV) of additional H_2O upon increasing d_{CC}. In fact, these results suggest that the highly charged, minimally hydrated interface becomes hydrophobic at values $d_{CC} < 6.7$ Å. This intriguing effect is due to strong interfacial correlations and the trigonal symmetry of H_3O^+ and ionized SGs (SO_3^- head groups). It is interesting to note in this context, that $d_{CC}\approx 7$ Å has been identified in experiment as a critical value for the occurrence of long-range proton conduction at lipid and stearic acid monolayers [96, 128, 129]. According to the theoretical results, this value represents a favorable trade-off between long-range correlations and flexibility of SGs, which is a prerequisite for long-range proton transport [82]. For longer SGs, resembling the sidechains in real PEM, 2D correlations and partial dissociation could persist up to $d_{CC} \simeq 15$ Å [130]. Current work explores proton transfer events in such brushes of acidic surface groups.

The MD simulations of proton transport in model pores employing empirical valence bond interaction models described the pore surface in a similar fashion as a regular array of static sulfonate groups, rotationally mobile or tethered sulfonate groups, or as entire Nafion sidechains [83, 84].

6
From Meso-to-Macroscale: Effective Transport Properties

Understanding the effects of membrane structure, water content and water distribution on proton conductivity has to invoke additional theoretical tools. They have to bridge many length scales from molecular environments to random heterogeneous structure at macroscale. This involves phenomenological concepts and homogenization methods in order to average over microscopic details [65].

The conductivity of a single pore is determined by the charge density of protons ρ^+ and the proton mobility μ^+. As discussed above, ρ^+ and μ^+ are functions of the position within the pore. In highly hydrated PEMs, μ^+ is highest in the bulk and smallest close to the surface. The proton distribution from simple Poisson–Boltzmann theory decreases monotonously from the interface towards the pore center. Refined calculations of the proton distribution take into account the finite size of protonated complexes and a repulsive part of the intermolecular potential near the pore surface. This modified PB approach predicts an excluded region for the hydrated protons close to the surface (mainly related to the finite size of proton complexes). A maximum in ρ^+ exists at about 1.5 Å away from the interface. From this position ρ^+ decreases continuously towards the center of the pore. This density profile is in reasonable agreement with results from MD simulations [84]. Further possible refinements, such as incorporating variations of dielectric constant with position within the pore and with pore size, were not included in these calculations.

In the absence of correlation effects in proton transport, the conductance of a single membrane pore can be calculated straightforwardly using the distributions of proton density and mobilities. The conductance of a slab-like pore is given by

$$\Sigma_p = \frac{L_x}{L} \int_{-z_0}^{z_0} dz \mu^+(z) \rho^+(z) , \qquad (2)$$

where L is the length of the pore and $\pm z_0$ denote the positions of the interfacial layers. The corresponding expression for a cylindrical pore of radius R is

$$\Sigma_p = \frac{2\pi}{L} \int_0^R r \, dr \mu^+(r) \rho^+(r) . \qquad (3)$$

The mobility in the pore includes molecular mechanisms of proton transport in bulk water and along the array of charged surface groups. An idealized two-state approach based on this distinction was considered in [82]. This simple model can reproduce a continuous transition from surface-like to bulk-like proton conductance upon increasing the water content in a pore. In calculations of pore conductances it was taken into account that the average separation between SO_3^- surface groups varies with increasing pore size and that the dielectric constant is a function of pore size. The latter effect arises due to the confinement of water in nm pores and the proximity to hydrophobic pore walls. On average, the number of hydrogen bonds per water molecule decreases below the value for bulk water. The remaining H-bonds have increased strengths. The reduced orientational flexibility in the stiffer H-bond network offers more resistance to water reorganization resulting in a decrease of the dielectric constant [131–133].

In the Poisson–Boltzmann approach, proton density and strength of electrostatic interactions of protons with surface groups determine the water content at which the transition between surface and bulk mechanisms occurs and the extent of the effect on activation energies. The earlier work in [82, 107] exaggerated the electrostatic contributions to these effects due to oversimplified model assumptions (point like surface charges, no fine structure and no flexibility). Pronounced effects of pore size on activation energies of proton transfer and pore conductances were found in good agreement with experimental conductivity data [50].

The refined Poisson–Boltzmann theory in [84], based on a more realistic electrostatic model of surface charges, showed activation energies that were almost independent of pore sizes. While results of this modified electrostatic theory were consistent with findings of classical MD simulations, they could not explain the experimentally observed increase of activation energies from 0.10–0.13 eV in "wet" PEMs to 0.22–0.36 eV in "dry" PEMs. Seemingly, the transition from bulk to surface mechanism of proton transport upon pore deswelling is not simply due to a rescaling of electrostatic interactions but it involves more dramatic reorganizations of the interfacial conformations which should invoke full ab initio studies. In this context, ab initio study of the TAM solid [112] and more recent ab initio simulations for a minimally hydrated interfacial layer of acidic surface groups [122] could provide insight into relevant correlations and proton transport mechanisms in the interfacial layer at molecular scale.

MD simulations based on empirical interaction functions are able to overcome some of the statistical limitations of the ab initio MD scheme. It is possible, at least for single pore environments to calculate proton mobilities in a statistically accurate way. The chemical nature of proton transfer, i.e., the structural diffusion from one hydrated cluster to the next, is efficiently taken into account by the use of empirical valence bond (EVB) models, which have been introduced by Warshel [134] and later extensively used for aqueous proton transport in bulk water by Vuilleuimier and Borgis [92, 135, 136] and the group of Voth [93, 137–141]. In the simplest version of such a model, a two-state EVB model [102], the proton can be regarded as being in a superposition state between two different valence bond states, the first one corresponding to $H_3O^+...H_2O$, and the second one corresponding to $H_2O...H_3O^+$. In the first valence bond state, the proton is localized on the first water molecule, in the second one it is located on the second molecule. Using a configuration-dependent coupling between those zero-order states, one can describe proton transfer as a sequence of $H_5O_2^+ \rightarrow H_3O^+ \rightarrow H_5O_2^+ \rightarrow H_3O^+$ transitions, where the H_3O^+ forms the central ion of an Eigen complex $H_9O_4^+$.

This EVB model was employed to the systematic study of model pores or channels. Taking a simplistic view, the polymer was regarded as a rigid framework in which slab or cylinder pores of constant thickness or radius, respectively, are formed. Within this approach, proton transport in pores has been studied as a function of a variety of generic structural and dynamical features of the polymer and operational parameters of the working fuel cell (such as temperature and humidity). These studies revealed a number of factors determining the proton mobility, such as the width of the channel, distance between the sidechains, and their flexibility. The main lessons of the simulations and the theoretical analysis were:

- The denser the sidechains, the lower the activation energy of proton mobility.
- The narrower the channels, the lower the mobility, and, in principle the higher the activation energy. However, MD simulations as well as PB simulations that allow for Debye–Waller fluctuations of SO_3^- groups showed only a minor increase in activation energy for the single pore in the range of nanometer pores, supporting the idea that much narrower water channels presumably control the proton transport at low water content.
- The more movable are the SO_3^- groups about the sidechain anchoring points, the higher the proton mobility, but the weaker are the previous two effects.
- Longer, flexible sidechains facilitating proton conductance, impede methanol diffusion [142]. This could be a very practical lesson.

The conclusions reflect certain physics. Whereas each sidechain group donates a proton to the aqueous phase, it also contributes to the Coulomb energy landscape which protons experience due to the electrostatic potential of the array of SO_3^- groups, screened by all other protons. The farther a proton moves from the surfaces of SO_3^- groups, the stronger the screening, and the smaller the role of non-zero harmonics in a Fourier expansion of the field, which decay exponentially from the plane where the charge groups are localized. This correlation between flat density distribution and overall large proton mobility in individual pores is clearly visible in the simulation results [109, 143]. In a wide channel there are such remote regions. Because of the attraction of protons to the sidechains, there will be less protons in those regions, but the mobility will be higher there. The overall channel conductance, a trade-off between mobility and concentration, will thus be higher in a wider channel. In narrow channels there will be no such remote regions.

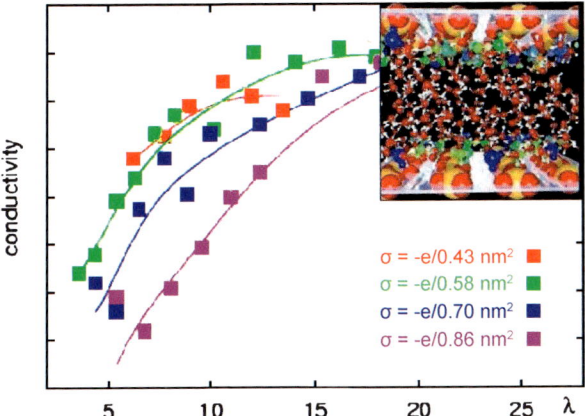

Fig. 6 Slab pore conductance as a function of water content λ for various densities of sulfonate groups indicated in terms of surface charge density σ

Figure 6 demonstrates that the total conductance of a single pore (modeled as the product of mobility and proton density in the pore) increases with increasing water content and also with increasing surface density of sulfonate groups. The conductivity increase is due to the increasing proton concentration in the pores with higher surface charge density. The self diffusion coefficient of protons actually *decreases* in the simulations with increasing surface charge density. Figure 7 (top) shows the simulated proton densities inside 17 Å wide slab pores, with sulfonate charge densities on the pore surface as indicated. With increasing surface charge density the proton density distribution shifts towards the pore surface and becomes narrower, in accordance with the expectations of simple Gouy–Chapman theory. For the narrower proton distributions repulsive proton–proton interactions increase. The net result is that the proton mobility decreases, contrary to the PB results which predict lower Coulomb barriers with increasing sulfonate density. The effect is, however, overcompensated by the increase in proton concentration in the more densely charged pores.

Figure 7 also shows in the bottom a comparison of the proton densities obtained by MD simulations, the simple PB approach and the modified PB approach of [84]. MD results show stronger proton-sulfonate correlations than the PB results. Taking into account the finite size of the proton cluster in the MPB approach leads to density distributions closer to the MD results. Other features of the MPB approach such as replacing point charges by extended charge distribution and accounting for thermal motion lead to an additional slight broadening of the density maximum relative to the PB results.

The temperature dependence of the proton mobility in these model pores showed Arrhenius behavior over the investigated range of temperatures

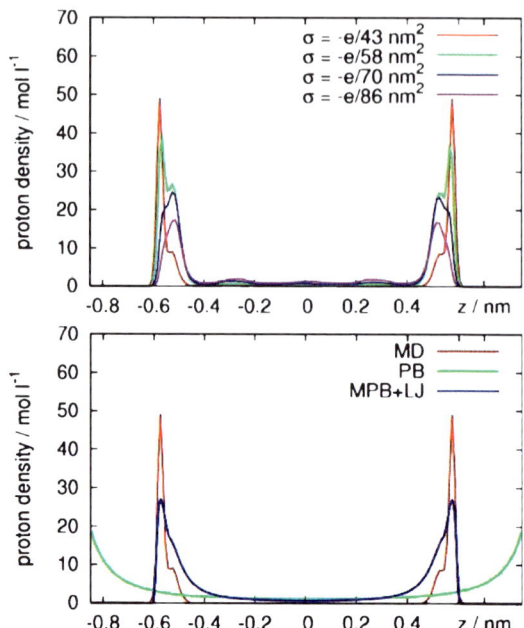

Fig. 7 *Top*: Proton density distributions in 17 Å wide slab pores with varying sulfonate surface charge densities as indicated. Bottom: Proton density distributions in the 17 Å wide slab pore with $\sigma = -e/0.70\,\text{nm}^2$ from MD, PB, and MPB approach

between 330 and 400 K. Calculated activation energies were in the range 0.1–0.15 eV and did not depend on humidity in rigid pores. They could thus not explain the experimental data in Fig. 2. Only a very small (significantly smaller than the experimental) trend of the activation energies to become higher at lower water content was observed in the case of pores with flexible pore surfaces. This nurtures speculation that at low humidity the Cappadonia results could be the consequence of the very narrow necks in the aqueous network and a dominant role of polymer dynamics on both the formation of these necks and the proton transport through the necks. In other words, at low humidities, the model of elongated pores as the basic proton conducting unit is likely to be inapt.

A possible alternative realization of such a process was given in [31] as considered above (see Fig. 1). The joining of pores as discussed above (Fig. 1) may explain the experimental results more naturally. Obviously, swelling or deswelling are not uniformly occurring processes in the membrane. A single pore cannot represent the whole network. It could just be a small number of critical pore connections that control the overall proton transport, e.g., via the dynamic formation of temporary bridges between disconnected aqueous regions, as suggested in [53], or that cause the apparent transition from bulk to interfacial mechanisms of proton transport. It is, thus, important to understand, how the conductance of single pores can be converted into the effective conductivity of a membrane.

Several other attempts have been made to model the humidified Nafion nano-phase-separated structure and the temperature dependence of proton transport by atomistic MD simulations [53, 59–64]. It was observed that more filamentous aqueous regions at low humidity change into clusters of more micellar shape at intermediate water content, which connect into channels at high water content [60]. Other studies noted a certain effect of sidechain arrangement (statistical vs. blocks) on the size of the phase-separated regions [59]. These calculations frequently suffer from an ergodicity problem due to the different characteristic time scales of water and polymer.

7
Macroscopic Network Models

Effective conductivity of the membrane is related to its macroscopic morphology, viz. the random heterogeneous domain structure of polymer and solvent phases. On the basis of Gierke's cluster network model, a random network model of microporous PEMs was developed in [22]. This approach highlighted the importance of connectivity and swelling properties of pores. Random distributions of pores and channels as their interconnections were assumed. The connectivity between pores was considered as a phenomenological parameter.

In general, pores swell non-uniformly. As a simplification, the random network was assumed to consist of two types of pores. In this two-state model, non-swollen pores ("red" pores) permit only a small residual conductance due to tightly bound water molecules solvating the surface groups. Swelling pores ("blue" pores) contain extra water in the bulk and thereby promote the high bulk-like conductance. Water uptake by the membrane is accomplished by the swelling of "blue" pores and by the increase of their relative fraction. Proton transport in the membrane is mapped on a site percolation problem, wherein randomly distributed sites represent pores of variable sizes and, thus, conductances. The assignment of pores of different color corresponds to distinct proton transport mechanisms, dominated by interfacial processes or by bulk water-like transport. Water uptake by "blue" pores promotes the transition between these mechanisms, as corroborated in the previous subsections. Chemical structure of the membrane is factored in at these subordinate structural levels, i.e. the molecular scale and the single pore level.

The relative portion of "blue" pores as a function of the water content w is

$$x(w) = \frac{\text{number of blue pores}}{\text{total number of pores}}.$$

It determines bond probabilities

$$p_{bb}(w) = x(w)^2, \quad p_{br}(w) = 2x(w)\left(1 - x(w)\right), \quad p_{rr} = \left(1 - x(w)\right)^2$$

between two "blue" pores, a "blue" and a "red" pore and two "red" pores, respectively. Channels connecting pores are assumed to be "red" (connecting two "blue" pores).

The model involves a phenomenological law of pore swelling. It utilizes Gierke's experimental data for the structural reorganization of the membrane upon water uptake. The following dependencies of the number of SO_3^- groups in an average pore, $n(w)$, and of the average volume of water-filled pores, $v(w)$ were used

$$n(w) = n_0(1 + \alpha w) \quad \text{and} \quad v(w) = v_0(1 + \beta w)^3, \tag{4}$$

where n_0 is the number of SO_3^- groups in a dry ("red") pore and v_0 the volume of a red pore. Invoking the conservation of the total number of dissociated SO_3^- groups in the membrane and the proportionality between total water content and the volume increase of "blue" pores, the swelling law could be derived

$$x(w) = \frac{\gamma w}{(1 + \beta w)^3 - \alpha \gamma w^2}. \tag{5}$$

The parameters α, β, and γ can be adjusted in order to reproduce the amount of swelling and the extent of reorganization of the membrane upon water uptake. This law accounts for the possibility of merging of smaller pores into larger pores upon swelling. It could represent distinct elasticity of the polymeric membrane matrix that leads to distinct types of water distributions. In a soft polymer matrix, water would be distributed rather heterogeneously with individual pores swelling to a large equilibrium radius and thus taking up a lot of water. In a more elastic polymer matrix, pores swell more homogeneously.

A rigid microporous morphology, which does not reorganize upon water uptake, corresponds to a simple linear relation $x(w) = \gamma w$. In this limiting case, the model resembles the archetypal problem of percolation in bi-continuous random media [105]. Because of swelling, universal percolation exponents in relations between conductivity and water content are not warranted.

Pore-size dependent conductances are assigned to individual pores and channels, determined by the overall water content and the law of swelling. Three possible types of bonds between pores exist. The corresponding bond conductances, viz. $\sigma_{bb}(w)$, $\sigma_{br}(w)$, and $\sigma_{rr}(w)$, can be established straightforwardly. The model was extended towards calculation of the complex admittance of the membrane by assigning capacitances in parallel to conductances to individual pores.

The simplest approach to solve the Kirchhoff equations that correspond to the random network of conductance elements is obtained in single-bond effective medium approximation (SB-EMA), wherein a single effective bond between two pores is considered in an effective medium of surrounding bonds. The effective bond conductivity is obtained from

$$p_{bb}(w)\frac{\sigma - \sigma_{bb}(w)}{\sigma_{bb}(w) + q\sigma} + p_{br}(w)\frac{\sigma - \sigma_{br}(w)}{\sigma_{br}(w) + q\sigma} + p_{rr}(w)\frac{\sigma - \sigma_{rr}(w)}{\sigma_{rr}(w) + q\sigma} = 0, \quad (6)$$

where q represents the connectivity of the pore network.

If the conductivities of red pores and channels vanish, the true percolation behavior is obtained,

$$\sigma(w) = \frac{1}{q}\left[(1 + q)x^2(w) - 1\right]\sigma_{bb}(w), \quad (7)$$

with percolation threshold

$$x_c = \sqrt{\frac{1}{1+q}}. \quad (8)$$

A value $q = 24$ reproduces well the quasi-percolation behavior of EW 1100 Nafion.

This random network theory could explain differences in $\sigma(w)$ relations for various sulfonated ionomer membranes [22]. It is useful in rationalizing the effects of membrane elasticity and swelling behavior on performance under varying degrees of hydration. The EMA solution of the random network model correctly reproduces the percolation behavior observed in Nafion-type membranes and Nafion-composite membranes. A highly elastic membrane matrix and a high connectivity of pores provide small values of the percolation threshold and, thus, favorable $\sigma(w)$ relations. In a soft polymer matrix, on the other hand, the fraction of water-filled pores $x(w)$ increases slowly at low to intermediate water contents, indicating that pores swell rather heterogeneously. In this respect Nafion membranes of equivalent weight 1100, seemingly possesses a favorable elasticity of the polymer matrix.

In [82] different model variants of pore-space evolution (random network, serial and parallel pore models) were compared to each other. A morphology of equally swelling parallel pores gives the most favorable $\sigma(w)$ relations with steepest increase of proton conductivity at small water contents. Results obtained for such a morphology are in good agreement with conductivity data of Dow membranes, which possess shorter pendant sidechains than Nafion.

8
Membrane Operation in the Cell

After all, any well-performing membrane should be a good water retainer. The excellent proton mobility in currently used PEMs depends, however, in particular on the existence of a network of loosely bound water molecules. Because of their weak interactions with the polymer and acidic residues, those water molecules are highly mobile themselves, and they easily evaporate at temperatures approaching 100 °C.

Under ideal operation of a PEM in a Polymer Electrolyte Fuel Cell the membrane would perform like a linear Ohmic resistance. This implies that the membrane is uniformly hydrated at the value of water content that gives the highest proton conductivity. Because of the strongly non-linear coupling between proton and water fluxes inside the membrane, caused primarily by the so-called electro-osmotic effect, this ideal performance is observed only in the limit of small current densities. The electro-osmotic coefficient n, which under usual conditions lies between one and two [6], is sufficient to cause a dehydration zone near the anode, if currents are high, membrane thickness L is large enough, and no special water management measures are taken. Non-uniformity of the water distribution near the anode will increase with current density. Deviations of the membrane's current-voltage characteristics from the simple Ohmic behavior result in a critical current density at which the non-linear increase of the membrane resistance could cause the failure of the complete cell. The origin of this dramatic effect is very low proton conductivities in locally dehydrated membrane parts on the anode side of the membrane. As theory shows [11] the critical current density scales as $\propto n^{-1}L^{-1}$.

Even if such dramatic effects do not occur in state-of-the-art fuel cells under feasible operating conditions, the membrane is pivotal in regulating water fluxes through the complete cell. It controls the amounts of water that leave the cell through anode and cathode compartments and determines the operating conditions that should be provided in electrodes. Throughout the recent history of Polymer Electrolyte Fuel Cells, water management has been one of the major issues for materials design and system optimization, including all components [6, 11–17, 144–147].

Membrane operation in the fuel cell is affected by structural characteristics and detailed microscopic mechanisms or proton transport, discussed above. However, at the level of macroscopic membrane performance in an operating fuel cell with fluxes of protons and water, only phenomenological approaches are feasible. Essentially, in this context, the membrane is considered as an effective, macrohomogeneous medium. All structures and processes are averaged over micro-to-mesoscopic domains, referred to as representative elementary volume elements (REVs). At the same time, these REVs are small compared to membrane thickness so that non-uniform distributions of water content and proton conductivities across the membrane could be studied.

The basic mechanisms of membrane operation in an operating fuel cell are shown in Fig. 8. Proton flow, as the primary membrane process, induces water transport from anode to cathode by electro-osmosis. Stationary operation implies that the electro-osmotic water flux has to be balanced by an internal backflux. This requires an appropriate gradient in water content across the membrane as a driving force. The stationary balance between electro-osmotic flux and backflux, thus, establishes to a profile of water across the membrane,

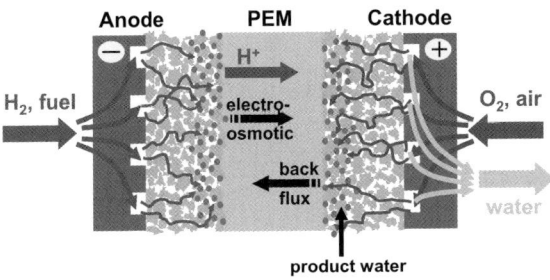

Fig. 8 The membrane inside the PEFC configuration. Proton current, electro-osmotic drag and backflux are indicated.

decreasing from cathode toward anode. The balance of water fluxes through the membrane has to be consistent with water fluxes prevailing in anode and cathode, taking into consideration also the water produced in electrochemical reactions at the cathode. It, moreover, is subject to gas pressures in both electrode compartments.

Modeling approaches that explore membrane water management have been reviewed in [16]. Overall, the complex coupling between proton and water mobility at microscopic scale is replaced by a continuum description involving electro-osmotic drag, proton conductivity and water transport by diffusion or hydraulic permeation. Essential components in every model are the two balance equations for proton flux (Ohm's law) and for the net water flux. Since local proton concentration is constant due to local electroneutrality of the membrane, only one variable remains that has to be solved for, the local water content.

Why has this seemingly simple and straightforward problem led to so many different approaches and ambiguous results [16]? The required effective transport coefficients (proton conductivity, electro-osmotic drag coefficient, diffusivity and permeability of water) are functions of the water content. Since it is impossible to calculate them from the first principles, they have to be determined from experiments (see discussion of Figs. 15 and 17 in [6]). Because of the dependence of these coefficients on water content, the system of equations becomes highly non-linear. The ambiguity in models of water management is in part due to the difficulties in determining electro-osmotic drag coefficients, diffusivity and permeability independently. Moreover, the unclear situation with respect to the membrane microstructure and the structure of water inside the membrane has led to two major types of models, viz. diffusion and hydraulic permeation models.

In simplest approximation the membrane is considered as a continuous non-porous phase in which the water of hydration is dissolved. The prevailing mode of water transport is, therefore, molecular diffusion [145]. In view of what has been written in previous sections about water pathways in a network of pores, the corresponding so-called diffusion approaches evidently overestimate the role of polymer-water correlations in well-hydrated membranes.

Structural models, on the other hand, emerge from the notion of the membrane as a heterogeneous porous medium, which is supported by a wealth of experimental data on structure and mobility. They focus on the role of percolating water networks as the active medium inside the membrane. This view clearly promotes hydraulic permeation as the main mechanism of water backflux. The relative humidity primarily controls the capillary pressure, which determines the water-filling of the membrane. Gradients in capillary pressure, related to gradients in pore radii, are, thus, the primary internal driving forces of water backflux in the membrane [11]. An additional external gas pressure gradient between cathode and anode sides may be superimposed on these internal gradients. Knowledge of the pore size distribution and of the wetting properties of pores are evidently of key importance in these hydraulic permeation models. It was found for instance, that the critical current density of membrane dehydration is directly proportional to the first moment r_1 of the pore size distribution [11, 16]. Using for mathematical simplicity a δ-function-like pore size distribution in the model, i.e. an idealized monodisperse distribution that exhibits a sharp peak at r_1, gives the following explicit dependence of the critical current density

$$j_{pc} \sim \frac{(w_s - w_c) r_1}{nL},$$

where w_s is the saturation water content and w_c is the percolation water content.

The hydraulic permeation model is appropriate for well-hydrated membranes. However, it cannot appropriately describe water transport in lowly hydrated membranes since it underestimates polymer-water correlations. Both model variants are mathematically similar and complementary in their range of applicability. Indeed, it is a straightforward task to merge them into a unified approach, as suggested in [11, 16, 144, 147].

The major membrane performance guidelines are as follows. During fuel cell operation a non-uniform water distribution emerges inside PEMs. Strong dehydration arises in the interfacial regions close to the anode, whereas the other parts of the membrane remain in a well-hydrated state, as shown in Fig. 9. The type of highly non-uniform water distribution, predicted in the hydraulic permeation model, is consistent with experimental data for water profiles and membrane resistance as a function of current density [148–150]. Estimated critical current densities at which the membrane resistance increases strongly due to local dehydration, are typically not reached for standard Nafion-type membranes in the feasible range of fuel cell operation; however this is not necessarily so for poorer membranes. Although we will not reach this limit in top performing membranes, the signature of this effect is felt earlier. Electro-osmotic coefficients themselves may depend on water

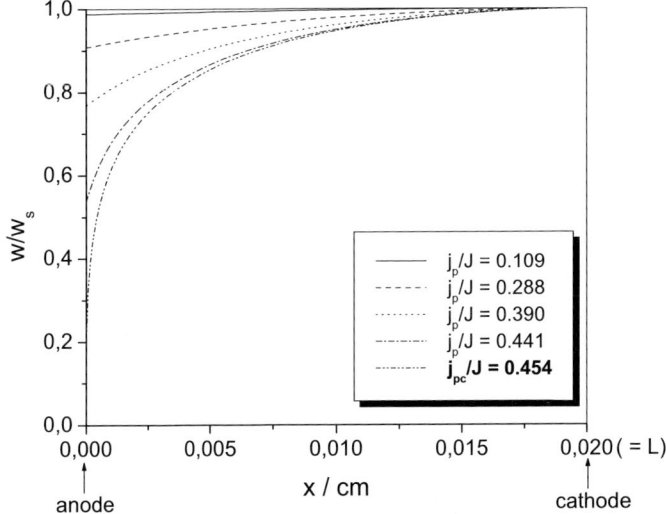

Fig. 9 Water content profiles in the membrane, calculated in the hydraulic permeation model, at various fuel-cell current densities. A typical value of the parameter J that determines the onset of membrane dehydration near the anode was estimated as $J \sim 5\text{--}10 \text{ A cm}^{-2}$ for Nafion 117 [11, 16]

content [6]; being typically smaller for smaller water content [6] they somewhat "smoothen" the dehydration effects. Generally membranes with higher water uptake are less prone to dehydration. The usage of thinner membranes is also beneficial (unless they rupture—one of the reasons for cell degradation), since, as mentioned, the critical current density increases inversely proportional to thickness. Two modes of membrane water management could be identified. Either the water removed by electro-osmosis has to be replenished by a matching water flux at the anode or a gas pressure gradient should be applied between cathode and anode, that supports the internal hydraulic mechanism of water backflow.

9
Concluding Remarks

The majority of "solid" proton-conducting membranes, most commonly used in contemporary fuel cell technology, are hydrated perfluorinated sulfonic acid ionomers. In recent years, enormous programs in membrane research have explored empirically how various modifications of the benchmark material, viz. Nafion, affect the physical membrane properties. The main modifications include (1) varying the hydrophobic/hydrophilic composition of the polymer, (2) controlling the grafting density and lengths of the sidechains,

(3) exchanging the acid functional groups, (4) incorporating inorganic particles into the membrane, (5) blending of ionomer with other stabilizing polymers, (6) replacing water by other anhydrous, possibly immobilized, solvents. A plethora of effects of these modifications on membrane morphology, stability, swelling behavior with water uptake, transport of protons, water and methanol, have been demonstrated.

Looking back, the only unequivocal membrane improvement, in spite of all these efforts, has been the reduction of thickness from ~200 μm in 1995 to <50 μm in 2005. In terms of chemical or morphological modifications at the microstructural level, no definite recommendations could be discerned so far. The focus of the works reviewed herein has been exploring the fundamental relations between micromorphology and transport from micro- to macroscales for prototypical polymer electrolyte membranes and the understanding of their major principles of operation.

Understanding the performance of these state-of-the-art membranes on the microscopic level has progressed quite a lot during the last two decades due to a combination of systematic experimental and theoretical studies. As discussed herein, the understanding of membrane performance in the well-humidified state seems to have been established. Overall, the picture of the effects which control proton and water motion, namely the formation of a percolating aqueous subphase in the swollen membrane in conjunction with microscopic transport mechanisms is quite obvious here.

The nature of the bottlenecks for proton conductance in the dry membrane state or "on the way to it" is, however, still the subject of debates. This will only be resolved after more detailed experimental studies (of macroscopic transport parameters such as proton conductance and electro-osmotic coefficients as a function of water content, or gas and liquid permeability before and after operation, and of microscopic structural probes such as small-angle neutron and X-ray scattering) will have discriminated between competing models. By and large, the direction of effects that go with dehydration is obvious enough to be introduced into phenomenological models of overall cell performance.

The main challenge for developing the theory of membrane operation can be easily illustrated. Proton transfer as the main membrane process is largely determined by molecular fluctuations on an fs time scale. On the other hand, membrane stability, which is a key benchmark of fuel cell operation, has to be demonstrated for 1000s of operating hours. This spans more than 21 orders of magnitude. Approximation schemes a la "Born–Oppenheimer" justify decoupling processes at different scales of resolution and studying them separately. However, when it comes to membrane operation and optimized design, we need to consider a coherent picture that incorporates all processes at different scales.

In the major part of this article, we have dwelt on different proton transport mechanisms. The two limiting views consider proton transport either as a bulk-like mechanism in well-hydrated membranes or as an interfacial phenomenon in lowly hydrated membranes. The transition from bulk-like proton transport to proton transport along the minimally hydrated array of charged surface groups upon dehydration could be triggered by a morphological transition of the membrane microstructure that encompasses the emergence of extremely narrow necks. We then saw how this dual mechanism scenario could be incorporated into a random network model of membrane conductivity. This model could, moreover, account for the specific swelling behavior of membranes due to distinct polymer elasticity. It manages to rationalize the percolation transition observed upon membrane dehydration as well as the saturation of conductivity at high water contents. It can explain conductivity relations of prototypical PEMs.

At the level of membrane operation in the fuel cell, considering bulk-like water as the active medium in well-hydrated membranes implies that hydraulic permeation should be regarded as the major mode of water back transport that counterbalances the electro-osmotic drag. This balance between electro-osmotic drag and water backflux determines the degree of membrane dehydration at the interface between membrane and anode as well as the critical fuel cell current density at which voltage losses in the membrane increase dramatically, i.e. in a highly non-ohmic fashion.

Overall, the distinction of different aqueous environments is a common thread for explaining membrane operation at distinct relevant scales (microscopic mechanisms of PT, conductance in the single pore environment, random network model of membrane conductivity and membrane operation in the cell).

The membrane is a highly non-linear medium; water uptake is a non-linear phenomenon, as water transport in the dry state is much different from that in the wet state. The three-dimensional water and current distribution in a fuel cell is highly inhomogeneous [151]. For instance, the existence of "torches" of current in front of the edges of the channels of bipolar plates as seen in quasi-3D simulations [152] (and avoidable by alternative designs [153]) suggests a highly inhomogeneous release of heat, leading first to extreme local dehydration and then possibly pinhole formation in membranes. Membrane rupture, a likely source of performance degradation, is thus ultimately affected by the details of atomic transport. Microscopic theory can only rationalize these effects. However, incorporating them into phenomenological models of cell performance can lead to the observation of new effects with huge practical relevance.

The most challenging task, to suggest chemical modifications of membranes, which minimize electro-osmosis and/or methanol diffusion, maximize proton conductance and membrane stability at a reduced price is only indirectly affected by theory and simulation. Most conclusions, once made, are rather obvious (such as higher channel connectivity, homogeneity of swelling, higher sidechain density, etc). Others are less trivial, such as the facilitation of proton transport by flexibility and charge smearing, or the opposite directions of the effect of sidechain length on proton and methanol mobility.

Today, the development of membranes moves rapidly away from simple membranes, which are amenable to such ideal model constructions, to composite materials, in which not one compound serves as a panacea for all fuel cell illnesses, but where different functions are assigned to different chemical compounds. Nevertheless, even in these more complex systems the same issues such as water structure, state of water, molecular interactions and proton transfer mechanisms will govern the control of chemical architectures yet to be developed.

After all our efforts, membrane research is still challenging and in need of fresh and innovative ideas. It is a highly interdisciplinary field, based on molecular chemistry, polymer physics, interfacial science and the science of random heterogeneous media. Could it be possible that the future lies in ordered nanostructured materials such as, for example, ordered polyelectrolyte brushes? In such materials, studying the role of the sidechains (length, separation, controlled flexibility, hydrophobicity) and mechanisms of self-assembly, which will determine proton distribution at the mesoscopic scale, will be central for design and optimization.

Will the currently used water-saturated polymers be able to overcome the difficulties in membrane operation? Or is a completely new design needed? It certainly looks that way, and many laboratories in Europe, North America, Japan and China put tremendous efforts into the development of such a "dream membrane". Possible candidates for highly proton-conductive materials might be polymer brushes or solid crystal hydrates, in which water is, if not frozen, at least substantially immobilized. Because of limits in their chemical and mechanical stability and their inability to form good contacts with the catalyst layer, such materials are not yet available for applications.

The lessons we have learned in understanding proton and water transport in hydrated polymer electrolytes are stimulating and have demonstrated some practical relevance but, paraphrasing the conclusion of Steven Hawking's second millennium lecture in the White House "*the greatest* inventions *may still lie where we don't expect them!*"

References

1. Agmon N (1995) Chem Phys Lett 244:456
2. Savadogo O (1998) J New Mater Electrochem Sys 1:47
3. Doyle M, Choi SK, Proulx G (2000) J Electrochem Soc 147:34
4. Li Q, Jensen JO, Bjerrum NJ (2003) J Chem Mater 15:4896
5. Rozière J, Jones DJ (2003) Annu Rev Mater Res 33:503

6. Kreuer K-D, Paddison SJ, Spohr E, Schuster M (2004) Chem Rev 104:4637
7. De Grotthuss CJ (1806) Ann Chim (Paris) LVIII:54
8. Marx D, Tuckerman ME, Hutter J, Parrinello M (1999) Nature 397:601
9. Mosdale R, Srinivasan S (1995) Electrochim Acta 40:413
10. Paganin VA, Ticianelli EA, Gonzales ER (1996) J Appl Electrochem 26:297
11. Eikerling M, Kharkats YI, Kornyshev AA, Volfkovich YM (1998) J Electrochem Soc 145:2684
12. Kulikovsky AA (2002) Electrochem Commun 4:845
13. Weber AZ, Darling RM, Newman J (2004) J Electrochem Soc 151:A1715
14. Ihonen J, Mikkola M, Lindbergh G (2004) J Electrochem Soc 151:A1152
15. Berg P, Promislow K, St Pierre J, Stumper J, Wetton B (2004) J Electrochem Soc 151:A341
16. Eikerling M, Kornyshev AA, Kulikovsky AA (2007) In: Bard AJ, Stratmann M (eds) Encyclopedia of Electrochemistry, vol 5, Electrochemical Engineering, vol ed by Digby Macdonald D and Schmuki P, Chapt 8.2, pp 447–543, Wiley, Weinheim
17. Eikerling M (2006) J Electrochem Soc 153:E58
18. Mathias MF, Makharia R, Gasteiger HA et al (2005) Interface. The Electrochem Soc 14:24
19. Kuznetsov AM, Ulstrup J (1999) Electron Transfer in Chemistry and Biology. Wiley, Chichester
20. Eikerling M, Ioselevich AS, Kornyshev AA (2004) Fuel Cells 4:131
21. Hsu WY, Barkley JR, Meakin P (1980) Macromolecules 13:198
22. Eikerling M, Kornyshev AA, Stimming U (1997) J Phys Chem B 101:10807
23. Smitha B, Sridhar S, Khan AA (2005) J Membrane Sci 259:10–26
24. Haile SM (2003) Acta Materialia 51:5981–6000
25. Savadogo O (2004) Power J Sources 127:135–161
26. Hickner MA, Ghassemi H, Yu Seung K, Einsla BR, McGrath JE (2004) Chem Rev 104:4587–4612
27. Schuster M, Rager T, Noda A, Kreuer KD, Maier J (2005) Fuel Cells 5:355
28. Yang Y, Holdcroft S (2005) Fuel Cells 5:171
29. Hickner MA, Pivovar BS (2005) Fuel Cells 5:213
30. Herz HG, Kreuer KD, Maier J, Scharfenbergr G, Schuster MFH, Meyer WH (2003) Electrochim Acta 48:2165
31. Ioselevich AS, Kornyshev AA, Steinke JHG (2003) J Phys Chem B 108:11953
32. Mauritz KA, Moore RB (2004) Chem Rev 104:4535
33. Gierke TD, Munn GE, Wilson FC (1981) J Polym Sci, Polym Phys 19:1687
34. Hsu WY, Gierke TD (1982) Macromolecules 15:101
35. Fuijimura M, Hashimoto R, Kawai H (1981) Macromolecules 14:1309
36. Divisek JJ, Eikerling M, Mazin V, Schmitz H, Stimming U, Volfkovich YI (1998) J Electrochem Soc 145:2677
37. Gruger A, Regis A, Schmatko T, Colomban P (2001) Vib Spectrosc 26:215
38. Elliott JA, Hanna S, Elliott AMS, Cooley GE (2000) Macromolecules 33:4161
39. Gebel G (2000) Polymer 41:5829
40. Rubatat L, Rollet AL, Gebel G, Diat O (2002) Macromolecules 35:4050
41. Ding J, Chuy C, Holdcroft S (2001) Chem Mat 13:2231
42. Ding J, Chuy C, Holdcroft S (2002) Macromol 35:1348
43. Falk M (1980) Can J Chem 58:1495
44. Cable KM, Mauritz KA, Moore RB (1995) J Polym Sci, Part B: Polym Phys 33:1065
45. Eisenberg A, Yeager HL (eds) (1982) Perfluorinated Ionomer Membranes. ACS Symposium Series 180. Am Chem Soc, Washington, DC

46. Yoshida H, Miura Y (1992) J Membrane Sci 68:1
47. Pineri M, Volino F, Escoubes M (1985) J Polym Sci, Polym Phys Ed 23:2009
48. Zawodzinski TA, Neeman M, Sillerud LO, Gottesfeld S (1991) J Chem Phys 95:6040
49. Cappadonia M, Erning JW, Stimming U (1994) J Electroanal Chem 376:189
50. Cappadonia M, Erning JW, Niake SM, Stimming U (1995) Solid State Ionics 77:65
51. Yang Y, Shi Z, Holdcroft S (2004) Macromolecules 37:1678
52. Yang Y, Holdcroft S (2005) Fuel Cells 5:171
53. Vishnyakov A, Neimark AV (2001) J Phys Chem B 105:9586
54. Netz RR, Andelmann D (2003) Phys Rep 380:1
55. DeGennes P-G (1979) Scaling Concepts in Polymer Physics. Cornell University, Ithaca
56. Limberger RE, Potemkin II, Khokhlov AR (2003) J Chem Phys 119:12023
57. Khalatur PG, Talitskikh SK, Khokhlov AR (2002) Macromol Theory Simul 11:566
58. Chitanvis SM (2003) Phys Rev E 68:061802
59. Yang SS, Molinero V, Cağin T, Goddard WA III (2004) J Phys Chem B 108:3149
60. Seeliger D, Hartnig C, Spohr E (2005) Electrochim Acta 50:4234
61. Petersen MK, Wang F, Blake NP, Metiu H, Voth GA (2005) J Phys Chem B Lett 109:3727
62. Petersen MK, Voth GA (2006) J Phys Chem B 110:18594
63. Venkatnathan A, Devanathan R, Dupuis M (2007) J Phys Chem B 111:7234
64. Devanathan R, Venkatnathan A, Dupuis M (2007) J Phys Chem B 111:8069
65. Torquato S (2002) Random Heterogeneous Materials, Springer Series in Interdisciplinary Applied Mathematics, vol 16. Springer, Berlin Heidelberg New York
66. Oosawa F (1971) Polyelectrolytes. Marcel Dekker, New York
67. Daniel RM, Finney JL, Stoneham M (2004) Is life possible without water? Philos T Roy Soc B 359:1448
68. Erdey-Gruz T (1974) Transport Phenomena in Electrolyte Solutions. Adam Higler, London
69. Johnston J (1909) J Am Chem Soc 31:987
70. Noyes A (1910) J Chim Phys 6:505
71. Noyes A (1910) Z Phys Chem Phys 70:356
72. Eucken A (1948) Z Elektrochem 52:6
73. Eucken A (1948) Z Elektrochem 52:255
74. Gierer A, Wirtz K (1949) J Chem Phys, Annalen der Physik 6:257
75. Gierer A (1950) Z Naturforsch 5a:581
76. Franck EU, Hartmann D, Hensel F (1965) Discuss Faraday Soc 39:200
77. Gierer A, Wirtz K (1949) Annalen der Physik Folge 6:17
78. Kohlrausch F, Heydweiller A (1894) Annalen der Physik und Chemie (Neue Folge) 53:14
79. Uosaki K, Okazakai K, Kita H (1990) J Electroanal Chem 287:163
80. Davydov AS (1991) Solitons in molecular systems, 2nd edn. Kluwer, Dordrecht
81. Bountis T (ed) (1992) Proton transfer in hydrogen bonded systems. NATO ASI Series B: Physics, Vol 291, Plenum, New York
82. Eikerling M, Kornyshev AA, Kuznetsov AM, Ulstrup J, Walbran S (2001) J Phys Chem B 105:3646
83. Spohr E, Commer P, Kornyshev AA (2002) J Phys Chem 106:10560
84. Commer P, Cherstvy AG, Spohr E, Kornyshev AA (2002) Fuel Cells 2:127
85. Bernal JD, Fowler RH (1933) J Chem Phys 1:515
86. Eigen M, De Maeyer L (1958) Proc Roy Soc (London) A247:505
87. Conway BE, Bockris JO'M, Linton H (1956) J Chem Phys 31:834

88. Zundel G, Metzger H (1968) Z Phys Chem 58:225
89. Kreuer K-D (1992) In: Colomban P (ed) Proton Conductors. Cambridge University Press, Cambridge, p 474
90. Tuckerman ME, Marx D, Klein ML, Parinello M (1997) Science 275:817
91. Zundel G, Fritsch J (1986) In: Dogonadze RR, Kálmán E, Kornyshev AA, Ulstrup J (eds) The Chemical Physics of Solvation. Part B: Spectroscopy of Solvation. Elsevier, Amsterdam, p 21
92. Vuilleumier R, Borgis D (1999) J Chem Phys 111:4251
93. Schmidt UW, Voth GA (1999) J Chem Phys 111:9361
94. Holm RH, Solomon EI (1996) Special Issue on Bioinorganic Enzymology. Chem Rev 96(7):2237–3042
95. Kuznetsov AM, Ulstrup J (1999) Can J Chem 77:1085
96. Mulkidjanian AY, Heberle J, Cherepanov DA (2006) Biochim Biophys Acta 1757:913
97. Kato M, Pisliakov AV, Warshel A (2006) Proteins 64:829
98. Braun-Sand S, Strajbl M, Warshel A (2004) Biophys J 87:2221
99. Hummer G, Dellago C (2006) Phys Rev Lett 97:245901
100. Gottesfeld S, Fuller TF (eds) (1999) Proton Conducting Membrane Fuel Cells. Proc vol 98–27, The Electrochem Soc Inc, Pennington, NJ
101. Colomban P (ed) (1992) Proton Conductors. Cambridge University Press, Cambridge
102. Walbran S, Kornyshev AA (2001) J Chem Phys 114:10039
103. Kornyshev AA, Kuznetsov AM, Spohr E, Ulstrup J (2003) J Phys Chem B 107:3351
104. Kirkpatrick S (1973) Percolation and Conduction, Rev Mod Phys 45:574
105. Stauffer D, Aharony A (1992) Introduction to Percolation Theory. Taylor and Francis, London
106. Kuznetsov AM (1995) Charge Transfer in Physics, Chemistry and Biology. Gordon and Breach, Amsterdam
107. Eikerling M, Kornyshev AA (2001) J Electroanal Chem 502:1
108. Paddison SJ, Paul R, Zawodzinski TA (2000) J Electrochem Soc 147:617
109. Spohr E (2004) Mol Simulation 30:107
110. Paddison SJ (2001) J New Mater Electrochem Sys 4:197
111. Eikerling M, Paddison SJ, Zawodzinski TA Jr (2002) J New Mater Electrochem Sys 5:1
112. Eikerling M, Paddison SJ, Pratt LW, Zawodzinski TA (2003) Chem Phys Lett 368:108
113. Elliott JA, Paddison SJ (2007) Phys Chem Chem Phys 9:2602
114. Paddison SJ, Elliott JA (2006) Phys Chem Chem Phys 8:2193
115. Kreuer KD (2001) J Membrane Sci 185:29
116. Paddison SJ, Reagor DW, Zawodzinski TA Jr (1998) J Electroanal Chem 459:91
117. Paul R, Paddison SJ (2001) J Chem Phys 115:7762
118. Lau AWC, Lukatsky DB, Pincus P, Safran SA (2002) Phys Rev E 65:051502
119. Krishtalik LI (1986) Charge Transfer Reactions in Electrochemical and Chemical Processes. Plenum, New York
120. Spencer JB, Lundgren J-O (1973) Acta Cryst B 29:1923
121. Barthel J, Meier R, Conway BE (1999) J Chem Eng Data 44:155
122. Roudgar A, Narasimachary SP, Eikerling M (2006) J Phys Chem B 110:20469
123. Kresse G, Furthmüller J (1996) Comp Mat Sci 6:15
124. Kresse G, Hafner J (1993) Phys Rev B 47:RC558
125. Kresse G, Hafner J (1993) Phys Rev B 48:13115
126. Kresse G, Hafner J (1994) Phys Rev B 49:14251
127. Kresse G, Hafner J (1994) J Phys: Condens Matter 6:8245
128. Zhang J, Unwin PR (2002) Phys Chem Chem Phys 4:3814
129. Leite VBP, Cavalli A, Oliveira ON Jr (1998) Phys Rev E 57:6835

130. Narasimachary SP, Roudgar A, Eikerling M, Electrochim. Acta, in press
131. Booth F (1951) J Chem Phys 19:391
132. Kornyshev AA, Kuznetsov AM, Ulstrup J, Stimming U (1997) J Phys Chem B 101:5917
133. Paul R, Paddison SJ (2001) J Chem Phys 115:7762
134. Warshel A, Weiss RM (1980) J Am Chem Soc 102:6218
135. Vuilleumier R, Borgis D (1998) J Phys Chem B 102:4261
136. Vuilleumier R, Borgis D (1998) Chem Phys Lett 284:71
137. Schmitt UW, Voth GA (1998) J Phys Chem B 102:5547
138. Schmitt UW, Voth GA (2000) Chem Phys Lett 329:36
139. Day TJF, Soudackov A, Cuma M, Schmitt UW, Voth GA (2002) J Chem Phys 117:5839
140. Petersen MK, Voth GA (2006) J Phys Chem B 110:18594
141. Petersen MK, Wang F, Blake NP, Metiu H, Voth GA (2005) J Phys Chem B 109:3727
142. Commer P (2003) PhD thesis. Heinrich Heine Universität, Düsseldorf
143. Commer P, Hartnig C, Seeliger D, Spohr E (2004) Mol Simulation 30:755
144. Weber AZ, Darling R, Meyers J, Newman J (2003) In: Vielstich W, Lamm A, Gasteiger HA (eds) Handbook of Fuel Cells: Fundamentals, Technology and Applications, vol 1. Wiley, Weinheim, p 47
145. Springer TE, Zawodzinski TA, Gottesfeld S (1991) J Electrochem Soc 138:2334
146. Nguyen N, White RE (1993) J Electrochem Soc 140:2178
147. Weber AZ, Newman J (2004) J Electrochem Soc 151:A311, A326
148. Büchi FN, Scherer GG (2001) J Electrochem Soc 148:A183
149. Büchi FN, Huslage J, Scherer GG (1997) PSI—Annual Report, Annex V. General Energy Research, p 48
150. Mosdale R, Gebel G, Pineri M (1996) J Membrane Sci 118:269
151. Kulikovsky AA (2003) J Electrochem Soc 150:A1432
152. Kulikovsky AA, Divisek J, Kornyshev AA (1999) J Electrochem Soc 146:3981
153. Kulikovsky AA, Divisek J, Kornyshev AA (2000) J Electrochem Soc 147:953

Structural and Morphological Features of Acid-Bearing Polymers for PEM Fuel Cells

Yunsong Yang[1,3] · Ana Siu[1,4] · Timothy J. Peckham[1,2] · Steven Holdcroft[1,2] (✉)

[1] *Present address (Steven Holdcroft):*
Department of Chemistry, Simon Fraser University, 8888 University Drive, Burnaby, BC V5A 1S6, Canada
holdcrof@sfu.ca

[2] *Present address (Timothy J. Peckham):*
Institute for Fuel Cell Innovation, National Research Council Canada, 3250 East Mall, Vancouver, BC V6T 1W5, Canada

[3] Ballard Power Systems Inc., 9000 Glenlyon Parkway, Burnaby, BC V5J 5J8, Canada

[4] Risø National Laboratory, The Danish Polymer Centre Risø, Building 111, P.O. Box 49, 4000 Risø, Denmark

1	Introduction	57
2	Factors Determining the Properties of Acid-Bearing Polymers	61
3	Studies of Acid-Bearing Polymers and Membranes	63
3.1	Morphological Studies of PEMs	69
3.1.1	Randomly Sulfonated PEMs	69
3.1.2	Block Copolymer PEMs	78
3.1.3	Graft Copolymer PEMs	85
3.2	Proton Conductivity	87
3.2.1	Effect of Polymer Microstructure and Morphology	94
3.2.2	Effect of Pretreatment and Casting Solvents	97
3.2.3	Proton Transport	100
3.3	Water Management	103
3.3.1	The Nature of Water in PEMs	104
3.3.2	Water Transport	106
3.4	MEA and Fuel Cell Performance	111
4	Summary and Future Prospects	114
	References	118

Abstract Chemical structure, polymer microstructure, sequence distribution, and morphology of acid-bearing polymers are important factors in the design of polymer electrolyte membranes (PEMs) for fuel cells. The roles of ion aggregation and phase separation in vinylic- and aromatic-based polymers in proton conductivity and water transport are described. The formation, dimensions, and connectivity of ionic pathways are consistently found to play an important role in determining the physicochemical properties of PEMs. For polymers that possess low water content, phase separation and ionic channel formation significantly enhance the transport of water and protons. For membranes that contain a high content of water, phase separation is less influential. Continuity of ionic aggregates is influential on the diffusion of water and electroosmotic drag within a membrane. A balance of these properties must be considered in the design of the next generation of PEMs.

Keywords Acid-bearing polymers · Chemical microstructure · Fuel cells · Ion-containing polymers · Membranes · Morphology · Proton conductivity

Abbreviations
AFC	Alkaline fuel cell
AFM	Atomic force microscopy
ATRP	Atom transfer radical polymerization
BDSA	4,4′-Diaminobiphenyl 2,2′-disulfonic acid
BPSH	Disulfonated biphenol based poly(arylene ether sulfone) copolymer
BVPE	1,2-Bis(vinylphenyl)ethane
CL	Catalyst layer
DMF	N,N-Dimethylformamide
DMFC	Direct methanol fuel cell
DMSO	Dimethyl sulfoxide
DS	Degree of sulfonation
DSC	Differential scanning calorimetry
DTA	Differential thermal analysis
DVB	Divinylbenzene
EG	Ethylene glycol
EIS	Electrochemical impedance spectroscopy
EOD	Electroosmotic drag
ETFE-g-PSSA	Poly(ethylene-co-tetrafluoroethylene) grafted with poly(styrene sulfonic acid)
EW	Equivalent weight
FEP-g-PSSA	Poly(tetrafluoroethylene-co-perfluoropropyl vinyl ether) grafted with poly(styrene sulfonic acid)
GDE	Gas diffusion electrode
GDL	Gas diffusion layer
HMEA	Half-MEA
IEC	Ion-exchange capacity
MCFC	Molten carbonate fuel cell
MEA	Membrane electrode assembly
NTDA	Naphthalene-1,4,5,8-tetracarboxylic dianhydride
ODA	4,4-Oxydianiline
PAFC	Phosphoric acid fuel cell
PAN-g-mac PSSA	Polyacrylonitrile grafted with poly(styrene sulfonic acid) macromonomer
PEM	Polymer electrolyte membrane
PEMFC	Polymer electrolyte membrane fuel cell
PFG-NMR	Pulse field gradient NMR spectroscopy
PFSI	Perfluorosulfonic acid ionomer
PS-g-mac PSSA	Polystyrene grafted with poly(styrene sulfonic acid) macromonomer
PSSA	Poly(styrene sulfonic acid)
PTFE	Polytetrafluoroethylene

PTFSSA	Sulfonated trifluorostyrene–trifluorostyrene copolymer
PVDF	Poly(vinylidene fluoride)
P(VDF-co-HFP)	Vinylidene fluoride–hexafluoropropylene copolymer
P(VDF-co-HFP)-b-SPS	Vinylidene fluoride–hexafluoropropylene copolymer-block-sulfonated polystyrene
PVDF-g-PSSA	Poly(vinylidene fluoride) grafted with poly(styrene sulfonic acid)
SANS	Small-angle neutron scattering
SAXS	Small-angle X-ray scattering
SOFC	Solid oxide fuel cell
SPEEK	Sulfonated poly(ether ether ketone)
SPEEKK	Sulfonated poly(ether ether ketone ketone)
SPI	Sulfonated polyimide
SPSU	Sulfonated bisphenol A polysulfone
SPSU-b-PVDF	Sulfonated bisphenol A polysulfone-block-poly(vinylidene fluoride)
SSEBS	Sulfonated polystyrene-block-(ethylene-co-butylene)-block-sulfonated polystyrene
SSIBS	Sulfonated polystyrene-block-(isobutylene)-block-sulfonated polystyrene
TEM	Transmission electron microscopy
THF	Tetrahydrofuran
TM-AFM	Tapping mode atomic force microscopy
λ	Water molecules per acid site, H_2O/SO_3H
D_{Chem}	Fickian diffusion coefficient of water
D_{H_2O}	Self-diffusion coefficient of water
D_σ	Proton mobility
n_{drag}	Electroosmotic drag coefficient
R_m	Bulk membrane proton resistance
R_u	Uncompensated resistance
T_g	Glass transition temperature
X_v	Water volume fraction

1
Introduction

Fuel cell systems are distinguished by the type of electrolyte used in the cells and include polymer electrolyte membrane (PEMFC), alkaline (AFC), phosphoric acid (PAFC), molten carbonate (MCFC), and solid oxide (SOFC) fuel cells. PEMFC technology differentiates itself from other fuel cell technologies in that a solid polymer membrane is used as the electrolyte. The membrane serves as a separator to prevent mixing of reactant gases and as an electrolyte for transporting protons from anode to cathode. Because the cell separator is a flexible polymer film, issues such as sealing, assembly, and handling are often less complex than with other fuel cell systems. The need to handle corrosive acids and bases is mitigated in this system. PEMFCs can operate at very high current density and typically operate at lower temperatures ($\sim$80 °C) than other fuel cell systems, allowing for faster startup times and a rapid response to changes in demand for power. The PEMFC is attractive for transportation applications, and it is a major competitor for low power stationary applications [1] and a technology leader in portable power (<1 kW).

Polymer electrolyte membranes (PEMs) for PEMFC applications should possess: high protonic conductivity; good chemical, electrochemical, and thermal stability; adequate mechanical properties; low permeability to reactants; and low swelling in water. They should be readily processable into membrane electrode assemblies (MEAs). Reproducibility of preparation and fabrication costs should be appropriate for the application. Membranes based on perfluorosulfonic acid, e.g., Nafion®, have been the material of choice and the technology standard. Nafion® consists of a perfluorinated backbone that bears pendent vinyl ether side chains, terminating with SO_3H (Scheme 1). Nafion® membranes show good operation under normal conditions but have several critical drawbacks: they are limited in their temperature of operation (usually to below 80 °C) and are prone to dehydration. They are relatively expensive and difficult to synthesize and process. For direct methanol fuel cell (DMFC) applications, they offer a poor barrier to methanol. In H_2/O_2 applications, they exhibit a high osmotic drag coefficient that leads to dehydration of the anode and/or flooding of the cathode, unless physical precautions are taken to prevent it. Recycling of the fluorinated polymer is another concern. It is difficult to process and cannot be easily dissolved, melted, or extruded in its ionic form, which precludes its use in novel cell designs and nonplanar geometries. For these and other reasons, alternative proton conducting polymers are actively sought after. Several classes of polymer are under intense investigation. These include polyarylenes [2–11], polyimides [12–15], polyphosphazenes [16–19], radiation-grafted polystyrene [20, 21], organic–inorganic composites and hybrids [22–24], polystyrene di- and triblock copolymers [25–28], and acid complexes of basic polymers [29, 30]. Comprehensive reviews of specific PEMs can be found in a number of articles [31–51].

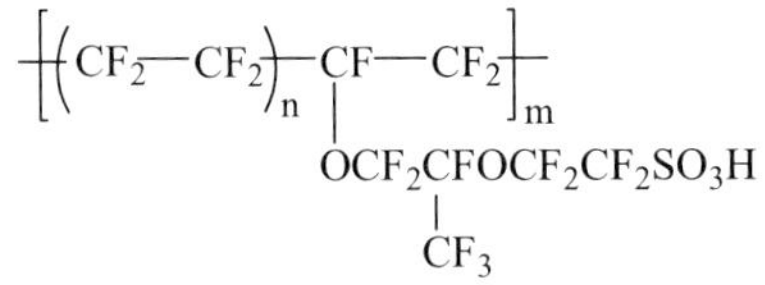

$$\text{-}\!\!\left[\!\left(CF_2\text{—}CF_2\right)_n\text{—}\underset{\underset{CF_3}{|}}{\underset{|}{CF}}\text{—}CF_2\right]_m\text{-}$$
$$OCF_2CFOCF_2CF_2SO_3H$$

Scheme 1 Chemical structure of Nafion®

A characteristic feature of alternative membranes is that they nearly always exhibit a lower proton conductivity compared to Nafion® for a similar ion content. The ionic conductivity can be improved by increasing the ion-exchange capacity (IEC) of the constituent polymer(s) but mechanical strength is frequently sacrificed; firstly, in the dehydrated state because of the high ionic content, and secondly in the hydrated state due to excessive swelling [43]. Moreover, virtually all alternatives to perfluorosulfonic acid membranes are less chemically and/or electrochemically robust under "real" fuel cell conditions, which limits their lifetime to a few thousand hours in the best cases [52]. Therefore, in order to obtain both satisfactory fuel cell performance and long term stability, it is necessary to design thermally, chemically, and mechanically stable *acid-bearing polymers* with sufficient ionic conductivity and low water uptake.

The importance of the *microstructure*—here we use the classical description referring to the arrangement of monomer units along a polymer chain—and *morphology* of the constituent acid-bearing polymers on the properties of PEMs has been widely demonstrated. For example, sulfonated polyimide block copolymers exhibit higher proton conductivity than analogous random copolymers [53, 54], and Nafion® when dissolved and recast exhibits an ionic conductivity up to several orders of magnitude lower than "as-received" membranes [55]. In order to optimize the chemical structure, microstructure, and morphology of acid-bearing polymers for application as PEMs, it is crucial to understand the relationship between structure, morphology, and physicochemical properties. Even though there are several excellent and detailed reviews on the alternative PEMs, and detailed reviews on the morphology of Nafion® [56], only a few reviews are available that expose relationships between the chemical structure, microstructure, and morphology of emerging and experimental PEMs [57, 58].

Proton conductivity is critical for operational PEMs and is often the first characteristic considered when evaluating membranes for fuel cell applications. High conductivity is essential to obtain high current densities. Proton conductivity is controlled by the structure of the constituent polymer(s) and the morphology of the resulting membrane; it is a strong function of water content and temperature. Proton conductivity may be measured ex situ using electrochemical impedance spectroscopy (EIS), or in situ during fuel cell operation using current interruption [59, 60] or high-frequency EIS methods [61].

Chemical/electrochemical stability and thermal stability are largely dependent on the chemical structure of polymers. For fuel cell applications they should be hydrothermally stable under high humidity and at high temperature (80–100 °C) and resistant to oxidation, reduction, and attack by free radicals. The membrane functions both as the electrolyte and separator of reactant gases. Membrane degradation causes the proton conductivity to drop and the mechanical properties to deteriorate, resulting in membrane rupture, pinhole formation, and mixing of anode and cathode reactant gases. For transportation applications, a PEM should be able to operate for more than 6000 h. Chemical stability is often investigated using the Fenton's reagent test [62]; electrochemical stability can be measured by cyclic voltammetry [63], and thermal stability is often estimated by thermogravimetric analysis, although these ex situ measurements may not accurately simulate fuel cell conditions.

Mechanical stability under operating fuel cell conditions (heating–cooling cycles, high temperature, humidity, and temperature/water gradients) is necessary for reliable operation in fuel cells. Resilient mechanical properties are essential for the fabrication of ultrathin membranes, which reduce proton resistance in a fuel cell and improve hydration at the anode through back diffusion. Water in the membrane acts as plasticizer for the constituent polymers, lowering both the glass transition temperature (T_g) and the modulus of the polymer. Mechanical properties may be determined using the tensile test, while T_g can be obtained by dynamic mechanical analysis or differential scanning calorimetry (DSC).

At the heart of the PEMFC is the MEA, comprising a PEM sandwiched between two gas diffusion electrodes (GDEs). The intrinsic electrocatalytic properties of the electrocatalyst, the electrode/electrolyte interface, the amount of water in the GDEs, water transport through both the PEM and the GDEs, and the rates at which gases flow through the GDEs all govern the rate of electrochemical reaction. Poor compatibility between the PEM and GDEs will lead to an excessively large interfacial resistance, accelerated degradation of performance during fuel cell operation, or even delamination of the PEM/GDE—with catastrophic failure of the fuel cell.

The power density and fuel efficiency of a fuel cell depend strongly on the mass transport properties in the PEM. Reactant crossover through the membrane results in low open circuit voltages and reduced lifetime. In the DMFC, methanol crossover leads to significant performance losses, and impacts cathode kinetics and overall fuel efficiency. Water management in the membrane is critical for efficient performance: the fuel cell must operate under conditions where the by-product water does not evaporate faster than it is produced because the membrane must remain hydrated. However, too little water removal results in "flooding" of the fuel cell. Primarily, water management is an engineering-based issue, but it is also related to the characteristics of the membrane.

The technical issues directly relevant to fuel cell performance are many and interdependent. This review will focus on the current literature and is intended to provide a deeper understanding of fundamental relationships between the chemical structure of an acid-bearing polymer and the morphology of the resultant PEM, and between morphology and fuel cell pertinent properties, such as proton conductivity and water transport. The microstructural, morphological, and macroscopic properties of various acid-bearing polymers and PEMs, as well as examples of specific acid-bearing polymers that have similar chemical structure but different microstructure and morphology—or have similar microstructure but different chemical structure—are described. Properties other than proton conductivity and water transport should also be considered in the design of PEMs for fuel cells, such as the effect of chemical structure, microstructure, and morphology on reactant crossover, mechanical properties, and chemical stability, but in order to limit the scope of this review these aspects are not included.

The review is organized as follows. Section 1 provides an overview of the issues associated with PEM design. Section 2 gives an overview of factors that determine the physicochemical properties of acid-bearing polymers. Section 3 reviews studies of microstructure, morphology, and transport properties of different classes of PEMs; it addresses aspects of proton conductivity and compares the properties of acid-bearing polymers that possess similar microstructure but different chemical backbones; it compares acid-bearing polymers that possess similar chemical composition but different chain microstructure; and it addresses issues associated with water transport. The body of work is then summarized and prospects for future considerations are suggested. The intention of these discussions is to provide a fundamental framework, based on microstructure and morphology, from which the next generation of PEMs may be designed.

2
Factors Determining the Properties of Acid-Bearing Polymers

Before detailing the relationship between the structure, morphology, and properties of PEMs, it is worth illustrating the factors that determine the properties of acid-bearing polymers as they relate to fuel cell applications. Generally, such polymers are amphiphilic, consisting of hydrophobic and hydrophilic subunits. Hydrophobic domains are responsible for the membrane possessing suitable mechanical properties, and they prevent the polymer dissolving in water; hydrophilic domains are responsible for providing a pathway for protons and for back transport of water from cathode to anode. Hydrophilic networks may be realized by doping a polymer matrix with acid, or incorporating protogenic groups (sulfonic acid, bis(sulfonyl) imide, phosphonic acid, or carboxylic acid functional groups) in the polymer chain. In comparison to sulfonated polymers, bis(sulfonyl) imides are stronger acids but their syntheses are more complicated. Phosphonic acid and carboxylic acid groups are much weaker acids than sulfonic acid and do not provide a sufficient concentration of dissociated protons at normal fuel cell operating temperatures. The large majority of experimental membranes are based on sulfonated polymers. Sulfonic acids may be grafted onto the polymer structure by post-sulfonation of pristine polymers or incorporated directly by polymerization of sulfonated monomers.

The properties of acid-bearing polymers are determined by: (a) the chemical nature of the monomers; (b) the microstructure of the polymer; (c) their molecular weight and molecular weight distribution; and (d) their morphology in the solid state. Sulfonated polymers usually exhibit a much higher proton conductivity than phosphonylated or carboxylated polymers for a given ion content, and generally aromatic polymers exhibit better chemical and thermal stability than aliphatic polymers. Moreover, physicochemical properties are affected by secondary bonding forces between polymer chains, e.g., dipole–dipole, hydrogen-bonding, and van der Waals interactions, which influence both the mechanical and transport properties of the solid polymer. The polymer's microstructure includes aspects of composition, architecture, and chain conformation. Polymers are often copolymers, and may be alternating, random, block, or graft. Examples of copolymers having different microstructure are shown in Fig. 1. A more detailed description of the relationship between copolymer composition and morphology can be found in the literature [57, 64, 65].

The most important property dependent on molecular weight and molecular weight distribution is mechanical strength and physical integrity. Properties such as melting point transitions, glass transition temperatures, and solubility are also related to the molecular weight of the polymer. However, for high molecular weight polymers the mechanical properties and thermal transitions vary little with varying chain length.

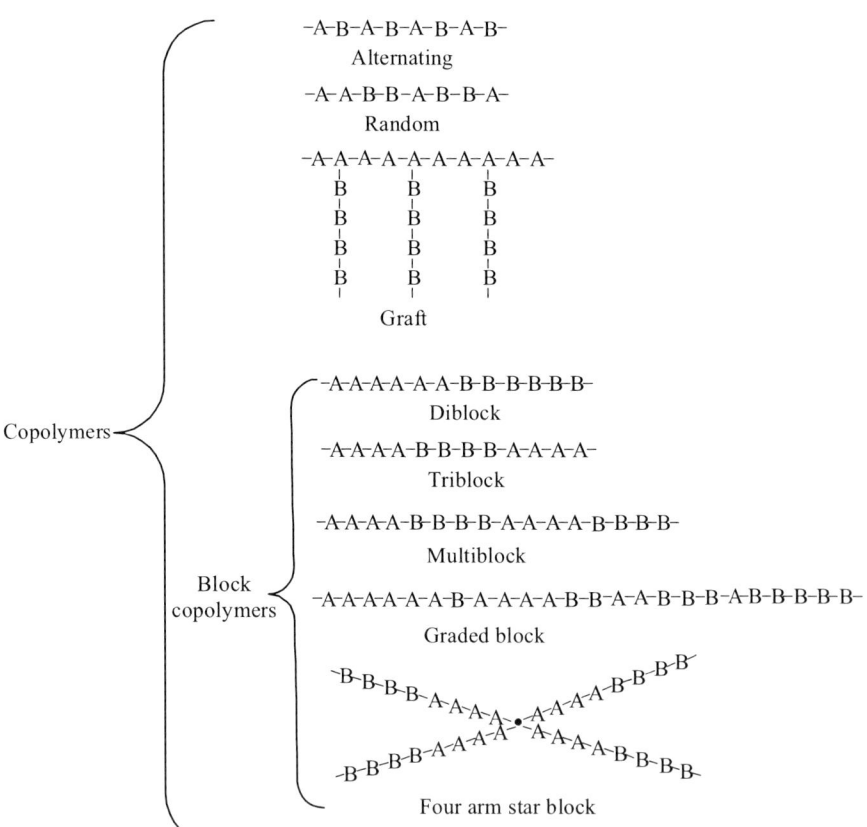

Fig. 1 Microstructures of copolymers

The morphology of an acid-bearing polymer is dependent on the chemical nature and microstructure of the polymer, but it is also affected by external conditions. The method of sample preparation—solution casting, extrusion, hot pressing, etc.—can have a profound influence. Additives such as fillers, reinforcing agents, and plasticizers often modify the physical and thermal properties of polymer membranes. Furthermore, a polymer's morphology may significantly change upon varying the temperature of annealing, and different morphologies for the same material may be observed in response to shear forces [66, 67], stretching [68], casting from different solvents [69, 70], rates of solvent evaporation [71], and film thickness [72]. For example, Nafion® films cast from ethanol/water solutions at room temperature are reported to be "mud-cracked", and soluble in a variety of polar, organic solvents at room temperature. In contrast, as-received Nafion® membranes are flexible, tough, and insoluble in virtually all solvents at temperatures below $\sim$200 °C. However, adding high boiling point polar solvents, such as N,N-dimethylformamide (DMF) or ethylene glycol (EG), to a commercially available dispersion and casting the mixture above 160 °C yields a membrane with similar properties to the extruded film [73]. Similarly, annealing the uncracked film, obtained from evaporation of the dispersing solvent at room temperature to >150 °C, also yields a film with similar properties to the extruded Nafion® [74].

3
Studies of Acid-Bearing Polymers and Membranes

Numerous polymers have been studied for their potential application in PEMFCs. Based on their chemical structure, these polymers can be categorized into (a) vinylic polymers, (b) aromatic polymers, and (c) polymer blends and composite/hybrid polymers. Generally, vinylic polymers are synthesized by addition polymerization, while aromatic polymers are synthesized by step-growth polymerization. The most studied vinylic polymers for PEMFC applications are perfluorosulfonic acid ionomers (PFSIs), in particular Nafion®, and styrene sulfonic acid-based polymers. Chemical structures of representative vinylic PEMs are shown in Scheme 2.

While a number of alternative polymer membranes have been developed, Nafion® is still considered the benchmark of proton conducting polymer membranes, and has the largest body of research literature devoted to its study. Alternative polymer membranes are almost invariably compared to Nafion®. Nafion® is a free radical initiated copolymer consisting of crystallizable, hydrophobic tetrafluoroethylene and a perfluorinated vinyl ether terminated by perfluorosulfonic acid. Nafion® 117 possesses an equivalent weight of 1100 (EW = mass of dry ionized polymer (g) in the protonic acid form that would neutralize one equivalent of base). Thus, there are $\sim$13 perfluoromethylene groups ($-CF_2-$) ($n = 6.5$) between pendent ionic side chains.

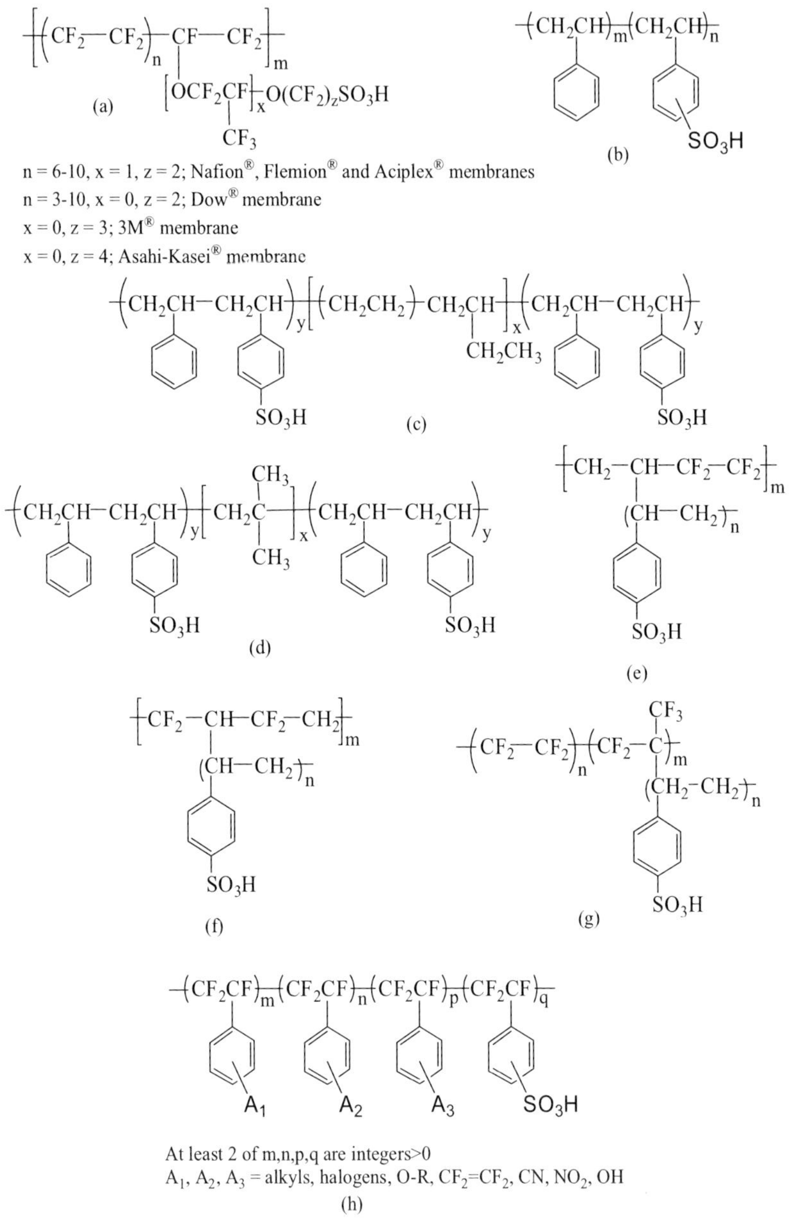

(a) $+(CF_2-CF_2)_n-CF-CF_2+_m$
   $[OCF_2CF]_x-O(CF_2)_zSO_3H$
   CF_3

n = 6-10, x = 1, z = 2; Nafion®, Flemion® and Aciplex® membranes
n = 3-10, x = 0, z = 2; Dow® membrane
x = 0, z = 3; 3M® membrane
x = 0, z = 4; Asahi-Kasei® membrane

(b) $+(CH_2CH)_m(CH_2CH)_n$ with phenyl and phenyl-SO_3H

(c) $+(CH_2CH-CH_2CH)_y+((CH_2CH_2)-CH_2CH)_x+(CH_2CH-CH_2CH)_y$ with phenyl-SO_3H, CH_2CH_3, phenyl, phenyl-SO_3H

(d) $+(CH_2CH-CH_2CH)_y+(CH_2C(CH_3)_2)_x+(CH_2CH-CH_2CH)_y$ with phenyl-SO_3H and phenyl-SO_3H

(e) $+(CH_2-CH-CF_2-CF_2)_m$
    $(CH-CH_2)_n$
    phenyl-SO_3H

(f) $+(CF_2-CH-CF_2-CH_2)_m$
    $(CH-CH_2)_n$
    phenyl-SO_3H

(g) $+(CF_2-CF_2)_n(CF_2-C(CF_3))_m$
    $(CH_2-CH_2)_n$
    phenyl-SO_3H

(h) $+(CF_2CF)_m(CF_2CF)_n(CF_2CF)_p(CF_2CF)_q$
    with A_1, A_2, A_3, SO_3H substituents

At least 2 of m,n,p,q are integers>0
A_1, A_2, A_3 = alkyls, halogens, O-R, $CF_2=CF_2$, CN, NO_2, OH

Scheme 2 Chemical structures of: **a** poly(perfluorosulfonic acid); **b** sulfonated polystyrene (SPS); **c** sulfonated polystyrene-*b*-(ethylene-*co*-butylene)-*b*-sulfonated polystyrene (SSEBS); **d** sulfonated polystyrene-*b*-(isobutylene)-*b*-sulfonated polystyrene (SSIBS); **e** poly(ethylene-*co*-tetrafluoroethylene)-*g*-polystyrene sulfonic acid (ETFE-*g*-PSSA); **f** poly(vinylidene fluoride)-*g*-polystyrene sulfonic acid (PVDF-*g*-PSSA); **g** poly(tetrafluoroethylene-*co*-perfluoropropyl vinyl ether)-*g*-polystyrene sulfonic acid (FEP-*g*-PSSA); **h** sulfonated trifluorostyrene–trifluorostyrene copolymer (PTFSSA)

Currently, Nafion® is available in two forms: extrusion-cast membranes and polymer dispersions. Nafion® polymer dispersions are used in the formulation of GDEs and reinforced membranes. Nafion® 117 is ~7 mils thick (1 mil = 25.4 μm) but can be obtained thinner. Table 1 illustrates the current offering of Nafion® from DuPont [75]. A number of excellent review articles exist that describe the structure, morphology, physical properties, and applications of Nafion® [56, 76–81].

Perfluorinated sulfonic acid-based ionomers other than Nafion® and Nafion®-like materials (e.g., Aciplex®, Flemion®) have also been developed. One of these is the DOW® membrane in which the side chain is reduced to only two CF_2 units in addition to the ethereal oxygen [82]. Although its performance was said to exceed that of Nafion® [83], work on this material in the past was limited due to the high cost of its manufacture. However, more recently, Solvay-Solexis has developed a cheaper route and is now marketing the membrane as Hyflon-Ion® [84]. Variations on this structure have also been extended to three (3M) [85] and four (Asahi-Kasei) [86] CF_2 units in the side chain. Asahi Glass has also reported a DOW-like material in which there is an intermediary CF_2 spacer between the polymer main chain and the ethereal oxygen of the side chain unit [87]. Finally, perfluorinated ionomers in which sulfonic acid groups have been replaced with sulfonamide groups for proton conduction have also been synthesized [88]. Conductivity tests on these materials at RH <70% have shown better values than analogous Nafion® derivatives, suggesting that sulfonamide groups are less sensitive to water content than sulfonic acid groups [89].

Very recently, an alternative route to perfluorinated ionomers has been demonstrated by De Simone et al. [90]. These materials were synthesized by copolymerizing a multifunctional, perfluorinated vinyl ether with styrene to yield a cross-linked, partially fluorinated membrane. Sulfonation of the aryl rings, post-fluorination to replace C–H bonds with C–F bonds, and subsequent hydrolysis of the sulfonyl fluoride groups yielded the proton conducting membrane. With its cross-linked structure, IECs in excess of those of Nafion® and similar materials are achieved along with higher conductivity values but without the membrane being water-soluble. These materials were also able to demonstrate significantly higher power densities

Table 1 Current Nafion® membranes

NR111 and NR112 (1100EW):
1 and 2 mil thickness, roll format, solution cast film

N112, N1135, N115, and N117 (1100EW):
2–7 mil thickness, pieces or rolls, extrusion cast film

N1035, N105 (1000EW):
3.5 and 5.0 mil thickness, pieces or rolls, extrusion cast film

than Nafion®, even with thicker membranes (190 versus 175 µm). Given that the monomers are liquids, it was also possible to cast the membranes into high surface area PEMs using micromolding/imprint lithography techniques. Higher performances were observed for patterned PEMs in comparison to their flat analogs.

Sulfonated styrene–divinylbenzene copolymers were used as proton conducting membranes in the early 1960s in the Gemini Space Program. Although only ∼500 h operation at 60 °C is reported due to their poor chemical stability, poly(styrene sulfonic acid)-based PEMs are still widely studied due to their ease of synthesis and the ability to control their microstructure. In order to modify the structure and improve their mechanical and chemical stability, and provide further understanding of structure–property relationships in these versatile systems, recent research on styrene sulfonic acid-based PEMs has focused on block and graft copolymers. Block and graft copolymers of styrene may be synthesized using living polymerization of appropriate monomers (including ionic, atom transfer radical polymerization (ATRP) and TEMPO-assisted radical polymerization) followed by post-sulfonation, or by the incorporation of macromonomers of polystyrene sulfonates into growing polymer chains. Another method to achieve styrene sulfonic acid-based polymers is radiation-induced grafting of polystyrene onto preformed polymer films followed by sulfonation. A variety of different fluoropolymer films have been investigated as the base polymer including polytetrafluoroethylene (PTFE), poly(vinylidene fluoride) (PVDF), poly(tetrafluoroethylene-*co*-perfluoropropyl vinyl ether) (PFA), poly(tetrafluoroethylene-*co*-hexafluoropropylene) (FEP), and poly(ethylene-*alt*-tetrafluoroethylene) (ETFE). Styrene and its derivatives, such as α-methylstyrene, α,β,β-trifluorostyrene, and substituted α,β,β-trifluorostyrene, have been employed as monomers and divinylbenzene (DVB), 1,2-bis(vinylphenyl)ethane (BVPE), and triallylcyanurate used as cross-linking agents. Acrylic monomers have also been incorporated but the resulting membranes have limited application in fuel cells due to the low acidity of the carboxylic acid unit. Thorough reviews of radiation-grafted membranes are available elsewhere [46, 91, 92].

Sulfonated aromatic polymers have been widely studied as alternatives to Nafion® due to potentially attractive mechanical properties, thermal and chemical stability, and commercial availability of the base aromatic polymers. Aromatic polymers studied in fuel cell applications include sulfonated poly(*p*-phenylene)s, sulfonated polysulfones, sulfonated poly(ether ether ketone)s (SPEEKs), sulfonated polyimides (SPIs), sulfonated polyphosphazenes, and sulfonated polybenzimidazoles. Representative chemical structures of sulfonated aromatic polymers are shown in Scheme 3. Aromatic polymers are readily sulfonated using concentrated sulfuric acid, fuming sulfuric acid, chlorosulfonic acid, or sulfur trioxide. Post-sulfonation reactions suffer from a lack of control over the degree and location of functionalization, and the

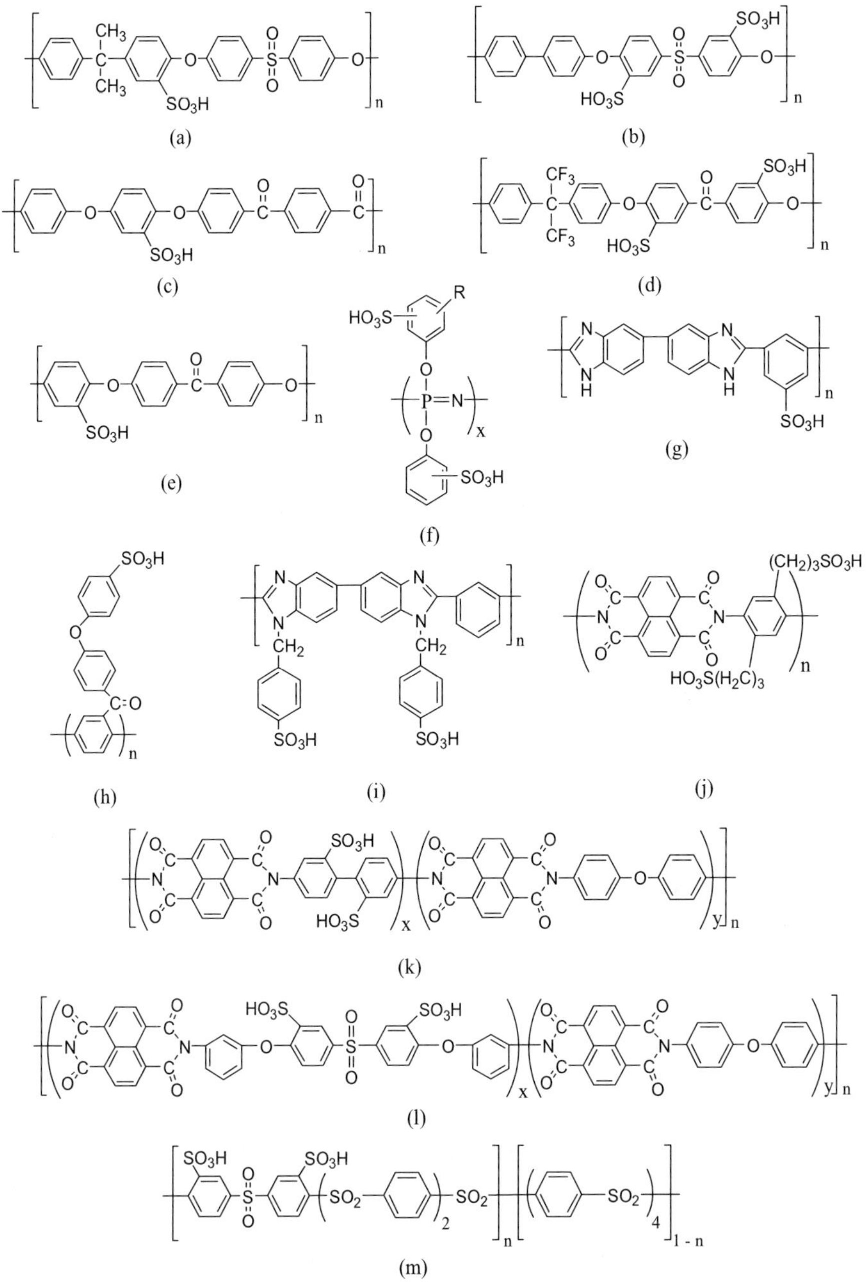

Structural and Morphological Features of Acid-Bearing Polymers

Scheme 3 Chemical structures of representative sulfonated aromatic polymers

possibility of side reactions (including cross-linking) and/or degradation of the polymer backbone [94, 95].

Recent efforts in the synthesis of sulfonated aromatic polymers are directed to the polymerization of sulfonated monomers (such as (b), (d), (g), (j), (k), and (l) shown in Scheme 3) [14, 15, 53, 54, 96–102] or coupling reactions of sulfonated compounds with functional groups attached to a polymer backbone [103, 104]. In post-sulfonation, attachment of the sulfonic acid group is restricted to the activated position *ortho* to the aromatic ether bond, as indicated in Scheme 4a, while in direct polymerization of sulfonated monomers, the sulfonic acid groups are attached to the deactivated site on the ring (Scheme 4b). An enhancement of stability toward desulfonation and a modestly higher acidity are expected for the structure shown in Scheme 4b. Recently, polymerization of sulfonated monomers was also used to obtain sulfonated polysulfone (m) via oxidation of a sulfonated polysulfide–polysulfone copolymer [105].

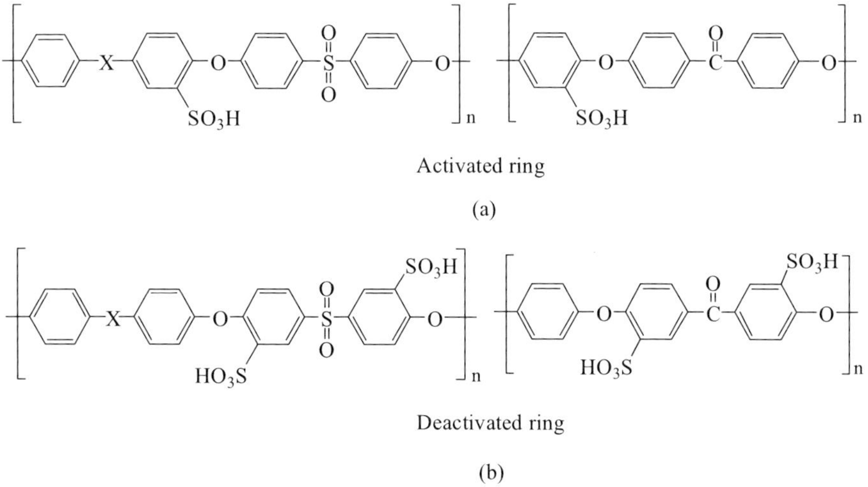

Activated ring

(a)

Deactivated ring

(b)

Scheme 4 Chemical structures of sulfonated polysulfones obtained by post-sulfonation and direct polymerization: **a** post-sulfonation; **b** direct polymerization

Since the material properties desired cannot always be obtained from a single homopolymer or copolymer, polymer blending and/or composite/hybrid formation is another strategy to improve the physical properties of polymer membranes. In the study of membranes for fuel cell application, these include cross-linked polymer blends, semi-interpenetrating networks (semi-IPNs) [106, 107], polymer–acid complexes, composite/hybrid polymers of Nafion® or sulfonated polymers with inorganic particles (e.g., TiO_2, SiO_2), inorganic acids (e.g., calcium phosphate, zirconium phosphate, heteropolyacids), and heterocyclic compounds (e.g., imidazoles). To limit the scope of this review, detailed discussions of polymer blends and composite/hybrid polymer membranes are not included here, and readers are directed to several excellent sources to learn more on these classes of PEMs [38, 41, 42].

3.1
Morphological Studies of PEMs

3.1.1
Randomly Sulfonated PEMs

As previously mentioned, proton conducting polymers combine, in a single macromolecule, a hydrophobic backbone and an ionic functional group (usually a sulfonic acid group). The sulfonic acid functionality aggregates to form hydrophilic domains that hydrate upon exposure to water, leading to a phase separated structure. The length scales of phase separation are typically nanometer in dimension. Hydrophobic domains impart mechanical stability to the membranes, whereas hydrated, hydrophilic domains serve to transport protons and water. The morphology of PEMs has chiefly been investigated through small-angle X-ray scattering (SAXS), small-angle neutron scattering (SANS) [108], transmission electron microscopy (TEM), and atomic force microscopy (AFM).

SAXS and SANS spectra of Nafion® membranes exhibit a broad maximum in the medium angular range (the so-called ionomer peak) and an upturn of scattered intensity at very low angles. A variety of models are proposed based on information gathered in order to fit calculated and observed scattering profiles, and to interpret transport properties and mechanical properties [109–115]. Central to each of these models is the recognition that ionic groups aggregate in the perfluorinated polymer matrix to form a network of clusters that allows for significant swelling by water and efficient ionic transport through nanometer-scale domains. Hsu and Gierke [109] established the formation of ionic clustering in Nafion®, and suggested the formation of inverted micelles with SO_3H groups forming hydrated clusters embedded in the fluorocarbon phase with diameters of 40–50 Å. More recently, Gebel et al. proposed a conceptual description for structural evolution that depends on the water content of Nafion® as depicted in Fig. 2 [114].

In dehydrated Nafion®, isolated spherical ionic clusters were calculated to have a 15 Å diameter and an intercluster spacing of 27 Å. This diameter is significantly smaller than the intercluster distance, which explains the low ionic conductivity observed at low water content. Absorption of water induces the formation of isolated spherical pools of water having a diameter of 20 Å, with the ionic groups located at the polymer/water interface. The interaggregate distance is ~30 Å, indicating that these spherical water pools are still isolated as evidenced by the low ionic conductivity of the membrane. With increasing water content, the diameter of the cluster increases to 40 Å while the inter-

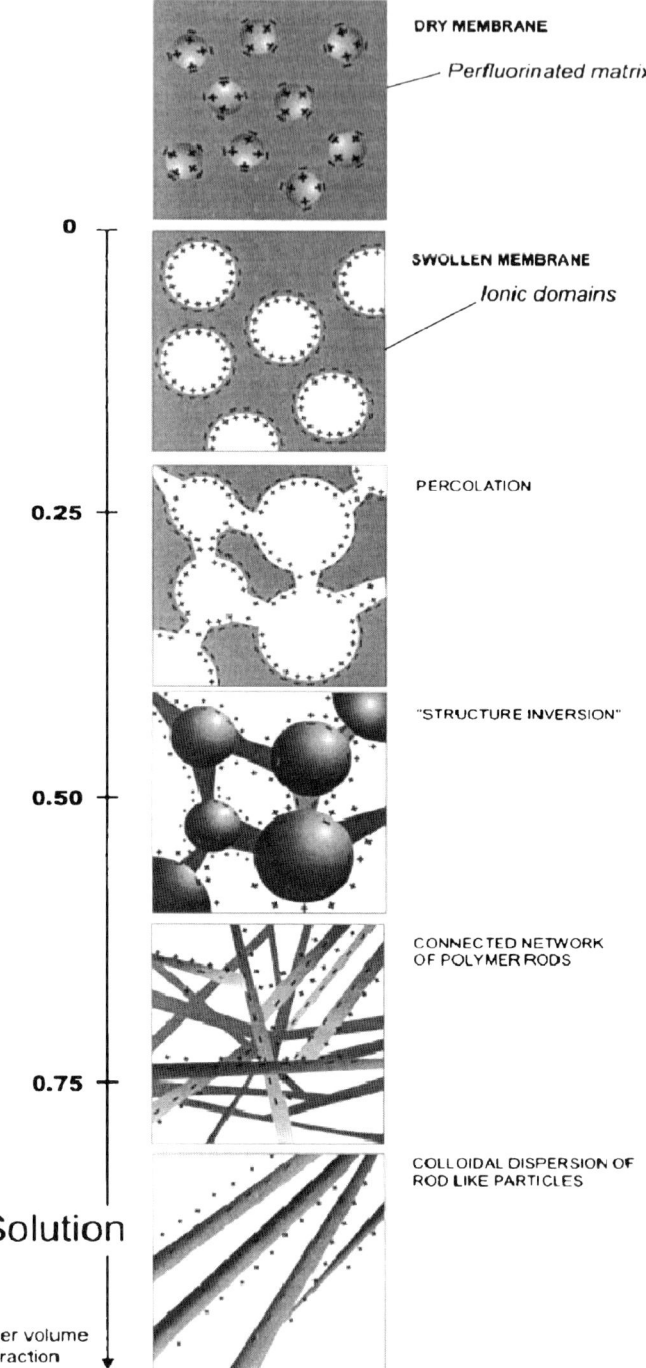

Fig. 2 Structural evolution of Nafion® depending on water content [114]

cluster distance increases only marginally. For a water volume fraction X_V larger than $X_V = 0.2$, a large increase in ionic conductivity is observed indicating a percolation threshold of ionic aggregates has been reached. Between $X_V = 0.3$ and 0.5, the structure is believed to be formed of spherical ionic domains connected with cylinders of water dispersed in the polymer matrix. The ionic domain diameter increases from 40 to 50 Å, and the increase of ionic conductivity as water content increases suggests that both the connectivity and the diameter increase. At X_V values larger than 0.5, an inversion of the morphology occurs and the membranes correspond to a connected network of rodlike polymer aggregates surrounded by water. Despite the fact that a spherical shape would be energetically favorable for the polymeric domains, apparently it is not possible to obtain such morphology at X_V values larger than 0.5 because the polymer volume associated with each ionic group is too large compared to the average distance between ionic groups along the polymer chain. The best compromise between the minimization of the interfacial energy and the packing constraints leads to the formation of connected cylindrical particles. Between $X_V = 0.5$ and 0.9, this connected rodlike network swells. The conductivity measurements indicate that the structure of the highly swollen membrane is close to the one observed for PFSI solutions.

In addition to investigations using SAXS and SANS, the morphologies of Nafion® membranes have been characterized by high-resolution TEM and AFM. These techniques provide direct visualization of phase separation and the presence of ionic clusters in the membrane. TEM images of dehydrated Nafion® membranes generally provide evidence for the existence of 3–10-nm-diameter ionic clusters, approximately spherical in shape [116–118]. Porat and coworkers carried out a TEM study on very thin, recast Nafion® solution-coated films cast from ethanol/water mixtures [118]. Zero-loss bright field images were obtained as well as Dage silicon intensified target (SIT) low light images with minimum specimen damage and specific elemental sulfur imaging. Single crystals of PTFE-like regions in the polymer with an average distance of several micrometers were observed to be randomly scattered across the film. The electron diffraction pattern of Nafion® films indicates an orthorhombic crystal structure which is similar to that of polyethylene (PE). The calculated d spacings of the corresponding lattice planes for the two polymers are also quite close suggesting similarity in their crystal structures. This result indicates that the fluorocarbon backbone of Nafion® is in the form of a linear zigzag chain as in PE and not a twisted chain as in PTFE, despite the similarity in the chemical composition of Nafion® and PTFE. Sulfonate groups were observed to be nonrandomly distributed in clusters of size 5 nm.

McLean et al. used tapping mode atomic force microscopy (TM-AFM) to characterize the fluorine-rich crystal aggregates within perfluorinated ionomers in addition to determining the relative morphological positions of the ionic species and domains [119]. Images of nonionic polymers including the Nafion® precursor polymer in its sulfonyl fluoride form served as controls to assign various phases. McLean et al. demonstrated that TM-AFM in phase-contrast mode can be used to characterize the positions in space of both ionic domains and crystalline lamellae or lamellar stacks in Nafion® membranes. Low oscillation amplitude tapping of Nafion® membranes generated remarkable images of ion clusters and their dependence on various conditions such as humidity. Figure 3 illustrates the low-energy phase images of Nafion® 117-K^+ after exposure to room temperature humidity (a) and to liquid water (b). The white regions in the images represent ionic domains. For the membrane exposed only to ambient humidity (a), the regions of ionic clusters exhibit a uniformity in spacing with an approximate size of 4–10 nm. Images taken of ionomers of varying EW show a relationship between the volume fraction of ionomer domains and the EW, although higher EW ionomers tend to appear less homogeneous in their distribution of ion-rich regions. The images of membranes exposed to water (b) indicate larger clumps of ionic regions 7–15 nm in size in the narrowest dimension and appear to have coalesced into channel-like shapes. The authors speculate that the technique provides direct images of ion clusters in their aggregated form, which when swollen with water are constrained by crystalline regions into the channel-like morphology. Notably, the analogous nonionic sulfonyl fluoride polymer under identical conditions yields no image contrast.

McLean et al. also examined the near-surface region of Nafion® membranes using a very light tapping force method. The surface of Nafion® membranes exposed to water vapor consists of a thin fluorine-rich skin region with low surface energy, ~ 1 nm thick. Based on results using different AFM techniques, the authors conclude that this region contains essentially no ionic species when

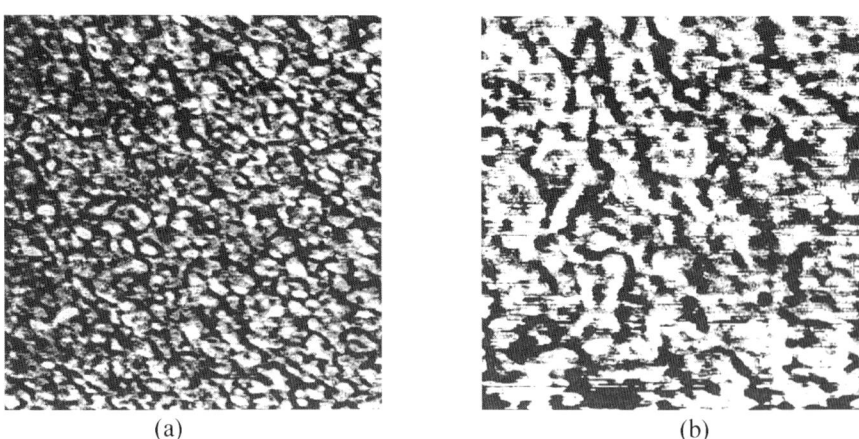

Fig. 3 TM-AFM phase images of Nafion® 117 (K^+) ionomer membrane after exposure to: **a** room temperature humidity; **b** deionized water. Scan boxes are 300×300 nm with a scale of 0–80° [119]

exposed to water vapor but rearranges rapidly when exposed to liquid water to allow diffusion of ionic groups to the surface. The rearrangement time for this process is fast (seconds) due to the high concentration of ionic groups as well as their relatively high mobility within the fluorocarbon matrix. This process is contrasted to the rearrangement process in poly(ethylene-co-methacrylic acid) ionomer neutralized with Zn metal ions, where the ionic species take substantially longer (hours) to migrate to the surface [119].

Dispersions of Nafion® are widely used in fuel cell materials such as catalyst coatings, gas diffusion media, and reinforced membranes. The properties and performance of Nafion® are directly related to its morphology, which in turn may be altered by the processing history. The morphology of recast Nafion® film has been studied in detail [73, 74, 120–122] and is directly related to the configuration of the polymer chains in the dispersions. Low EW Nafion® membranes (EW < 900) are soluble in many polar solvents, while high EW membranes (EW > 1000) are not. Dispersions of high EW Nafion® can be obtained by heating in a solvent mixture at 250 °C in an autoclave under pressure [123, 124]. As the temperature is increased, the fluorocarbon phase melts and reorganization of the ionic phase permits dissolution. Nafion® dispersions are characterized as containing relatively large colloidal aggregates of anisotropic structures [125–129]. Pinéri et al. suggest that the suspended particles resemble regular micelles with the charge sites extending out into the solution and the hydrophobic chain material buried in the interior [125]. Evaporation of the solvents at room temperature preserves the colloidal morphology. This film possesses ionic clusters but is amorphous, lacking crystallinity, mechanically weak, and soluble in a variety of polar organic solvents. When the Nafion® dispersion is annealed at high temperature (>150 °C) or cast at temperatures >160 °C, i.e., above the T_g for Nafion® (~150 °C) [120]), the ionic chains reorganize into a more entangled network in the solid film; this leads to an enhancement of crystallinity, and the films possess similar mechanical properties to as-received membranes [73, 74]. Cold cast films possess a micellar configuration with sulfonated exchange sites located on the outside of the micelle; annealing organizes the structure into an inverted micelle with the sulfonates located in the interior of the micelle.

Compared to fluoropolymers, aliphatic and aromatic polymers are less hydrophobic, and the sulfonic acid functional group is less acidic and less polar. Nafion® is a superacid, with a pK_a ~ -6 as estimated using "pK_a database 4.0" [130]. Ma et al. [131] determined the pK_a of Nafion® 117 and the sulfonated aromatic polymer, BPSH (Scheme 3b), to be -3.09 and -2.04, respectively, while the simplest aromatic sulfonic acid, benzene sulfonic acid, has a pK_a of 0.70 [132]. Aromatic polymers are more rigid than Nafion® and possess shorter ionic side chains, and therefore are expected to exhibit a lesser degree of separation between hydrophilic and hydrophobic domains. SAXS measurements on rigid aromatic polymers, such as SPIs based on naphthalenic dianhydride (Scheme 3k, $x = 5$, $x/y = 30/70$) indicate the absence of an ionomer peak, presumably due to the rigidity of the polymer chain [133]. Sulfonated polyetherketone exhibits an ionomer peak that is broadened and shifted toward higher scattering angles, and the scattering intensity at high scattering angles (the Porod regime) is higher. This indicates a smaller characteristic separation length with a wider distribution and a larger internal interface between hydrophobic and hydrophilic domains. Kreuer [134] proposed an illustration for the morphology of Nafion® and a sulfonated aromatic polymer polyetherketone (shown in Fig. 4) derived from SAXS experiments [135]. The illustration indicates the water filled channels in the sulfonated poly(ether ether ketone ketone) (SPEEKK) are narrower compared to those in Nafion®, and are less separated and more branched with more dead-end "pockets". These features correspond to a larger hy-

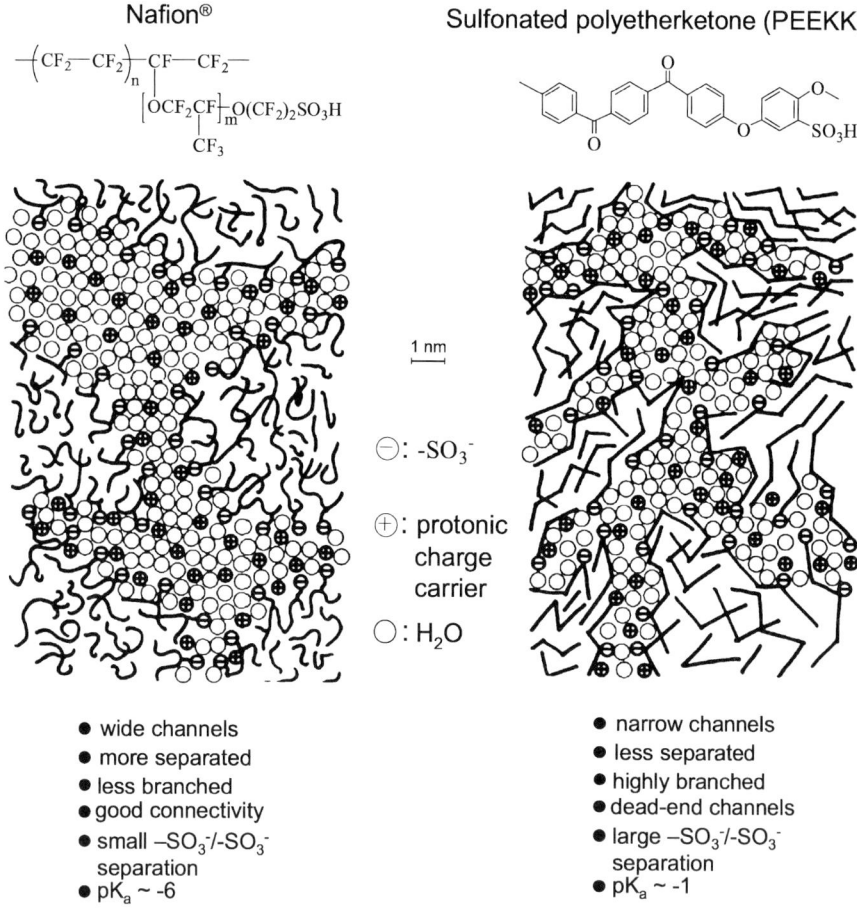

Fig. 4 Schematic representation of the morphology of Nafion® and a sulfonated polyetherketone [134]

drophobic/hydrophilic interface and, therefore, to a larger average separation of neighboring sulfonic acid functional groups.

The synthesis, film formation, and properties of random partially sulfonated polysulfones prepared by direct polymerization of disodium 3,3′-disulfonate-4,4′-dichlorodiphenyl sulfone with different biphenols is illustrated in Scheme 5. For BPSH membranes, DSC measurements show a single T_g that increases with disulfonated monomer content up to 40 mol % (corresponding to EW 588 g mol^{-1}). Membranes possessing 50 mol % (EW 500 g mol^{-1}) and 60 mol % (EW 417 g mol^{-1}) possess two T_g values, assigned to the nonionic matrix and ionic clusters, indicating that such materials are nominally phase separated [97]. TM-AFM images of different partially sulfonated polymers of BPSH are shown in Fig. 5. Non-sulfonated analogs (Fig. 5a) are featureless whereas the corresponding sulfonated polymers (Fig. 5b-e) show dark cluster-like regions 10–25 nm in dimension. A micrograph of Nafion® 117 is shown for comparison. The hydrophilic domain sizes of BPSH, and their connectivity, vary with sulfonated monomer content. For the polymer containing 20 mol % disulfonated monomer (EW 1087 g mol^{-1}), an isolated region of ionic clusters is found having diameters of 10–15 nm; for the 40 mol % polymer, the phase contrast of the hydrophilic ionic domains increases and is more readily distinguished from the nonionic matrix, but the domains are still segregated. The phase image undergoes a significant change when the sulfonate monomer content is increased to 60 mol %, and the hydrophilic ionic domains form a continuous network. With the cautionary note that these particular AFM images were obtained on dehydrated samples, it can be surmised that ionic regions in random copolymers of BPSH reach a percolation limit when the sulfonated monomer content is >50 mol %. The BPSH copolymers with a low degree of sulfonation, i.e., <50% (EW > 500 g mol^{-1}) generally show a closed domain structure denoted as *regime 1*, where isolated hydrophilic domains are surrounded by a hydrophobic matrix. Systems with a higher degree of disulfonation showed a more open or continuous hydrophilic domain structure, denoted as *regime 2*.

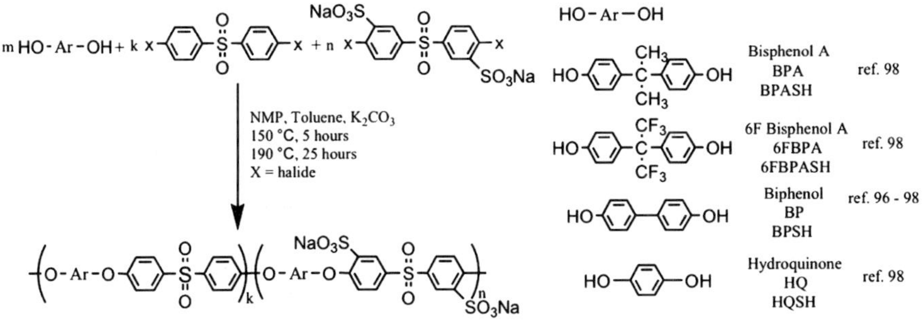

Scheme 5 Synthesis of random (statistical) disulfonated poly(arylene ether sulfone) copolymers

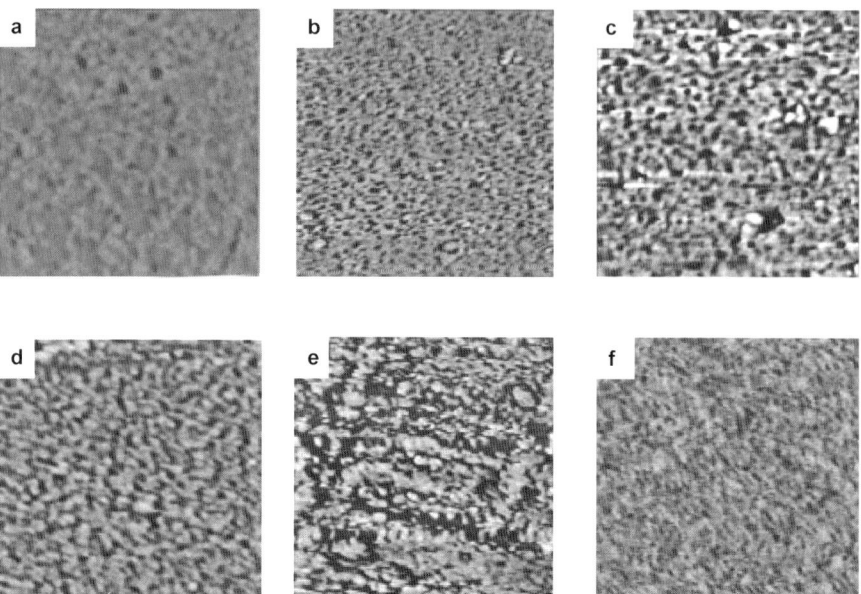

Fig. 5 TM-AFM images of random copolymer BPSH and Nafion® 117. **a** BPSH-00; **b** BPSH-20; **c** BPSH-40; **d** BPSH-50; **e** BPSH-60; **f** Nafion® 117 (acid form). Scan boxes are 700×700 nm, except **f** which is 350×350 nm [97]

The morphology of BPSH membranes was found to be dependent on the nature of hydrothermal treatment [136] and acidification [137]. Generally, low temperature conditioning of the membranes in water favors a closed hydrophilic domain structure, i.e., regime 1, while higher temperature conditioning produces a more open morphology, regime 2 [136]. The temperature of transition from regime 1 to regime 2 decreased with the temperature of hydrothermal treatment (Table 2). Acidification of solvent-cast membranes involved either immersion into sulfuric acid at 30 °C for 24 h and washing with water at 30 °C for 24 h (method 1, unboiled) or immersion in sulfuric acid at 100 °C for 2 h followed by a similar water treatment at 100 °C for 2 h (method 2, boiled). TM-AFM of BPSH containing 35 and 40 mol % of the disulfonated monomer showed that membranes boiled in acid possessed larger hydrophilic domains and a greater phase continuity than unboiled membranes (see Fig. 6). The morphology remains stable as long as the membrane is operated at a temperature below the post-treatment temperature. Heating the membranes above 120 °C at ~100% relative humidity causes an irreversible morphological change and a decrease in conductivity (Fig. 6c) [137].

SAXS data for poly(trifluorostyrene sulfonic acid) (PTFSSA) membranes possess ionomeric peaks that decreased in intensity with EW [138]. In comparison to SAXS studies of Nafion® 117, the scattering intensity is very weak.

Structural and Morphological Features of Acid-Bearing Polymers

Table 2 Transition temperature of BPSH and Nafion® 1135 [136]

Copolymer	Regime 1 to regime 2 transition temperature, T_p (°C)
BPSH-20	130
BPSH-30	100
BPSH-35	80
BPSH-40	70
BPSH-45	50
BPSH-50	30
BPSH-60	ND
Nafion® 1135	ND

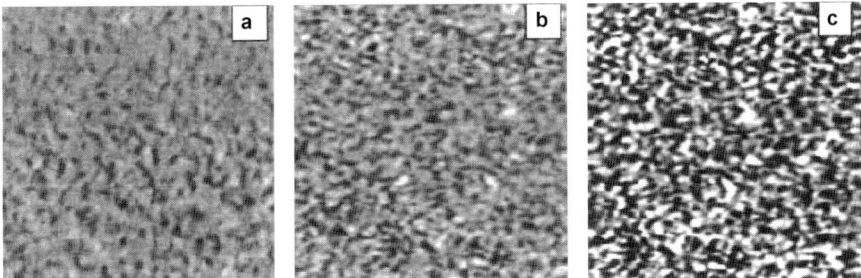

Fig. 6 TM-AFM images of random copolymer BPSH (35 mol% sulfonated monomer): **a** unboiled, **b** boiled, and **c** after high-temperature exposure (140 °C max.); scan size: 500 nm [137]

SAXS spectra for PTFSSA were analyzed using a model consisting of spherical domains randomly distributed in the polymeric matrix [139] and compared to that of Nafion® 117 [140]. The analysis suggested that these domains are very small ($R < 15$ Å). Even though evidence is presented for the existence of ionic aggregates, the large fraction of ions are dispersed homogeneously and do not give rise to ionomeric scattering [138]. The structure of water-swollen PTFSSA membranes characterized by low EW was also studied by SANS [141]. A contrast variation method based on both the deuteration of the counterions (TMA) and of the solvent was used to obtain different spectra for the same system. A local rodlike structure model, based on a perfluorinated core containing the CF_2–$CF(X)$ repeat units, a shell composed of the phenyl rings, and a second shell composed of ammonium ions and water (Fig. 7), was used for the fitting procedure. The radius of the perfluorinated core was calculated to be 4 Å and the external shell of substituted phenyl rings was 2.5 Å. The thicknesses of the phenyl ring and counterion shells are in agreement with their molecular size, and the ratio of the core volume to the shell

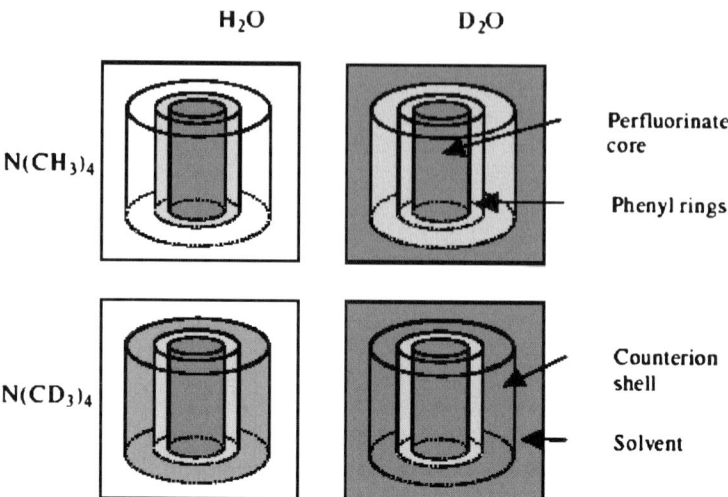

Fig. 7 Schematic representation of the scattering length density profile used for the fitting procedure for the different counterion–solvent combinations. The radius of the perfluorinated core is 4 Å and the thicknesses of the shells are 2.5 Å for the phenyl rings and 4.5 Å for the counterions [140]

correspond to the volume fraction of the fluorinated region. TEM analyses performed on dry Pb^{2+}-stained PTFSSA membranes [138] showed there is some evidence of a small degree of microphase separation, but it is not as well developed as in the case of Nafion® [116–118] and the distribution of ions is largely homogeneous.

3.1.2
Block Copolymer PEMs

Compared to random copolymers, block copolymers offer a diversity of morphologies and display a microphase-separated morphology in which the physicochemical properties of individual block components can be realized in a single polymer structure [142]. SAXS and SANS analyses of sulfonated polyimides indicate that the effect of charge content on SAXS and SANS profiles is not significant, while the effect of the charge distribution is very pronounced [133]. This also suggests that a polymer chain microstructure has a more significant impact on the morphology of PEMs than the copolymer composition. Recently, researchers have directed attention to preparing PEMs with controlled microstructure in order to study its effect on morphology, and ultimately on a PEM's physical properties.

Sequenced naphthalenic polyimides have been synthesized via a two-step condensation polymerization of aromatic diamines and dianhydrides as depicted in Scheme 6. The average length of the ionic sequence is controlled by

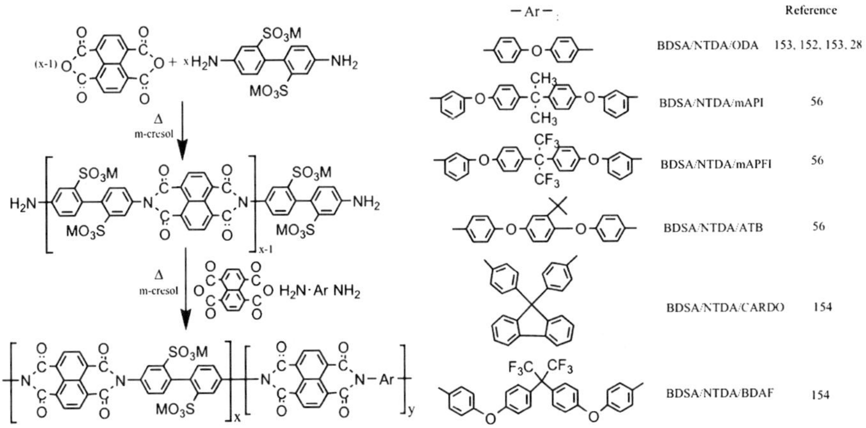

Structural and Morphological Features of Acid-Bearing Polymers 79

Scheme 6 Synthesis of sequenced sulfonated naphthalenic polyimides

the molar ratios of monomers. Gebel et al. [143] studied the effect of block length of 4,4′-diaminobiphenyl 2,2′-disulfonic acid (BDSA)/naphthalene-1,4,5,8-tetracarboxylic dianhydride (NTDA)/4,4-oxydianiline (ODA) copolymer (Scheme 3k) on its morphology using SANS (Fig. 8). For small sequences ($x = 3$), a broad maximum is clearly visible. Increasing the sequence lengths causes an increase in the upturn in intensity at very small angles, which indicates the development of large scale heterogeneities. SAXS profiles of more flexible phthalic sulfonated polyimides also indicated that as the ionic sequence length decreases, the ionomeric peak shifts to higher q; concurrently, the intensity $I(q_{max})$ and the intensity scattered at zero angle $I(q \to 0)$ decreases [133]. This behavior is in agreement with smaller and closer ionic domains as the ionic sequence length decreases.

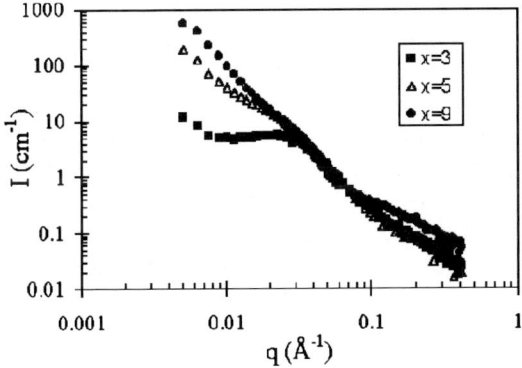

Fig. 8 SANS spectra of water-swollen $x/y = 30:70$ sulfonated polyimides possessing varying ionic block length [143]

SAXS profiles of sulfonated polystyrene-b-(ethylene-co-butylene)-b-sulfonated polystyrene (SSEBS) membranes (Scheme 2c) reveal cylindrical morphologies when the degree of sulfonation is <34 mol%. In comparison, hydrated Nafion® ionic clusters are reported to be irregular in shape and size, and randomly distributed [147]. However, different morphologies are also observed for SSEBS when films are cast from selected solvents [148, 149]. The morphology of SSEBS membranes cast from tetrahydrofuran (THF), methanol (MeOH), and THF/MeOH mixtures are reported. THF is an aprotic good solvent for EB and PS blocks, while MeOH is a protic good solvent for the SPS block. Membranes cast from THF and prepared from polymers having 27 mol% sulfonation exhibit a well-ordered lamellar morphology (Fig. 9a). The morphology of membranes cast from THF/MeOH is deformed with a diffusive phase boundary, and the phase domains are disorderly interconnected (Fig. 9b). The morphologies depend on the solvent compositions and the presence of residual solvent during membrane formation. As solvents evaporate the solids content in solution increases and the block copolymer membrane starts to phase-separate. In addition, the concentration of methanol in the mixed solvent increases with time because the vapor pressure of THF is higher than that of methanol (at 25 °C, 160.0 vs 129.2 mmHg). A high fraction of methanol partitions into SPS domains, but virtually none partitions into EB domains. Owing to selective swelling of SPS domains, the volume ratio of SPS domains to EB domains increases, and residual methanol plasticizes SPS domains to create a disorderly interconnected structure without changing the domain spacing of the EB blocks. The transformation of the morphologies was confirmed by SAXS. Sulfonated polystyrene-b-(isobutylene)-b-sulfonated polystyrene (SSIBS) with IEC be-

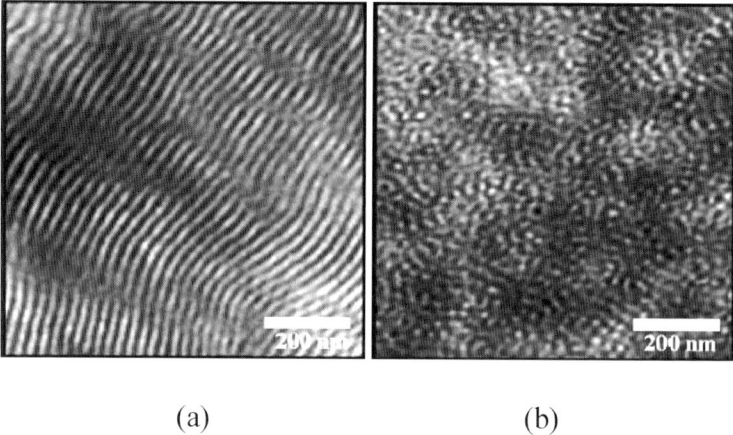

(a) (b)

Fig. 9 Cross-sectional TEM images of SSEBS (27 mol% sulfonation) cast from **a** THF and **b** a mixed solvent of MeOH/THF (20 : 80 v/v) [149]

tween 0.5 and 1.0 mmol/g exhibits an ordered lamellar structure [150]; the morphology of SSIBS-based membranes is also affected by the nature of the casting solvent [151].

Compared to aliphatic block copolymers, fluoro-block analogs are of special interest in fuel cell technology due to their enhanced hydrophobicity and thermo-oxidative stability. However, the number of main chain, fluoro-block copolymers are few because fluorous monomers cannot readily be polymerized by living ionic polymerization [152] or pseudo-living radical polymerization [153–156]. One approach, however, is to first synthesize halogen-terminated, low molecular weight fluoropolymers by means of telomerization, which are then used to initiate ATRP of a non-fluorinated monomer. For example, telomerization of vinylidene difluoride in the presence of $BrCF_2CF_2Br$ provides α,ω-dibrominated poly(vinylidene difluoride) (PVDF), which is subsequently used to initiate the ATRP of styrene to form PS-b-PVDF-b-PS triblock copolymers [156]. This α,ω-dibrominated PVDF oligomer (M_n = 1200 Da) was also used to prepare block copolymers of polysulfone-b-PVDF [158] by condensation polymerization, and subsequently sulfonated to prepare proton conducting membranes (Scheme 7) [159]. TEM analyses of sulfonated polysulfone homopolymer (SPSU) and sulfonated block copolymer (SPSU-b-PVDF) membranes (Fig. 10) show the presence of ionic aggregates. The size of the aggregates is smaller in the block copolymer (∼7 vs ∼11 nm) for the high IEC polymers. Ionic aggregates are also observed for low IEC polymers (Fig. 10c and d) but in addition to small aggregates the block copolymer possesses larger regions of ionic aggregation, similar to those reported for polymer blends of sulfonated polymers [160]. It appears that these 50–200 nm size domains are the result of gross phase separation of ionic and nonionic regions. However, it should be noted that the micrographs illustrated are measured under vacuum with the membrane in its dehydrated state, while transport properties such as conductivity are measured in their hydrated state.

PVDF, terminated with trichloromethyl groups and prepared by telomerization of VDF, has been used to initiate the ATRP of styrene, methyl methacrylate, methyl acrylate, and butyl acrylate to form various diblock copolymers [161]. The drawback of the telomerization approach is the low molecular weight of the fluoropolymer segments produced (2500 g mol^{-1}), which are too small in comparison to the non-fluorous segments for the block copolymers to take on fluoropolymer characteristics. In order to address this, chain transfer polymerization of fluoromonomers has been used to obtain higher molecular weight, halogen-terminated fluoropolymers. In one example, vinylidene difluoride (VDF) and hexafluoropropylene (HFP) were copolymerized by emulsion copolymerization in the presence of a halogenated chain transfer agent. This gave poly(VDF-co-HFP) possessing halogenated termini by chain transfer polymerization, which were subsequently used to initiate the ATRP of vinylic monomers [162]. Fluoropolymer

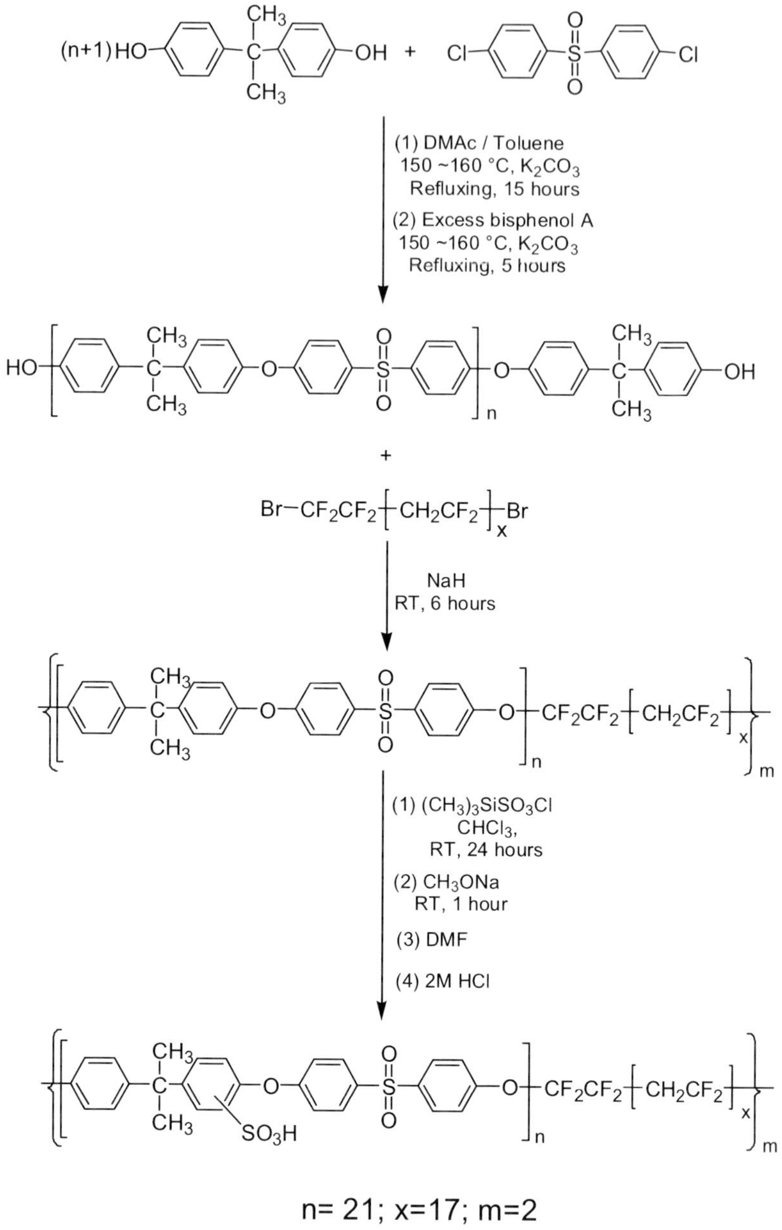

n= 21; x=17; m=2

Scheme 7 Synthesis of sulfonated polysulfone-b-PVDF

macroinitiators were used to initiate the ATRP of styrene and form P(VDF-co-HFP)-b-PS diblock copolymers. Post-sulfonation yields a proton conducting polymer (Scheme 8) [163]. Partially sulfonated polymers exhibited two T_g

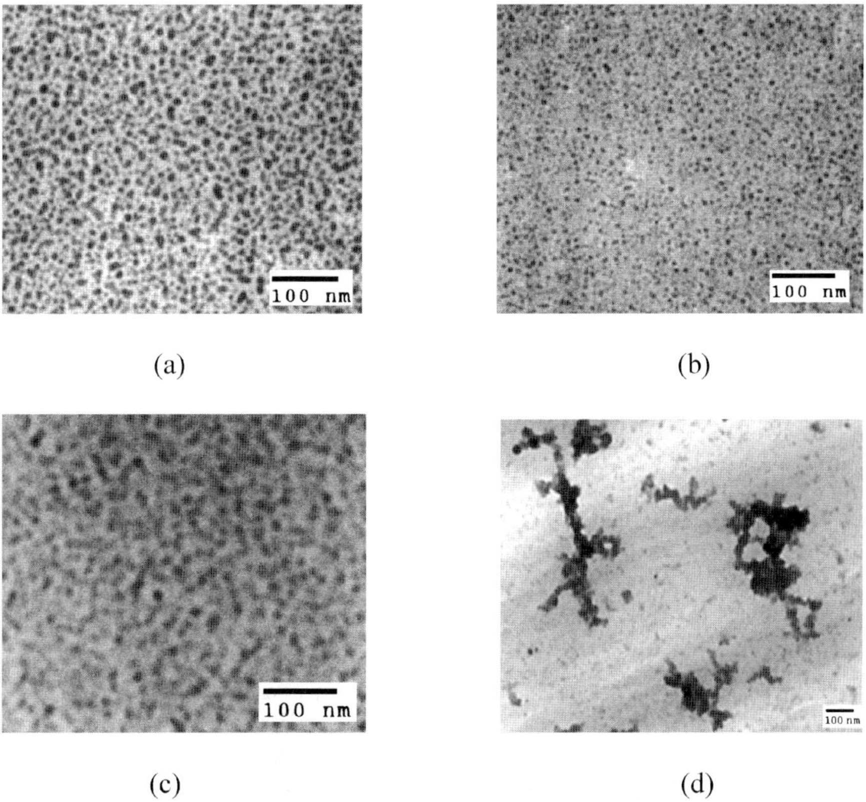

Structural and Morphological Features of Acid-Bearing Polymers

Fig. 10 TEM micrographs of polymer membranes. **a** SPSU (IEC = 1.55); **b** SPSU-*b*-PVDF (IEC = 1.62); **c** SPSU (IEC = 0.83); **d** SPSU-*b*-PVDF (IEC = 0.78) [159]

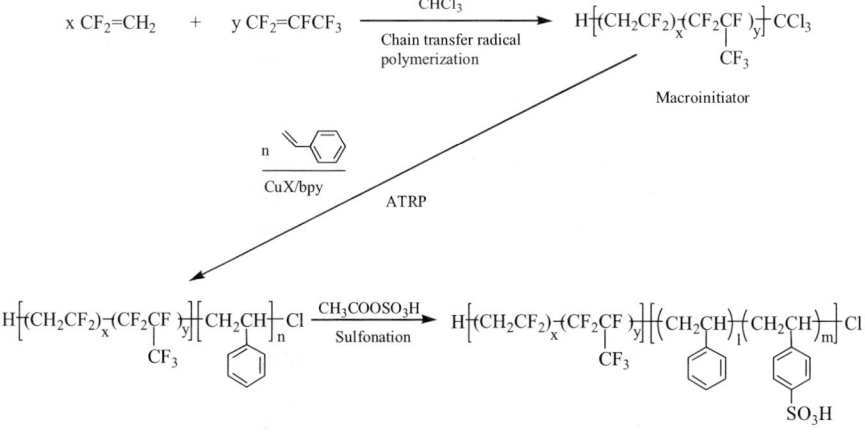

Scheme 8 Synthesis of P(VDF-*co*-HFP)-*b*-SPS diblock copolymer

values corresponding to the P(VDF-co-HFP) block (–34 to –43 °C) and the sulfonated polystyrene block (86 to 166 °C), respectively. The T_g of sulfonated polystyrene increases with increasing degree of sulfonation, while the T_g of the fluorous block decreases.

The morphology of ionic, fluorous block copolymers was characterized by TEM (Fig. 11). No distinct ion domains are observed for membranes possessing low degrees of sulfonation (DS) (12 mol%, IEC = 0.26 mmol g^{-1}). No clear phase separation is observed for a membrane possessing a degree of sulfonation of 17%. With increasing DS (20–40 mol%), a distinct morphology develops as observed in Fig. 11B–D. Ordered, connected ionic channels are observed that possess 20–40-nm interdomain spacings and width 8–15 nm. TEM indicates that the network is three-dimensional. Upon traversing the se-

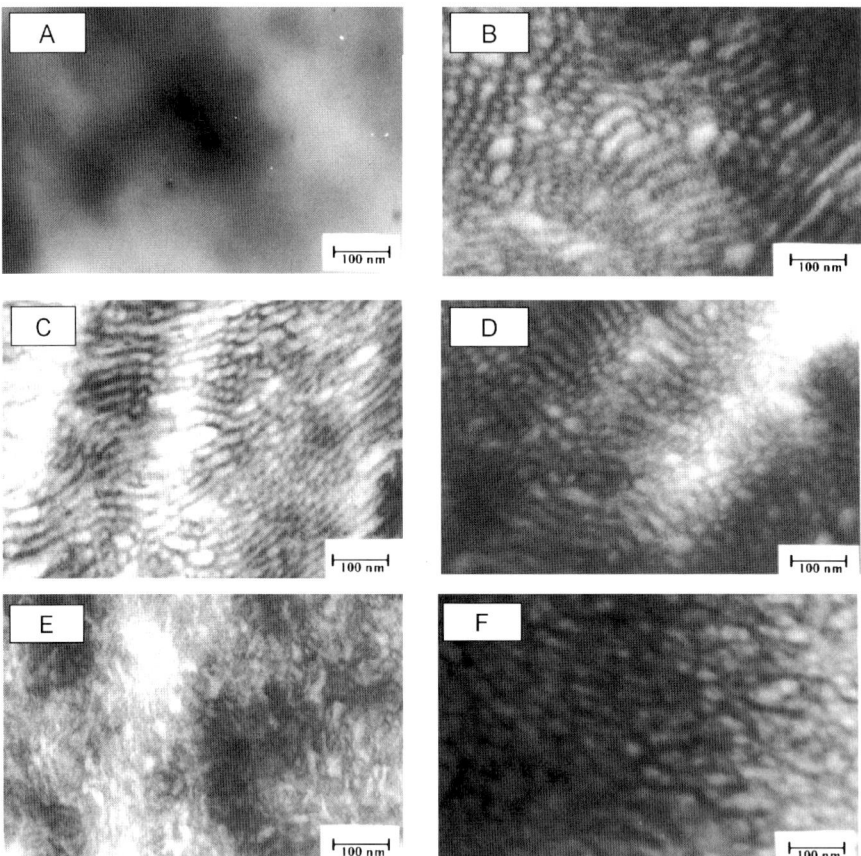

Fig. 11 TEM images of P(VDF-co-HFP)-b-SPS block copolymer membranes: **A** DS, 12%; IEC, 0.23 mmol g^{-1}; **B** DS, 22%; IEC, 0.62 mmol g^{-1}; **C** DS, 32%; IEC, 0.89 mmol g^{-1}; **D** DS, 40%; IEC, 1.08 mmol g^{-1}; **E** DS, 49%; IEC, 1.31 mmol g^{-1}; **F** DS, 100%

ries of membranes with increasing DS, a disruption in ordered morphology is observed (see Fig. 11E,F). For membranes with 100% DS, the interface between ion channels and the hydrophobic matrix is less sharp and the ionic domains aggregate, tending toward disordered structures.

3.1.3
Graft Copolymer PEMs

Graft copolymers are a special case of block copolymer and are characterized as possessing blocks of one or more species connected to the main chain. These blocks are derived from a different species of monomer than those that make up the main chain. Of the graft copolymers studied in the form of proton conductive membranes, the most extensively studied are those formed by radiation-grafting styrene monomer onto/into matrices with subsequent sulfonation of polystyrene [46]. The grafting and sulfonation processes introduce ionic conductivity while maintaining the desirable mechanical integrity associated with the base polymer. The process uses prefabricated commercial films and thus circumvents difficulties in obtaining thin films of uniform thickness. Parameters such as cross-link density and membrane thickness can also be controlled. Although ion clusters are observed by SAXS [164], and TEM and AFM images show that phase separation occurs in the order of 100–250 nm [165], systematic studies of the effect of molecular structure on morphology are not fully developed.

In the context of acid-bearing polymers, graft copolymers, in which ionic polymer grafts are attached to a hydrophobic backbone, are useful model systems for studying structure–property relationships, especially if the length of the graft and the number density of graft chains can be controlled. The length of the graft chain has the potential to influence the size of ionic domains, whereas the number density of graft chains determines the number of ionic domains per unit volume. Collectively, the size and number density of ionic aggregates/clusters are expected to control the degree of connectivity between ionic domains. Nevertheless, examples of controlled graft copolymers bearing an acid functionality are relatively few, because of the limited synthetic capabilities of preparing "living" sulfonic acid-based monomers and the lack of suitable strategies for their incorporation. Methodology has been reported to synthesize ion-containing polymers in which poly(styrene sulfonic acid) (PSSA) graft chains are attached to a hydrophobic polystyrene (PS) backbone [166, 167]. In this method, poly(sodium styrene sulfonate) (PSSNa) consisting of 32 repeat units and possessing a polydispersity index of 1.25 was prepared by stable free radical polymerization (SFRP). "Living" chains of PSSNa were terminated with DVB to form PSSNa macromonomers (*mac*PSSNa), which were subsequently copolymerized with styrene to form PS-*g*-*mac*PSSNa graft copolymers, as illustrated in Scheme 9.

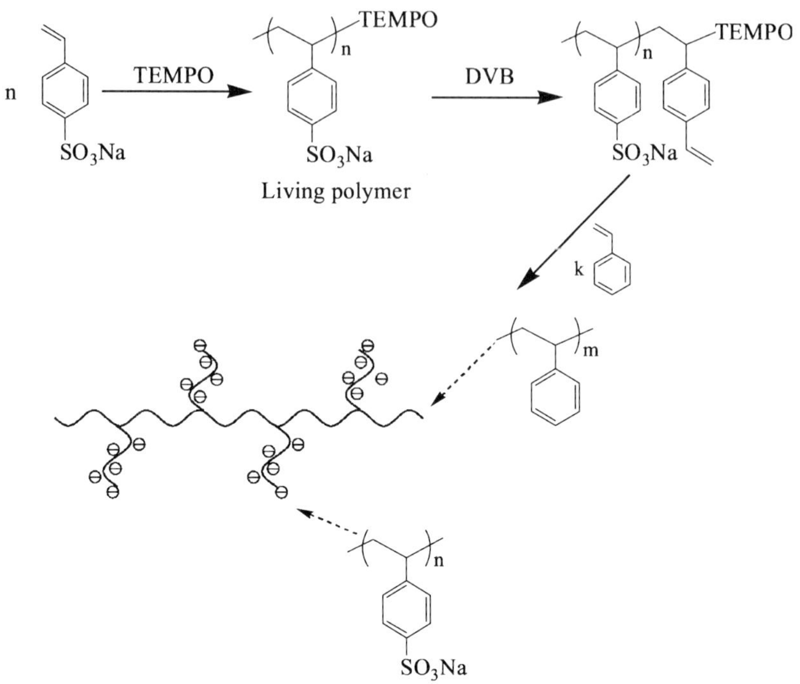

Scheme 9 Synthesis of PS-g-macPSSNa graft copolymers

TEM analyses of the graft copolymer membranes show ionic domains 5–10 nm wide that are visibly connected to yield a continuous network. Figure 12 shows that the ionic network is developed to a greater extent as the ion content is increased. Micrographs of random copolymer membranes show little evidence of microphase separation.

Several variations on this theme of polymer design have been reported. In one example, the length of the ionic graft in PS-g-macPSSA polymers was also shown to exert an influential role on the membranes' morphology [168]. Membranes incorporating longer ionic side chains (i.e., graft chains possessing degrees of polymerization of 17, 29, 62, and 102) phase separate to a greater extent for a given ionic content. Membranes possessing long graft chains and low ion contents, where there is insufficient ion content to form an extensive network, possess isolated ionic domains as shown in Fig. 13a. However, with increasing ionic content a continuous aqueous/ionic network is formed (Fig. 13b). A recent addition to the area of graft copolymer PEMs is composed of a partially fluorinated polyarylene backbone to which has been grafted polystyrene side chains [169]. TEM analysis of this material has shown a number of different morphologies. These morphologies range from wormlike at low grafting density to cylindrical at high grafting density.

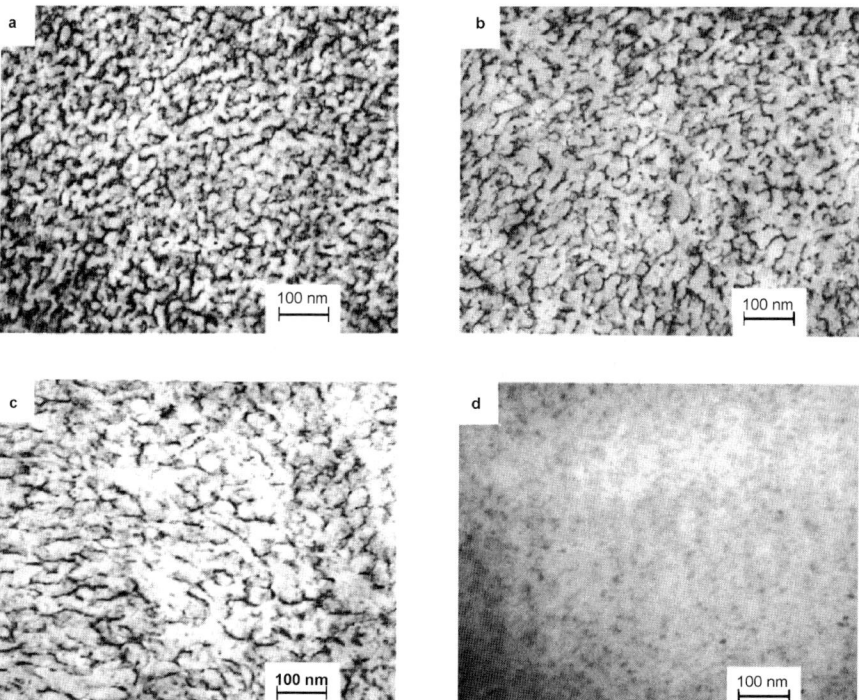

Fig. 12 TEM micrographs of PS-g-macPSSA graft copolymer membranes possessing ion contents of **a** 19.1 mol%, **b** 11.9 mol%, and **c** 8.1 mol%, and **d** a random copolymer membrane (PS-r-PSSA), ion content 12.0 mol% [166]

Not surprisingly, higher conductivity values are observed with increasing side chain content.

3.2
Proton Conductivity

Proton conductivity is often the first characteristic considered when evaluating membranes for fuel cell application. Proton conductivity is dependent on chemical structure, morphology, equivalent weight (EW = 1000/IEC, IEC: ion-exchange capacity), water content, and temperature. Conductivities and water content of fully hydrated ETFE-g-PSSA, PTFSSA, SSEBS, Nafion®, and BPSH membranes (see Schemes 2 and 3) as a function of EW are shown in Fig. 14a and b, respectively. Since EW and water content are correlated, plots of conductivity versus water content yield similar trends to plots of conductivity versus EW, as shown in Fig. 14c. Data for ETFE-g-PSSA, PTFSSA, SSEBS, and Nafion® membranes measured at room temperature are reported in references [138, 170] and [171], respectively. Data for BPSH membranes at 30 °C are taken from references [97] and [137]. Morphology, conductiv-

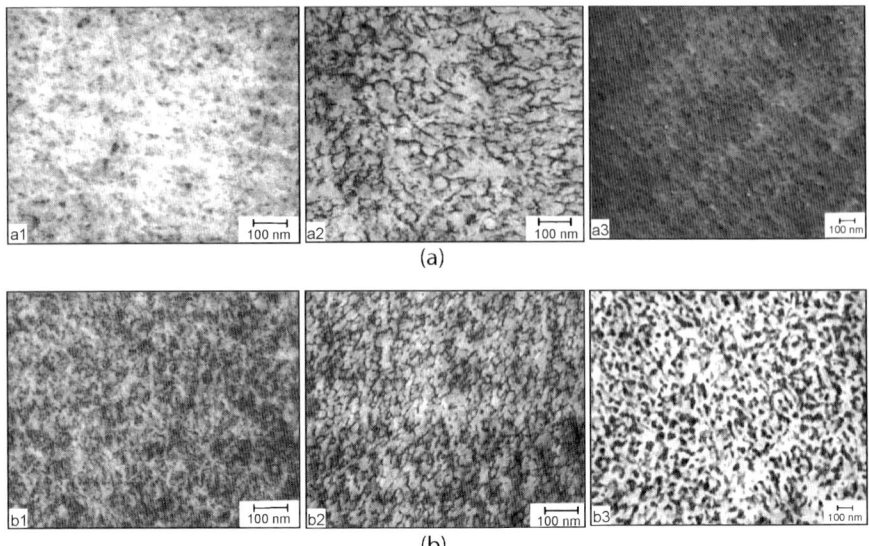

Fig. 13 TEM images of PS-*g*-*mac*PSSA membranes as a function of graft length (degrees of polymerization of *mac*PSSA provided in parentheses) for **a** low and **b** high ion content (given as mol%). (*a1*) PS-*g*-*mac*PSSA(21), 8.1 mol%; (*a2*) PS-*g*-*mac*PSSA(32), 8.1 mol%; (*a3*) PS-*g*-*mac*PSSA(102), 8.6 mol%; (*b1*) PS-*g*-*mac*PSSA(21), 16.6 mol%; (*b2*) PS-*g*-*mac*PSSA(32), 14.8 mol%; (*b3*) PS-*g*-*mac*PSSA(102), 13.2 mol% [57]

ity, and water uptake of BPSH membranes are reported to be dependent on the method of acidification, while these same properties for Nafion® 1135 membrane are independent of the method. For comparison with other membranes, proton conductivity and water uptake data for BPSH membranes treated using "method 1" are used, because this is similar to the method used to treat ETFE-*g*-PSSA, PTFSSA, and SSEBS membranes [138, 170]—i.e., soaked in sulfuric acid for 48 h to ensure complete protonation, and stored in Milli-Q water for at least 24 h prior to use.

Values of proton conductivity of ETFE-*g*-PSSA, Nafion®, PTFSSA, and BPSH membranes are strongly dependent on EW (Fig. 14a). The conductivity of Nafion® and PTFSSA membranes maximizes at intermediate EWs, while the conductivity of ETFE-*g*-PSSA and BPSH membranes increases with decreasing EW. ETFE-*g*-PSSA membranes possess an exceptionally high conductivity. Compared to ETFE-*g*-PSSA membranes, a moderate conductivity can be achieved for Nafion®, PTFSSA, and BPSH membranes by adjusting the EW of the membrane. However, based on the conductivity–EW relationship for SSEBS membranes, the maximum conductivity obtained is only ~0.05 S cm^{-1}, and this is much lower than that of other membranes.

In Fig. 14b, water contents are shown to range from 86 to 26%, 81 to 28%, 62 to 45%, 44 to 11%, and 60 to 11% for PTFSSA, SSEBS, ETFE-*g*-PSSA,

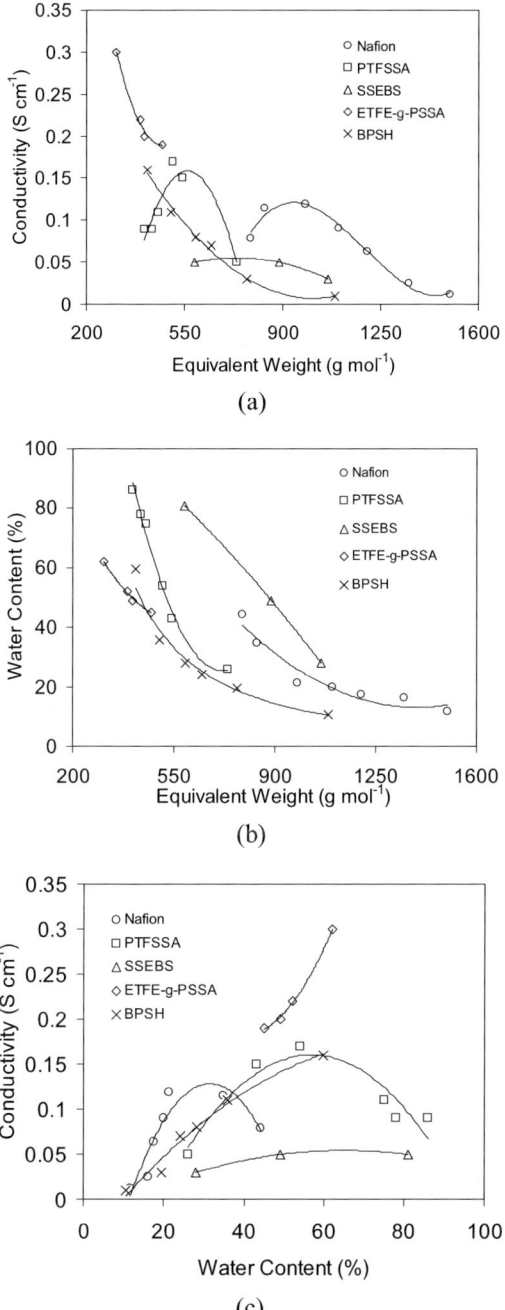

Fig. 14 Plots of proton conductivity, water content, and EW for various membranes. **a** Conductivity vs EW; **b** water content (wt%) vs EW; **c** conductivity vs water content (wt%)

Nafion®, and BPSH membranes, respectively. Water contents vary considerably but generally decrease with increasing EW as expected. Compared to Nafion® membranes, ETFE-g-PSSA, BPSH, PTFSSA, and SSEBS are able to support large contents of acid functionality (low EW) and take up large masses of water without dissolution. Even though the perfluoro structure is very hydrophobic, such high ion content and water content cannot be achieved with Nafion® without dissolution of the membranes. In the low EW range, water content in PTFSSA and BPSH increases rapidly with decreasing EW, while water content in ETFE-g-PSSA and SSEBS increases much more slowly with decreasing EW.

As illustrated in Fig. 14c, the conductivity of Nafion® and PTFSSA membranes maximizes at intermediate water contents, while the conductivity increases with increasing water content for ETFE-g-PSSA and BPSH membranes, and the conductivity is relatively independent of water content for SSEBS membranes. A plot of conductivity versus H_2O/SO_3^- ratio (λ, the number of water molecules per fixed ion site) is shown in Fig. 15. Large values of λ (>100) are exhibited by low EW PTFSSA and SSEBS membranes. These values are much larger than those of the ETFE-g-PSSA series membranes, even though low EW ETFE-g-PSSA membranes possess water contents up to 62%. Low EW PTFSSA and SSEBS membranes imbibe relatively large amounts of water such that the ion groups are significantly diluted. Nafion®, BPSH, and ETFE-g-PSSA membranes possess exceptionally low EW and hence the dry membranes possess high concentrations of sulfonic acid groups. Even though ETFE-g-PSSA membranes imbibe a significant mass of

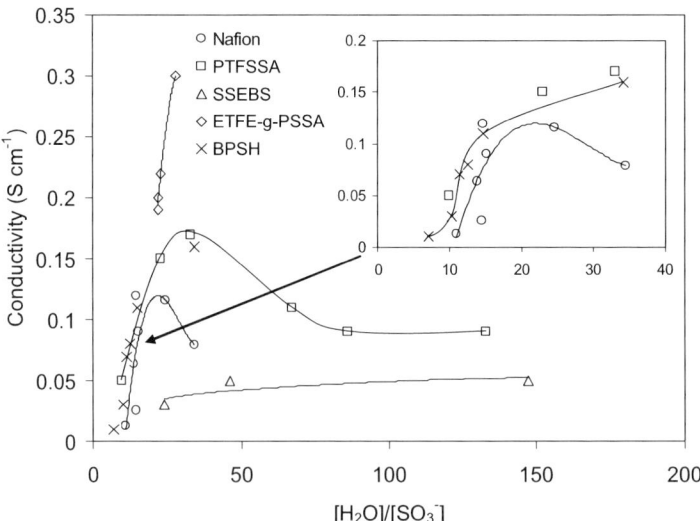

Fig. 15 Plot of conductivity versus H_2O/SO_3^- ratio for various membranes. *Inset*: expanded view

water (up 62%), this value is lower than anticipated based on water content–EW relationships observed for other membranes, e.g., PTFSSA, SSEBS, BPSH, and Nafion®. ETFE-g-PSSA membranes are prepared from preformed hydrophobic membranes (ETFE) by the incorporation of polystyrene and its subsequent sulfonation. The base ETFE membrane limits the extent to which water is imbibed. The volume of water imbibed by ETFE-g-PSSA is significant but offset by the initial high ionic content of the material, so that the λ ratios are very much lower than those of low EW PTFSSA and SSEBS membranes, and comparable to Nafion® and BPSH membranes.

The conductivity of ETFE-g-PSSA and BPSH membranes increases with λ indicating that additional water per ion group assists proton conduction. This is interpreted to mean [138] that the membranes are "water poor" in the context of ion conductivity (because λ values are low). The conductivities of the SSEBS membranes appear less sensitive to λ even though their values range from 24 to 147. The PTFSSA series yields a different trend: the conductivity increases with λ for 735, 542, and 509 g mol^{-1} EW membranes indicating that they can be classified as being water poor, i.e., an increase in water content (by virtue of lowering the EW) leads to an increase in conductivity. However, in traversing the series 509, 455, 436, and 407 g mol^{-1} EW, conductivity first decreases with λ and then becomes independent of this ratio in a manner exhibited by the SSEBS series of membranes. Nafion® membranes show a similar trend to PTFSSA membranes but the EWs available are not sufficiently low, and thus λ is not sufficiently high to observe this leveling off effect: the conductivity increases with λ for 1500, 1350, 1200, 1100, and 980 g mol^{-1} EW membranes, while in traversing the series 834 and 785 g mol^{-1} EW, conductivity decreases with λ.

There are difficulties in attempting to correlate proton conductivity with water content. These include: the very high conductivity of ETFE-g-PSSA membranes; the absence of an effect of λ on the conductivity of SSEBS membranes; and the volcano-type curves exhibited by Nafion® and PTFSSA membranes for plots of conductivity versus EW, water content, and λ. More insight into these anomalies is obtained when the overall proton concentration in the water-swollen membranes is taken into consideration (Fig. 16). PTFSSA membranes exhibit a volcano-type curve in which the maximum conductivity correlates to the intermediate EW membranes. Choosing PTFSSA 509 (54% H$_2$O) as a reference point, it is apparent that the lower EW PTFSSA membranes possess a lower conductivity even though their water content is higher. This is interpreted to mean that the low EW PTFSSA membranes are "water rich"; membranes of lower EW possess lower conductivity simply because their acid concentration is diluted. The leveling-off effect observed for plots of conductivity versus λ for low EW PTFSSA membranes (Fig. 15) is explained on the basis that increasing the acid content of the dry membrane by lowering the EW is offset by an increasing dilution of the acid. In contrast, upon traversing the series PTFSSA 509–735, the conductivity decreases

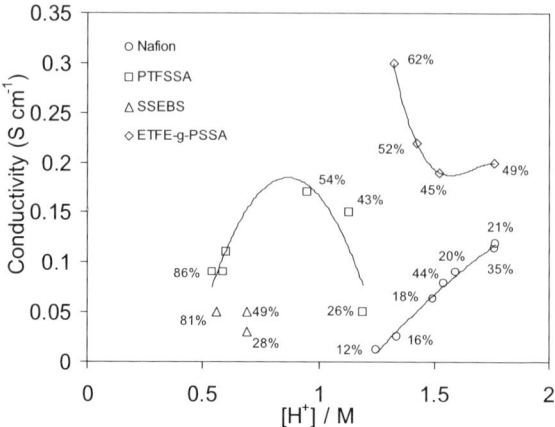

Fig. 16 Plot of conductivity versus [H$^+$] for various membranes

even though [H$^+$] increases; however, the conductivity is very sensitive to water content in this regime, suggesting that these membranes are water poor from the perspective of optimizing conductivity. As EW is increased, [H$^+$] increases due to a disproportionate decrease in water uptake. The conductivity is compromised due to low water content and the diminishment of a hydrophilic, ion conductive network.

The [H$^+$] of SSEBS membranes does not vary significantly even though their EWs and water contents do. This is presumably due to similar reasons given for PTFSSA membranes; the increased acid content of the dry membrane obtained for lower EW membranes is offset by an increasingly disproportionate water uptake that effectively dilutes the acid. Unlike PTFSSA, SSEBS membranes do not exhibit a maximum in conductivity. This may be due to the limited data set and the relatively lower elasticity of the base polymer, which enables the membranes to imbibe a large amount of water. The SSEBS 684 membrane is most likely water rich, as evidenced by its high λ, and its conductivity is likely determined largely by its acid concentration. SSEBS 887 and 1062 possess 49 and 28% water, respectively, and are judged to be water poor. From the structures of the base polymers, it is presumed that the SSEBS membranes possess a lower internal elastic force than the polymers constituting the PTFSSA membranes; as a result, for a given acid concentration in the dry membrane, SSEBS membranes swell to a greater extent when exposed to water. This would explain why the corresponding acid concentration and hence proton conductivity of the swollen SSEBS membranes is lower.

For Nafion® membranes, a linear relationship between conductivity and ionic concentration is realized. The [H$^+$] of the ETFE-g-PSSA membranes is comparatively larger than those of PTFSSA and SSEBS membranes, and similar to Nafion®. However, the water content and λ for ETFE-g-PSSA mem-

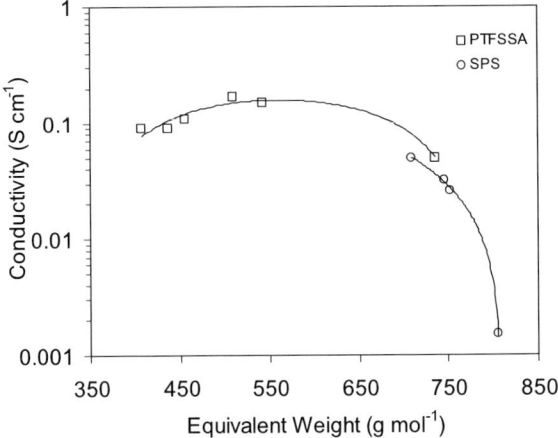

Fig. 17 Proton conductivity of SPS and PTFSSA

branes are larger than for Nafion® and thus proton conductivity is enhanced. The high [H^+] and adequate λ for ETFE-g-PSSA membranes explain their high conductivity. Even so, these membranes are water poor, as judged by the fact that conductivity increases with λ.

As described previously, Nafion® membranes exhibit much higher proton conductivity than any other aliphatic and aromatic PEMs bearing similar ion content due to the special chemical structure and morphology. Partially sulfonated polystyrene (SPS) and PTFSSA have the same backbone except PTFSSA possesses a fluoropolymer backbone. The dependence of proton conductivity on EW for SPS at 22 °C [172] and PTFSSA [138] membranes is shown in Fig. 17. The conductivities of the fluorinated block copolymer P(VDF-co-

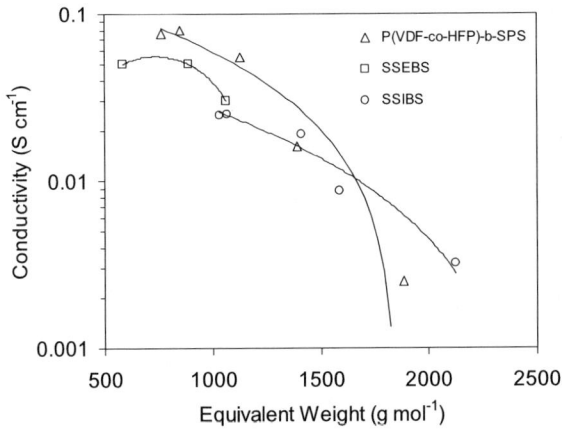

Fig. 18 Conductivity of P(VDF-co-HFP)-b-SPS, SSEBS, and SSIBS

HFP)-*b*-SPS [163] and non-fluorinated block copolymers SSEBS [138] and SSIBS [173] are shown in Fig. 18. It can be concluded that fluorous structures generally enhance the conductivity of PEMs. Also, as can be inferred from Fig. 17, fluorous polymers can be prepared with lower EWs without dissolution in water due to the increased hydrophobicity of the backbone. The higher conductivity is most likely due to the decrease in swelling, the concomitant increase in [H$^+$], and an increased hydrophilic network.

3.2.1
Effect of Polymer Microstructure and Morphology

SPSU-*b*-PVDF membranes (Scheme 7) may be compared with SPSU membranes in order to examine the effect of fluoropolymer blocks on membrane morphology and proton conductivity [159]. Proton conductivity is illustrated in Fig. 19. In the low IEC regime, SPSU-*b*-PVDF copolymers exhibit a conductivity up to four times greater than that of the corresponding homopolymers. As IEC is increased the conductivity of the homopolymer and block copolymer increases, but the difference in conductivity between the two series diminishes. At the highest IEC examined the effect of the fluoropolymer block is negligible. Water contents and λ values ($[H_2O]/[SO_3H]$) are similar for both polymers at a given IEC. The enhancement in conductivity of the block copolymers is suggested to be due to an enhancement of the ionic network. For higher IEC membranes, where both the concentration of acidic sites and λ values are also much higher and the network of ions more fully formed, the presence of the relatively small fluoropolymer segment is less influential.

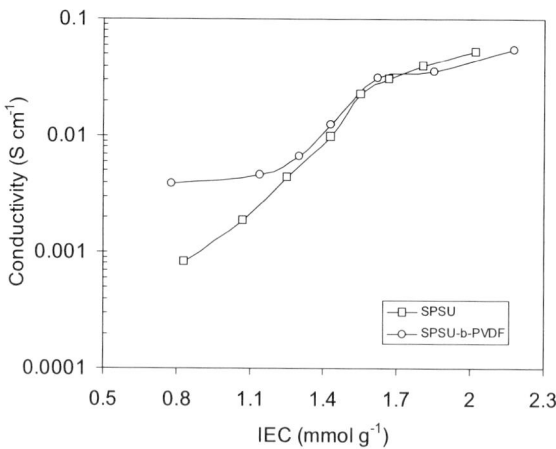

Fig. 19 Dependence of proton conductivity on IEC for SPSU and SPSU-*b*-PVDF membranes. 30 °C, 95% relative humidity

Structural and Morphological Features of Acid-Bearing Polymers 95

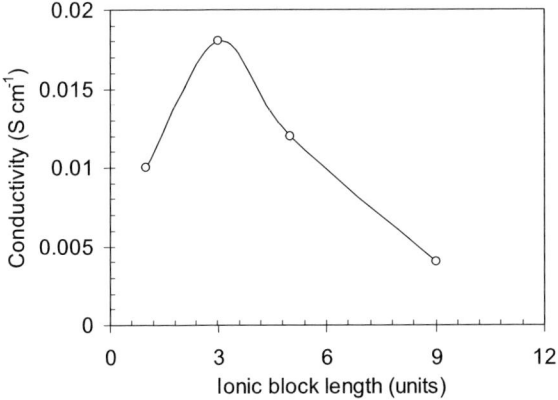

Fig. 20 Ionic conductivity versus ionic block length for BDSA/NTDA/ODA copolymers (Scheme 3k) with $x/y = 30:70$

In another example of the importance of polymer microstructure, the variation of proton conductivity of BDSA/NTDA/ODA copolymers (Scheme 3k) with ionic sequence length is shown in Fig. 20 [143], where an ionic length of "1" represents a statistical copolymer. Water uptake for copolymers with different ionic block lengths (block lengths in brackets) are: (1) 39.3%, (3) 40.6%, (5) 44.3%, and (9) 69.3%. Corresponding λ ($[H_2O]/[SO_3H]$) values are 17, 18, 20, and 30. Proton conductivity does not linearly relate to water uptake in the polymers, and this may reflect a secondary influence of proton concentration in the hydrated membranes.

Membranes consisting of sequenced sulfonated naphthalenic polyimides (ionic block length of 5) also exhibit higher conductivity and λ values

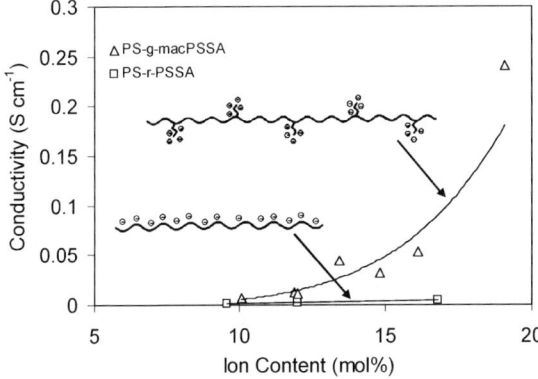

Fig. 21 Comparison of proton conductivity of PS-r-PSSA and PS-g-macPSSA copolymer membranes as a function of ion content

than membranes prepared from analogous random copolymers [53]. The difference in conductivity between sequenced and random polymers decreases as IEC increases [53], as observed for SPSU-*b*-PVDF block copolymers. In similar work, sequenced copolymers of NTDA, ODA, and 9,9-bis(4-aminophenyl)fluorine-2,7-disulfonic acid (BAPFDS), in which the ionic block length is 2, are reported to exhibit higher proton conductivity than random copolymers [54]. Increased phase separation is suggested as the explanation [53, 54].

The effect of incorporating ionic groups in graft chains, as opposed to randomly along the main chain, can be observed by comparing the properties of PS-*g*-*mac*PSSA membranes to those based on PS-*r*-PSSA membranes. As observed in Fig. 21, the graft structures exhibit a significantly higher conductivity as a function of ion content even though the water contents are much lower. PS-*g*-*mac*PSSA membranes use their associated water more effectively to transport protons [167].

Membranes based on PS-*g*-*mac*PSSA copolymers incorporating longer ionic side chains, i.e., graft chains possessing degrees of polymerization of 102, exhibit very poor proton conductivity when the ion contents are low (fewer graft chains per unit length of main chain) (Fig. 13a) because the polymers are below the percolation threshold for ion conduction. However, when the percolation threshold is reached and exceeded with increasing ionic content (Fig. 13b) both water uptake and proton conductivity are significantly enhanced compared to membranes prepared from short graft PS-*g*-*mac*PSSA polymers. The relationship between proton conductivity and IEC and the effect of the ionic chain length is illustrated by the plot shown in Fig. 22 [168].

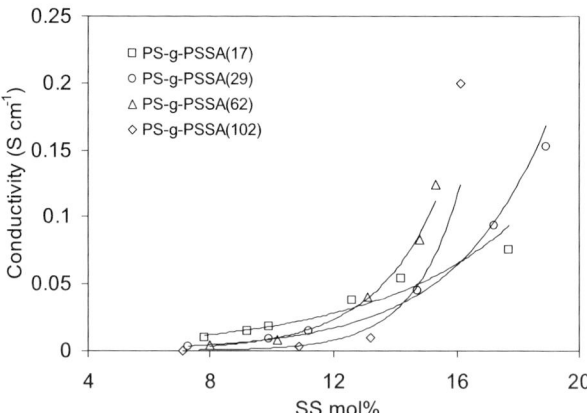

Fig. 22 Proton conductivity of PS-*g*-*mac*PSSA (DP) polymer membranes as a function of ion content (mol% styrene sulfonate, SS)

3.2.2
Effect of Pretreatment and Casting Solvents

Conductivity and water uptake of BPSH membranes (Scheme 3b) prepared using different treatments for acidification are shown in Fig. 23 (redrawn from references [137] and [174]). Membranes boiled in sulfuric acid exhibited an increase in connectivity of the hydrophilic domains, and consistently exhibited a higher water sorption capacity and greater proton conductivity than membranes possessing isolated hydrophilic domains (i.e., those treated with acid at 30 °C). From trends in water uptake, the percolation threshold of BPSH membranes treated by method 2 was shifted to a lower degree of disulfonation (from ∼50 down to 35%).

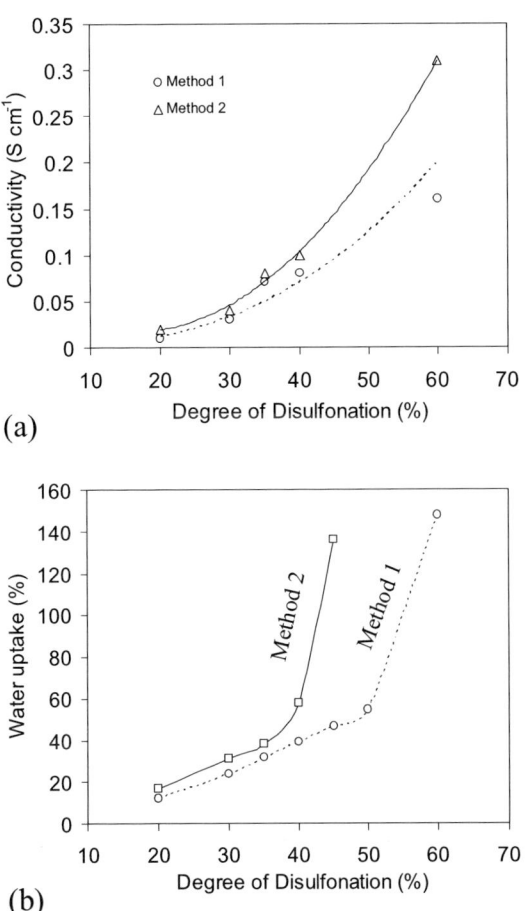

Fig. 23 Conductivity and water uptake of BPSH membranes treated by different methods. Method 1, boiling in sulfuric acid; method 2, treated with acid at 30 °C

Pozio and coworkers [175] studied the proton conductivity of solution-cast Nafion® membranes, and compared the data to those obtained from commercial Nafion® 112 and Nafion® 115 membranes. Membranes were cast under the conditions listed in Table 3. Process 01, process 02, and process 03 represent the different methods used to obtain Nafion® powder. In process 01, a commercial 5 wt% dispersion of Nafion® 117 in low aliphatic alcohols was used to cast a membrane. An alcohol-free Nafion® 117 dispersion was prepared using an azeotropic distillation, the concentrated dispersion was neutralized with 0.1 M NaOH solution, the solution was heated at 60 °C to remove solvent, and Nafion® powder was ball-milled to form a fine powder. In process 02, as-received Nafion® 112 membranes in their protonic form were heated in a mixture of ethanol/water at 250 °C to prepare a dispersion. An alcohol-free Nafion® dispersion was prepared by distillation, the dispersion was neutralized with 0.1 M NaOH solution, heated at 60 °C to remove solvent, and the resulting Nafion® powder was ball-milled. Process 03 consisted of the same procedure but, in contrast to process 02, membranes were neutralized with 0.1 M NaOH prior to dissolution. In order to form cast films, PFSI powder was mixed with the solvents listed in Table 3, sonicated, and the homogeneous dispersion placed into petri dishes. DMF–PFSI dispersions were heated at 165 °C; DMSO and EG dispersions were heated at 180 °C. Dried, cast films were re-acidified using boiling nitric acid (20 wt%) followed by two boiling water treatments.

The EW (g mol^{-1}) and hydrated thicknesses were 1020 and 55 μm for N112, 1079 and 149 μm for N115, 985 and 54 μm for EG01, 1057 and 60 μm for DMSO01, 1030 and 96 μm for DMF01, 1049 and 80 μm for DMF02, and 1000 and 96 μm for DMF03. Figure 24 shows a plot of proton conductivities at 25 °C. Although it was reported that an increase of thickness correlates with an increase in conductivity [176, 177], considering the conductivity data on

Table 3 Precursors, solvents, and temperature used for obtaining solution-cast Nafion® membranes

Sample	Nafion® precursor	Casting solvent	Heat treatment (°C, min)
N112	Nafion® 112 commercial membrane	–	–
N115	Nafion® 115 commercial membrane	–	–
EG01	Nafion® 117 commercial dispersion—process 01	EG	180, 90
DMSO01	Nafion® 117 commercial dispersion—process 01	DMSO	180, 90
DMF01	Nafion® 117 commercial dispersion—process 01	DMF	165, 90
DMF02	Nafion® process 02	DMF	165, 90
DMF02	Nafion® process 02	DMF	165, 90
DMF03	Nafion® process 03	DMF	165, 90

Structural and Morphological Features of Acid-Bearing Polymers

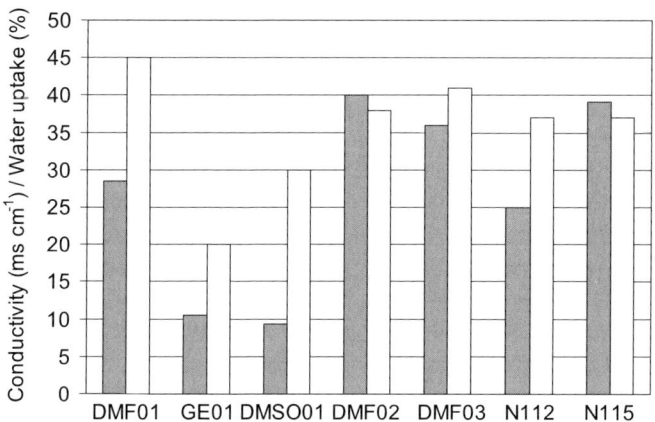

Fig. 24 Conductivities (*dark bars*) at 25 °C and water uptake (*open bars*) of solution-cast Nafion® membranes and Nafion® 112 and 115 (redrawn from [175])

a relative basis, the results indicate that different casting solvents promote different conductivity and water sorption values for recast Nafion® membranes. Data for membranes obtained using process 01 show the following trend in conductivity: DMF > EG > DMSO; only DMF provided membranes with a conductivity value comparable to that of commercial membranes. It was suggested that the Nafion®–solvent interaction plays a strong role in determining the outcome of the membranes' properties, and solvents with lower boiling point (i.e., near to the T_g of Nafion®) are preferred. Conductivity is correlated to water uptake, rather than simply EW. It was hypothesized that a continuum of ion clusters absorbs more water than isolated clusters. Samples DMF02 and DMF03, prepared by dispersing Nafion® 112 membranes, possess conductivities similar to that of Nafion® 115, which indicates that process 02 and process 03 lead to morphologically different recast membranes than those obtained from commercially available dispersions (process 01). In related research, Zanderighi et al. [121] report that aqueous dispersions of Nafion® provide membranes that absorb more water than commercial dispersions of Nafion® and have equal and/or higher proton conductivity.

The importance of the choice of casting solvent on membrane properties is also evident in studies of proton conductivity of sulfonated polystyrene-*block*-(ethylene-*co*-butylene)-*block*-sulfonated polystyrene (SSEBS) (Scheme 2c) membranes prepared from different compositions of mixed casting solvents (MeOH/THF) (Fig. 25). For example, the conductivity of SSEBS membranes possessing a degree of sulfonation (DS) of 27 mol % increases with an increasing fraction of methanol in the solvent mixture. For membranes with a DS of 42 mol %, the proton conductivity increases with concentration of methanol fraction and levels off at higher volume fraction.

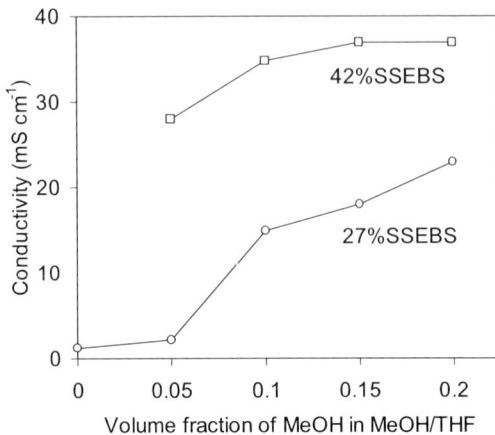

Fig. 25 Proton conductivities of SSEBS membranes with degrees of sulfonation of 27 and 42% cast from MeOH/THF mixtures (redrawn from [149])

In contrast, membranes possessing a DS of 8% had conductivities that did not vary with methanol content (not shown in this figure). These data are explained on the basis of a morphological change upon increasing the methanol content of the casting solvent.

3.2.3
Proton Transport

Rationalizing proton conductivity with membrane parameters such as EW (IEC), water content, λ, and [H$^+$] is not trivial, especially when considering such a diverse group of membranes. Proton conduction in ionic membranes has also to be viewed in the context of ionic clustering and phase separation between hydrophilic and hydrophobic domains, and issues of acidity (pK_a), as pointed out in a recent review [134]. In PEMs, ionic conductivity occurs within the continuity of hydrophilic domains: protons dissociate from their conjugate base, are solvated, and mobilized by water. The general picture of proton transport in PEMs is such that the majority of solvated protons are located in the central region of a hydrated hydrophilic nanochannel. Water in this region is bulk-like (except for low degrees of hydration), and mechanisms of local proton transport are similar to those found in bulk water [134, 178].

In bulk water, protons exist in interchangeable charged clusters of water: the hydronium ion H_3O^+; the hydrated hydronium ion (the Eigen cluster) $H_9O_4^+$ [179]; and the Zundel ion [179], $H_5O_2^+$. Many intermediate or more complex states of the "hydrated proton", $H^+(H_2O)_n$, can also be envisaged, but the three listed have been observed spectroscopically [181] and rationalized theoretically [182]. All have a finite lifetime and interconvert. Proton transfer occurs via different pathways dependent on the relative abundance of these three basic states [183].

In PEMs, proton transport properties are determined by the degree of confinement of water within the hydrophilic domain and the interaction of this water with the acidic functional groups. Confinement of water in PEMs can be considered on both a molecular and macroscopic scale. On the molecular level, for low water contents (small λ values), bulk-like water is largely absent and the water that exists is strongly bound to the ions present in the membrane. This leads to a decreased dielectric constant and reduced rate of hydrogen bond formation and bond breaking. Since the latter controls proton transport, the rate of proton transport is reduced by the confinement of water. With increasing degree of hydration, the properties of the water confined in the membrane approach those of bulk water. On a macroscopic scale, hydrated PEMs are heterogeneous systems. Since excess protons originate from immobile acidic functionality, they remain located in the vicinity of the counterion. The distribution of protonic charges in these regions and the corresponding electrostatic potential distribution is controlled by (1) the chemical interaction of the proton with the anion, (2) the local dielectric constant of the solvating water, and (3) the spatial separation of the immobile acid anions, which is usually on the order of 0.6–0.9 nm [184]. The association energy increases with decreasing acidity, decreasing dielectric constant, and increasing separation of the acidic functions. The restriction of proton transport to only the hydrophilic domains in a PEM also reduces protonic transport coefficients due to the reduced volume fraction.

Compared to Nafion®, a stronger confinement of water in the narrow channels of the sulfonated aromatic polymers leads to a significantly lower dielectric constant of the waters of hydration (20 compared to 64 in fully hydrated Nafion® [185, 186]). Of particular relevance to macroscopic models are the diffusion coefficients of water. As the amount of water sorbed by the membrane increases and molecular-scale effects are reduced, the properties approach those of bulk water on the molecular scale. Figure 26 shows the trend in proton mobility, D_σ, and water self-diffusion, D_{H_2O}, for Nafion® and the sulfonated polyetherketone membrane [134].

For high water volume fractions the extent of percolation in the hydrophilic domain appears similar for both polymer membrane systems, and since the water–polymer ion interaction is small in this regime, the water self-diffusion coefficients (D_{H_2O}) are comparable (Fig. 26). With decreasing water volume fraction, the water diffusion coefficient decreases more rapidly in sulfonated PEEKK (Scheme 3c) compared to Nafion®. The mobility of protonic charge carriers (D_σ), as obtained from conductivity data via the Nernst–Einstein relationship assuming full dissociation of the sulfonic acid functional groups, shows a similar behavior since this is approximately related to the water diffusion coefficient (Fig. 26). At very high water contents, however, D_σ is higher than D_{H_2O} indicating the influence of intermolecular

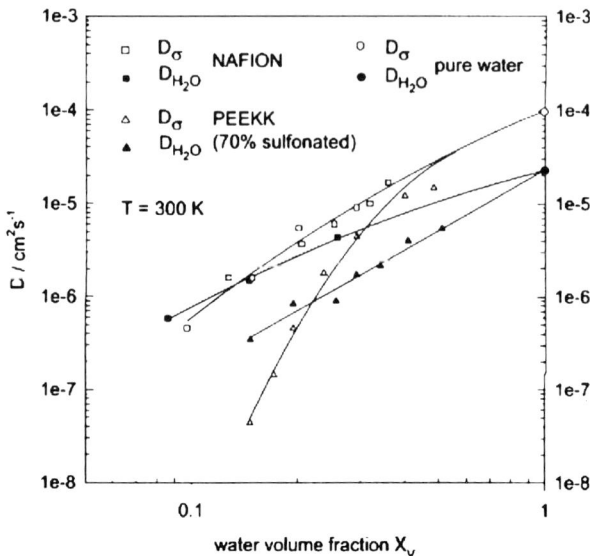

Fig. 26 Water self-diffusion coefficient (D_{H_2O}) and proton mobility (D_σ) as a function of the water volume fraction in Nafion® and sulfonated polyetherketone [134]

proton transfer on the mobility of protonic charge carriers—as in the case of dilute aqueous solutions of acids [183, 187, 188]. At very low water contents, the opposite is true. As a result of the decreasing degree of dissociation of the acidic functional group and the decreased dielectric screening of the counterion, the protons are more localized in the vicinity of the polymer-bound counterion, which leads to a larger decrease of D_σ compared to D_{H_2O} with decreasing water volume fraction. While this effect is negligible for Nafion®, it is quite pronounced in sulfonated PEEKK and is believed to be due to the higher pK_a of the acidic functional group and the lower dielectric constant of the water of hydration, which allows only for a weak dielectric screening of the negative charge of the sulfonic acid anion. The different extents of hydrophilic networks and differences in localization of protons are used to explain why the proton conductivity in sulfonated aromatic polymers decreases much more rapidly with decreasing hydration levels than in the case of Nafion® [178].

Proton diffusion coefficients for ETFE-g-PSSA membranes (Scheme 2e) have been calculated and compared to those of Nafion® 117 (Fig. 27) [189] as well as PTFSSA (Scheme 2h) [190]. Fully hydrated ETFE-g-PSSA and PTFSSA membranes exhibit much higher proton diffusion coefficients than hydrated Nafion® 117 due to their high water content. This is in agreement with the higher conductivities observed for fully hydrated ETFE-g-PSSA and, in general, PTFSSA membranes (Fig. 14). However, when proton diffusion coefficients are compared where the three polymer series overlap (e.g., $X_V = 0.1–$

Structural and Morphological Features of Acid-Bearing Polymers

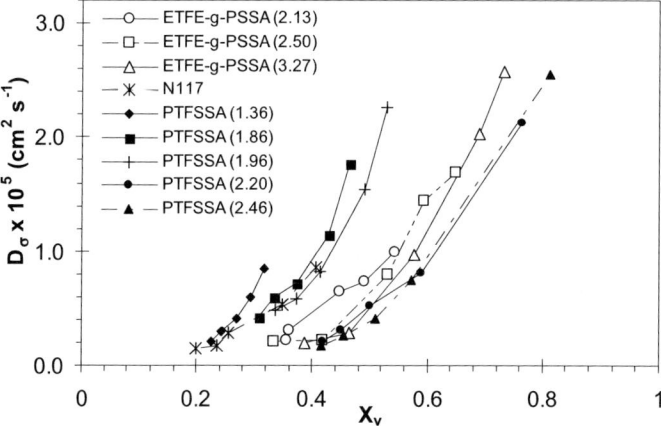

Fig. 27 Proton diffusion coefficients of ETFE-*g*-PSSA (Scheme 2e) and Nafion® 117 membranes as a function of water volume fraction, X_v

0.4), Nafion® 117 generally possesses a significantly higher diffusion coefficient from which it can be inferred that in this water volume fraction regime, the network of the hydrophilic domains in Nafion® 117 is more extensive than in either ETFE-*g*-PSSA or PTFSSA (1.96–2.46). Interestingly, however, the diffusion coefficients for PTFSSA (1.36 and 1.86) are actually higher than for Nafion® 117 although the origin for this is not clear. The transport of protonic charge carriers is often inherently correlated to the transport of water. The relationship between polymer structure, morphology, and water transport is discussed in the following section.

3.3
Water Management

Water management refers to the many different issues associated with maintaining an appropriate water balance in an operating fuel cell. From the perspective of PEMs, it includes preventing the anode and membrane from dehydrating, and the cathode from flooding. Most studies related to water management in fuel cells are primarily engineering-based and include (1) the design of new gas flow field channels with improved differential pressure loss, (2) the design of new gas diffusion layers (GDLs) to help remove excess water from the cathode to prevent flooding, (3) humidification of the anode to prevent it from drying out, and (4) using a differential pressure to help transport excess water from the cathode to the anode. An understanding of water transport through PEMs is essential in order to control water transport and fluxes and help maintain the membrane in a hydrated state.

3.3.1
The Nature of Water in PEMs

An understanding of water sorption behavior in membranes is an important area of research in PEMFCs. Water initially taken up by dehydrated membranes serves to solvate the SO_3^- and H^+ ions [191]. Water uptake in excess of $6H_2O/SO_3^-$ is believed to swell ion-rich clusters, thereby increasing the mobility of the protons [192]. Early research on hydrated Nafion® membranes by FTIR revealed three types of water: non-hydrogen-bonded water (3714 cm^{-1}), partially hydrogen-bonded water (3668 cm^{-1}), and hydrogen-bonded water (3524 cm^{-1}) [193]. More recent studies, based on calorimetry and gravimetry [194], categorize the state of water in PEMs as belonging to one of three types: (1) *free water*, water that is not intimately bound to the polymer chain through ion–water interactions and exhibits the same freezing and melting transitions as bulk water; (2) *freezable, loosely bound water*, water that is weakly bound to the polymer chain or interacts weakly with nonfreezable water and exhibits a shift in freezing/melting transitions with respect to bulk water; and (3) *nonfreezable water*, water that is strongly bound to the polymer chain through solvation effects, and which shows no detectable phase transition over the range of temperatures normally associated with bulk water. DSC is a useful method for evaluating different types of water [195, 196]. NMR spectroscopy can be used to provide additional quantitative information [197, 198].

Comparative data of λ for Nafion® 1135 and BPSH membranes measured using a combination of DSC and ^1H pulse NMR are shown in Table 4. For BPSH membranes with IEC < 1.3 mmol g^{-1}, no free water is found. Hydrated Nafion® 1135 membrane exhibits a very high free water content compared to BPSH membranes possessing IEC ≤ 2.0 mmol g^{-1}, even though in one instance BPSH membrane (IEC 2.0) has a much higher total λ than

Table 4 Comparative data of λ for Nafion® 1135 and BPSH membranes measured using a combination of DSC and ^1H pulse NMR [195]

PEMs	IEC (mmol g^{-1})	Water uptake (%)	λ Total	Non-freezable water	Loosely bound water	Free water	σ (S cm^{-1})
Nafion® 1135	0.9	33	20.2	2.2	13.1	4.9	0.11
BPSH	0.5	7	8.1	3.2	4.9	0	–
	0.9	18	10.8	3.6	7.2	0	0.01
	1.3	29	12.0	4.1	7.9	0	0.03
	1.7	56	18.0	5.1	10.8	2.2	0.08
	2.0	124	33.1	4.9	24.5	3.7	0.11

Nafion® 1135 membrane. Nafion® 1135 membranes possesses a much higher fraction of freezable water than BPSH membranes with similar total λ. This property affects not only the proton conductivity of membranes but also transport properties, such as the electroosmotic drag coefficient, and gas and methanol permeability.

In the absence of ^{1}H pulse NMR, DSC can only distinguish freezable water from nonfreezable water. Comparative data of λ (freezable and nonfreezable) determined by DSC are listed in Table 5 for various membranes. In complementary work [203], DTA measurements showed that below about six molecules of water per equivalent, ion-containing polymers

Table 5 Comparative data of λ (freezable and nonfreezable) determined by DSC for various membranes

PEMs	IEC (mmol g^{-1})	Water uptake (%)	λ Total	Non-freezable water	Freezable water	σ (S cm^{-1})	Refs.
Nafion® 117	0.91	–	20.8	12.5–13.1	7.7–8.3	–	[199]
Nafion® 117	0.97	36	20	11	9	0.09	[189]
ETFE-g-PSSA	2.13	69	22	13	9	0.15	[189]
	2.50	111	24	12	12	0.17	
	3.27	186	28	12	16	0.20	
PVDF-g-PSSA	0.30	6	9–11	9–10	–	< 0.0001	[200]
	0.90	22	9–15	9–10	–	< 0.0001	
	2.01	90	25	11	14	0.108	
	2.26	143	35	11	24	0.107	
	2.55	139	30	10	20	0.117	
PVDF-g-PSSA 5 mol % DVB	2.78	130	30	11	19	0.130	[200]
	0.66	7	15	10	5	< 0.0001	
	2.53	62	13–15	8–10	–	0.045	
	2.95	59	21	8–10	–	0.068	
PVDF-g-PSSA 5 mol % BVPE	2.29	114	28	11	17	0.100	[200]
	2.51	112	29	12	17	0.077	
Polysulfone	1.0	25	7	7	0	0.014	[201]
	1.1	30	9	9	0	0.020	
	1.3	48	16	13	3	0.032	
PAN-g-macPSSA (16)	0.71	14	20	20	0	0.003	[202]
	1.15	25	29	28.5	0.5	0.031	
	1.54	80	35	22	13	0.084	
PAN-g-macPSSA (106)	0.67	12	16	16	0	0.003	[202]
	1.13	23	22	21	1	0.013	
	1.51	91	31	21	10	0.049	
	1.85	100	36	19	17	0.098	

poly(ethylene sulfonic acid), poly(styrene sulfonic acid), poly-2-acrylamido-2-methylpropanesulfonic acid, and Nafion® do not exhibit a melting transition for water between $-40\,°C$ and ambient temperature.

Modeling and experimental validation indicate that membranes with a higher content of freezable water possess a higher proton conductivity [204]. The fraction of freezable water generally decreases with decreasing ionic content, as does proton conductivity. This phenomenon is a strong function of the polarity and flexibility of the polymer backbone, and membrane-swelling capacity. ETFE-g-PSSA, PVDF-g-PSSA, PVDF-g-PSSA cross-linked with 5 mol % BVPE and 5 mol % DVB, sulfonated polysulfone, and Nafion® 117 contain $\sim$11 molecules of nonfreezable water per sulfonic acid group, while polyacrylonitrile grafted with poly(styrene sulfonic acid) (PAN-g-macPSSA) with different graft chains contains $\sim$20. This is probably associated with the more hydrophilic backbone of the polyacrylonitrile structure. The swelling behavior of the polymers can be controlled with different cross-linkers. Polymers that have higher densities of cross-links exhibit lower water uptakes and possess smaller fractions of free water. Since water uptake, membrane conductivity, and freezable water content are lower for cross-linked membranes, a much higher grafting ratio (or ionic content) is required to reach proton conductivities comparable in magnitude to non-cross-linked membranes.

It is expected that mobile, freezable water significantly contributes to water transport across the membrane. Protons are also expected to be more mobile through free water than through highly polarized, nonfreezable water. The activation energies for proton conduction through free water and nonfreezable water are 0.15 and 0.5 eV, respectively [189, 205, 206]. Nonfreezable water may still participate in proton conduction, as has been shown in studies of low temperature proton conduction [189, 207]. No freezable water is detected for samples where the total uptake is low. Investigations of Nafion® and ETFE-g-PSSA membranes reveal that the fraction of freezable water approaches zero at 85% RH. Nonfreezable water varies between 14 and 9 H_2O/SO_3^- for ETFE-g-PSSA and $\sim$11 and 8 H_2O/SO_3^- for Nafion® 117 at humidities ranging from 100 to 85% RH. In circumstances where freezable water is absent, membrane conductivity corresponds to transport through nonfreezable water, for example, proton conductivities for Nafion® 117 and ETFE-g-PSSA at 85% RH are 0.049 and 0.069 S cm^{-1}, respectively [189].

3.3.2
Water Transport

Scheme 10 illustrates the different modes of water transport through PEM-FCs. Electroosmotic drag (EOD) is characterized by protons that drag waters of hydration as they traverse the membrane from anode to cathode. The production of water at the cathode results in a gradient of water content across

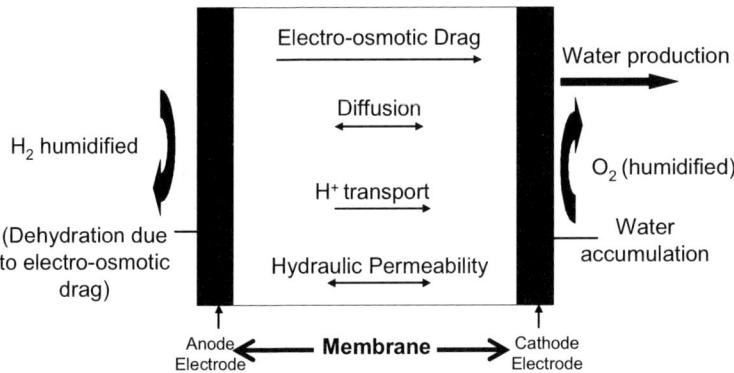

Scheme 10 Schematic drawing illustrating the modes of water transport in an operating H_2/O_2 PEMFC

the membrane that can result in the back diffusion of water from cathode to anode. The net water flux across the membrane is a combination of diffusion and EOD. The associated drag coefficient, n_{drag}, specifies the number of water molecules that are transported per proton from the anode to the cathode. Therefore, the water diffusion coefficient and the EOD coefficient, have a profound impact on fuel cell performance.

3.3.2.1
Diffusion of Water

Knowledge of the diffusion coefficient of water within a membrane is important for understanding the dynamics of water transport within PEMFCs. The water self-diffusion coefficient (D_{H_2O}) for Nafion® and other membranes has been determined by pulse field gradient NMR (PFG-NMR) [191, 199, 210–213]. Values of D_{H_2O} in Nafion® 117, BPSH, and SPEEKK membranes as a function of λ are shown in Fig. 28a, and as a function of IEC for BPSH membranes in Fig. 28b. D_{H_2O} increases with increasing water content and IEC. For high water contents (or high IEC), D_{H_2O} approaches the value of pure water (2.3×10^{-5} cm^2 s^{-1}) [214]. The increase of D_{H_2O} with increasing water content can be attributed to the higher volume fraction of free water. Nafion® 117 exhibits a higher D_{H_2O} than BPSH and SPEEKK at low water content, while at high water content, the D_{H_2O} values for Nafion® 117, BPSH, and SPEEKK membranes are similar. As described previously, aromatic polymers are reported to confine water to narrow channels, which leads to a significantly lower dielectric constant for water. As the amount of water sorbed by the membrane increases these effects are reduced, and the properties of water inside the membrane approach those of bulk water. For high water volume fractions the water self-diffusion coefficients (D_{H_2O}) are comparable (Fig. 28a). With decreasing water volume fraction,

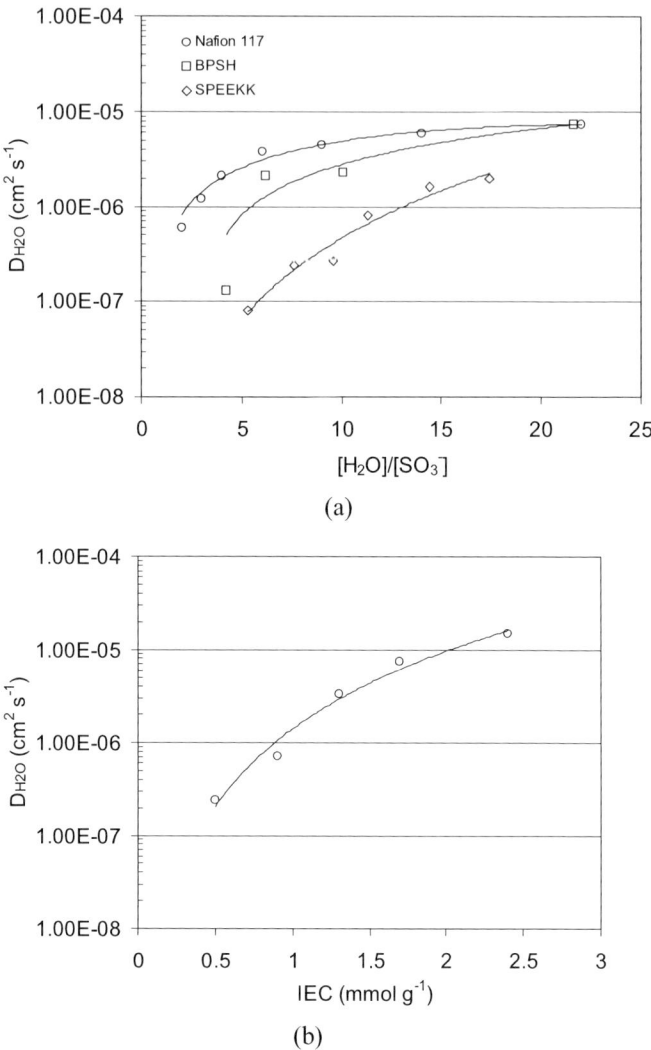

Fig. 28 D_{H_2O} in **a** Nafion® 117, BPSH, and SPEEKK membranes as a function of λ; **b** BPSH membranes as a function of IEC. Data are taken from refs. [191] (Nafion® 117), [213] (BPSH), and [178] (70% SPEEKK)

however, the water diffusion coefficient decreases more rapidly in BPSH and SPEEKK compared to Nafion®. SSEBS membranes are also reported to possess water diffusion coefficients lower than for Nafion® 117 membranes for the same water content [27, 211], but information on IEC or λ is not provided.

For the purpose of macroscopic transport of water, self-diffusion coefficients of water are converted to Fickian diffusion or the "chemical" diffusion coefficient D_{chem} using Eq. 1:

$$D_{chem} = \frac{d \ln a}{d \ln C} \times D_{H_2O}, \tag{1}$$

where a is the thermodynamic activity of water and C is the concentration of water. Techniques for measuring water diffusion coefficients also include steady-state permeation [215, 216], sorption kinetics [217–219], streaming potential measurements [220], and others [221–223]. Since D_{chem} is very sensitive to membrane history and the conditions under which measurements take place, even under similar conditions, e.g., a fully hydrated Nafion® 117-H^+ membrane, literature values vary by an order of magnitude [80]. A plot of water diffusion coefficient as a function of λ is shown in Fig. 29.

The variation in literature values of D_{chem} has been addressed in a study of water transport though Nafion® membranes [222]. Measurements of the diffusion of water were conducted wherein the membrane was exposed to liquid water on one side and dry flowing nitrogen on the other side, and water flux measurements were made across a Nafion® 115-H^+ membrane at 80 °C. Based on comparisons of their experimental data, the most accurate expression for the dependence of the D_{chem} on water content was exhibited using Fickian diffusion coefficient data obtained from [208]. The following expression was developed, where D_{chem} has units of cm^2 s^{-1}:

$$D_{chem} = 3.10 \times 10^{-3} \lambda (e^{0.28\lambda} - 1) \times \exp\left[\frac{2436}{T}\right] \quad \text{(for } 0 < \lambda \leq 3\text{)} \tag{2a}$$

$$D_{chem} = 4.17 \times 10^{-4} \lambda (161 e^{-\lambda} + 1) \times \exp\left[\frac{2436}{T}\right] \quad \text{(for } 3 \leq \lambda \leq 17\text{)} \tag{2b}$$

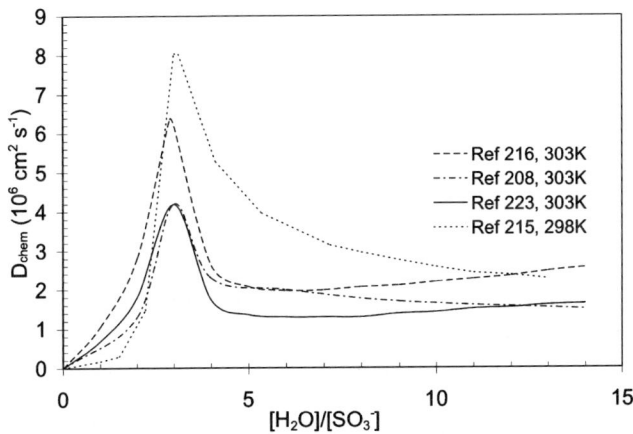

Fig. 29 Variation in D_{chem} as a function of λ for Nafion® 117

3.3.2.2
Electroosmotic Drag

The EOD coefficient (n_{drag}) is defined as the ratio of the flux of water through the membrane to the flux of protons in the absence of a concentration gradient of water [224]. EOD increases with current density and often exceeds the ability of the membrane to redistribute water by back diffusion. A hydrodynamic model for electroosmosis has been developed [225] which treats ions as spherical particles moving in a continuous viscous medium. The model describes the variation of the EOD within polystyrene-based model membranes quite well. Several methods have been utilized to measure EOD coefficients including streaming potential measurements [226, 227], the use of concentration cells [224, 228], water flux measurements [191], DMFC experiments [229, 230], and NMR spectroscopy [231]. Plots of n_{drag} values for selected membranes are shown in Fig. 30.

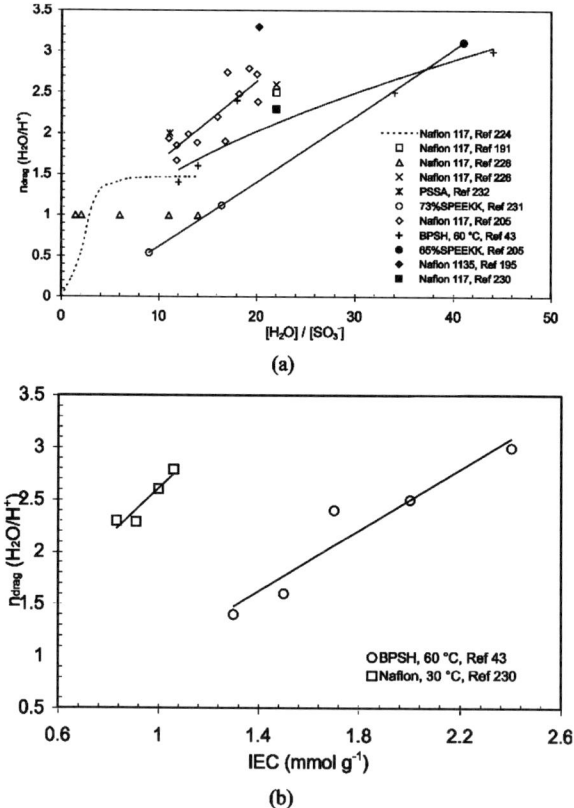

Fig. 30 Electroosmotic drag coefficient for different membranes as a function of **a** λ at ~300 K unless otherwise stated; **b** IEC

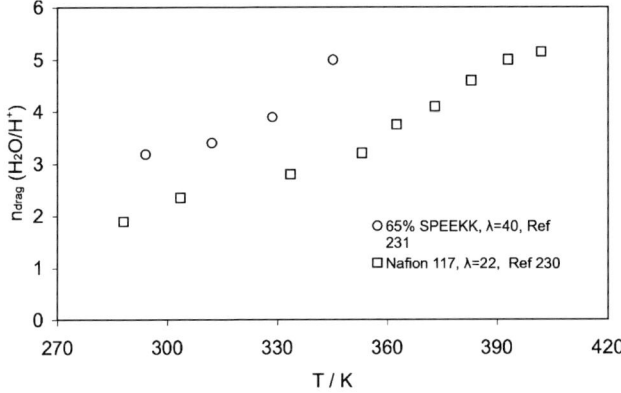

Fig. 31 Electroosmotic drag coefficient as a function of temperature for 65% SPEEKK (IEC = 1.46) and Nafion® 117 membranes

The drag coefficients vary, depending on the nature of the membrane and IEC, and, unfortunately, on the experimental method used. Generally, membranes possessing higher IECs within a series possess higher drag coefficients. Reducing the water content of the membrane by lowering the relative humidity reduces the EOD. Nafion® membranes yield higher n_{drag} values than aromatic polymers for similar water contents or IECs. The influence of channel size on EOD corroborates the importance of hydrodynamic radius. The EOD coefficient for SPEEKK is lower than that of Nafion®-type membranes due to the presence of smaller proton conducting channels in SPEEKK. Even for comparable channel dimensions, EOD values of SPEEKK are still lower, indicating that water–polymer interactions (and/or λ values) play a significant role in EOD. Increasing the temperature of the membrane increases the drag coefficient (Fig. 31)

3.4
MEA and Fuel Cell Performance

In Sects. 3.2 and 3.3, the essential properties of membranes for PEMFC are described. However, a true test of experimental membranes requires their integration and characterization in MEAs. An MEA comprises a PEM sandwiched between two GDEs. Detailed descriptions of structural and design considerations of GDEs and MEAs are captured in recent reviews [233–235]. Generally, a porous electrode structure is required, which usually consists of a Pt electrocatalyst dispersed on high surface area carbon black, held together into a cohesive coating with binding agents such as Nafion® and/or PTFE. The hydrophobic polymer backbone wet-proofs the porous electrode, thereby reducing the propensity to electrode flooding and promoting gas permeable pathways for rapid gas transport to catalytic sites [236]. It has been established that optimal electrochemical kinetics are achieved when a three-phase interface exists between polymer electrolyte, electrocatalyst, and reactant gas. The incorporation of proton conducting ionomer into the catalyst layer (CL) greatly enhances fuel cell performance [237, 238]. This gain in performance is attributed to increased catalyst utilization and optimization of the three-phase boundary.

The PEM indirectly affects the performance of a GDE through its influence on the properties of the membrane/GDE interface, and its role in determining the nature and extent of water transport between the cathode and anode. Both these effects can be observed in a recent report describing the characteristics of ETFE-g-PSSA membranes used in half-MEAs (HMEAs) in half-fuel cells [239] and complete MEAs in PEMFC systems [240]. In both systems, the GDEs contained Nafion® ionomer. Half-cell systems simulate fuel cells under flooded (or semi-flooded) conditions. In this study, the active catalyst areas of (EFTE-g-PSSA)-based HMEAs (having IEC 3.27, 2.56, 2.45, and 2.13 mmol g^{-1}, respectively) were found to increase with increasing IEC of the membrane, which is explained as being due to improved "wetting" of the CL. Active catalyst areas using the grafted membranes were higher than those based on Nafion® even though the composition of the GDEs was exactly the same. More specifically, it is shown using percolation theory that the membrane regulates the water content in the CL, and thereby the transport of reactant gases in the CL. Mass-transport properties in the half-cells were extracted by fitting data to the agglomerate model. C_0^{ref}, equivalent to the concentration of oxygen at the Pt/membrane interface, decreases with increasing IEC of the membrane; the same trend was observed in a Pt microelectrode investigation of PEMs [241]. The effective diffusion coefficient of oxygen in the CL and in the GDL decreases with increasing IEC of the membrane due to an increase in the water content of the HMEA, which reduces the electrode porosity and restricts gaseous O_2 mass transport within the CL and GDL. Consequently, the higher the IEC, the lower the rate of the oxygen reduction reaction.

In PEMFCs using full MEAs, the electrochemically active surface area of the CL reveals a slight dependence on IEC for MEAs prepared using ETFE-g-PSSA membranes having IEC 3.22, 2.73, 2.38, 1.95, and 0.75 mmol g^{-1}. The steady-state beginning-of-life polarization curves show an increase in fuel cell performance with increasing IEC for ETFE-g-PSSA membranes. The higher proton conductivity of the higher IEC membranes may partly explain this trend. However, MEAs based on Nafion® membranes yield a higher performance than most of the (ETFE-g-PSSA)-based MEAs despite the latter possessing a much higher proton conductivity. The membrane's IEC and molecular structure controls the water content within the membrane and regulates the water balance in MEAs during operation, thereby affecting the membrane/electrode interface. MEAs containing ETFE-g-PSSA membranes show an increase in uncompensated resistance (R_u) with decreasing IEC, because MEAs containing lower IEC membranes are less effective in promoting the back transport of water to the anode, and this leads to flooding of the cathode and dehydration of the anode. Compared to MEAs based on ETFE-g-PSSA, Nafion®-based MEAs exhibit a lower R_u, even though they possess a higher bulk membrane proton resistance (R_m) than the ETFE-g-PSSA membranes studied. This is due to the formation of a very good membrane/electrode interface and consequently negligible interfacial contact resistance. Bonding of Nafion® ionomer and Nafion® bulk membrane is chemically favorable. Hot pressing at 150 °C, which corresponds to the processing temperature of Nafion®, facilitates good interfacial adhesion. For ETFE-g-PSSA membranes, interfacial resistance decreases with increasing IEC of the membrane. It is believed that the increase in hydrophilicity of the higher IEC membranes provides better interfacial contact with the CL. The incompatibility between ETFE-g-PSSA membranes and GDEs impregnated with Nafion® ionomer has been reported by other research groups [242, 243].

The agglomerate model and percolation theory were used to analyze fuel cell data, and results were compared to those obtained using a half fuel cell setup. Under the fuel cell conditions employed, more facile gas transport in the cathode is observed as a result of more facile water removal via the GDL and/or back transport through the membrane. Flooding of the cathode CL in the fuel cell is not as extensive as in the case of the half fuel cell and the decrease in electrochemically active catalyst surface area, due to reduced wetting of the CL, is offset by increased O_2 mass-transport kinetics. Opposite trends in effects of IEC on the performance in fuel cell and half fuel cell systems were explained by considering the water balance in the membrane as shown in Scheme 11. In half fuel cells the effect of the increasing EOD with increasing IEC dominates water transport, whereas under fuel cell conditions this effect is overcompensated by the increasing permeability of water and increased back transport of water. Membranes with higher IEC improve fuel cell performance since they facilitate fuel cell water management. However, higher IEC membranes have a higher propensity to dehydrate and usually require extensive humidification.

In other studies of the electrode/membrane interfacial properties on MEA performance, Easton et al. [244] studied polypyrrole/Nafion® composite membranes and Nafion®-based GDEs for use in DMFCs. MEAs fabricated from membranes with high polypyrrole content on their outer surface performed poorly and possessed poor interfacial properties. However, composite membranes having less polypyrrole and more Nafion® character at their outer surface performed significantly better and exhibited better interfacial properties. A similar phenomenon is reported for Nafion®/silicon oxide composite membranes [245]. Kim et al. [246] also report that the performance of DMFCs was greatly improved using a partially fluorinated sulfonated poly(arylene ether benzonitrile) copolymer compared to non-fluorinated analogs, due to a better compatibility of the polymer electrolyte with the Nafion®-containing electrode.

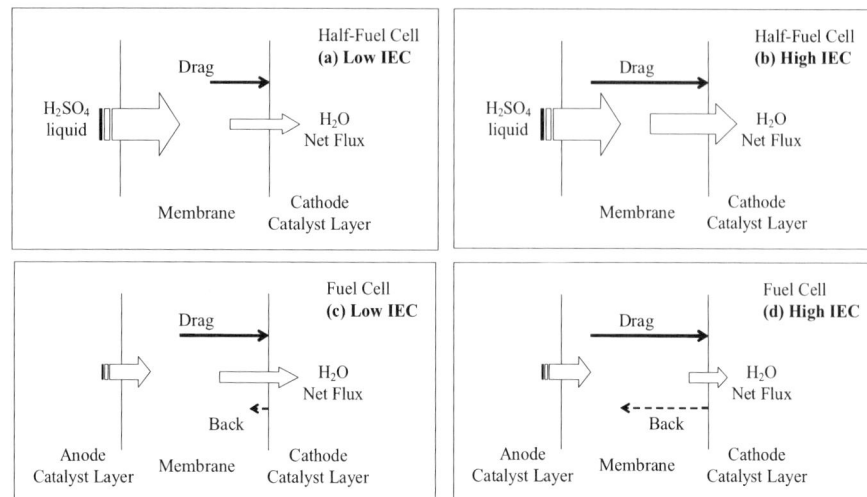

Scheme 11 Schematic diagrams illustrating different scenarios of water transport in half-cell conditions using **a** low IEC and **b** high IEC, and in fuel cells using **c** low IEC and **d** high IEC [240]

Substituting other ionomers for Nafion® in the CL is a growing area of research [43, 247]. For this purpose, formation of aqueous/alcohol dispersions of the acid-bearing polymers, and an understanding of the solution and dispersion properties of the polymers, is critical [43]. MEAs have been prepared using conventional Nafion®-containing GDEs in combination with SPEEK membranes as well as MEAs prepared from (SPEEK/PTFE)-containing GDEs in combination with SPEEK membranes [248]. The latter exhibits a significantly lower interfacial resistance. Despite this, a higher performance was still achieved with Nafion® 117 membranes due to more efficient water management and/or enhanced oxygen permeability at the interface. Further work is needed to understand the role of ionomer dispersions on CL formation, and even more is required in order to be able to correlate the role of ionomer morphology within the CL on its performance.

Fuel cell performance is the interplay of the properties of the GDEs, GDLs, and the PEMs that constitute a MEA. Fuel cell data vary widely because of different materials, fabrication procedures, and operating conditions employed. Representative data are listed in Table 6.

4
Summary and Future Prospects

The ability of a membrane to support fuel cell electrochemistry depends on many interconnected factors. Each class of polymer membrane requires in-

Table 6 Fuel cell performance data for representative PEMs

Item	PEM	IEC (mmol g^{-1})	σ (S cm^{-1})	λ	Operating conditions	I^a @0.6 V (mA cm^{-2})	Durability (h)	Refs.
1	Nafion® 117				7 cm^2 cell, 55 °C, H$_2$/O$_2$, 2 bar abs, dry gases	132		[242, 248]
	Nafion® 115				240 cm^2 cell, 80 °C, H$_2$/O$_2$, 3 bar abs, humidified gases	732		
2	Nafion® 112		0.086		30 cm^2 active area, 80 °C, H$_2$/O$_2$, ambient pressure	440		[249]
3	Nafion® 112		0.082			805		[250]
4	PTFSSA	1.4–2.5		10–133	Mark IV (50 cm^2 active area), 70 °C, H$_2$/air, 2.6 bar abs	730	14 000^b 3300^c 2100^d	[170, 251]
5	BPSH	1.3	0.03		H$_2$/air, 80 °C, RH 100%, 30 psig	840	>800	[43]
6	SPEEK (18 μm)	1.55	0.05	12	16 cm^2 active area, H$_2$/air, p_{H_2} = 3.5 and p_{air} = 4 bar abs, cell temperature 90 °C, humidifiers: 90 °C	950		[252]
7	SPEEK (50 μm)				16 cm^2 active area, H$_2$/air, p_{H_2} = 3.5 and p_{air} = 4 bar abs, cell temperature 110 °C, humidifiers: 120 °C	730		[253]
8	Sulfonated polyimide (SPI)	1.98	0.13	13	16 cm^2 active area, 70 °C, H$_2$/O$_2$, 1 bar abs, humidified gases, film cast from m-cresol	410 (at 0.7 V) 225	4300	[254]

Table 6 (continued)

Item	PEM	IEC (mmol g^{-1})	σ (S cm^{-1})	λ	Operating conditions	I^a 0.6 V (mA cm^{-2})	Durability (h)	Refs.
9	SPI (Scheme 3k)	1.98			50 cm^2 active area, 70 °C, H$_2$/O$_2$, humidified gases, 4 bars abs	500		[255]
10	SSEBS (60 μm)	1.78			45 °C, H$_2$/air, 1 bar	135 (at 0.7 V)		[256]
11	SPSU (Scheme 3a)	1.1	0.008	15	16 cm^2, 80 °C, H$_2$/O$_2$, p_{H_2} = 3 bar abs; p_{O_2} = 3 bar abs	400		[3]
12	ETFE-g-PSSA	3.22	0.19	28	H$_2$/O$_2$, 25 °C, RH 95%, ambient pressure	190		[240]
13	FEP25-g-(PSSA + 10%DVB)		0.040	~8	30 cm^2 active area, 80 °C, H$_2$/O$_2$, ambient pressure	305–665	>4150	[250]
14	FEP-g-(PSSA + 9%DVB)		0.104	10		450	6000 (60 °C) 1400 (80 °C)	[249]
15	PVDF-g-PSSA	1.83	0.050	23	5 cm^2 active area, 60 °C, atmospheric pressure, full humidification	135	130–150	[257, 258]

^{a} Current density
^{b} Membrane IEC = 2.5 mmol g^{-1}, Mark 513 (240 cm^2 active area), 80 °C, H$_2$/air, 3 bar abs
^{c} Membrane IEC = 2.5 mmol g^{-1}, Mark 513 (240 cm^2 active area), 8-cell stack, 85 °C, H$_2$/air, 3 bar abs
^{d} Membrane IEC = 2.6 mmol g^{-1}, Mark 513 (240 cm^2 active area), 75-cell stack, 85 °C, H$_2$/air, 3 bar abs

depth consideration in order to extract trends in physicochemical properties. Extending trends across different classes of membrane is difficult and must be carried out with caution, requiring extensive data sets. The main factors determining the performance of a fuel cell are proton conductivity, water transport, oxygen permeability, mechanical properties, and chemical stability.

The magnitude of proton conduction is determined by the proton concentration and the mobility of protons within the membrane. Proton mobility is enhanced by high water contents (high lambda values) and a well-formed, continuous network of aqueous domains; high water contents are favored by high IECs, whereas continuity of hydrophilic networks is favored by higher water contents and phase-separated morphologies. Proton concentration generally increases with IEC up to the point where the cohesive forces of the polymer are overcome by forces of osmotic pressure, after which additional increases in IEC result in a lowering of the proton concentration of hydrated membranes due to dilution effects.

Water transport is surprisingly the least studied property of experimental PEMs yet it has a profound effect on fuel cell performance. Membranes that yield large n_{drag} coefficients but do not allow sufficient back flow of water from cathode to anode are prone to flooding of the cathode and/or de-hydration of the anode. Membranes possessing low λ values generally yield lower n_{drag} but suffer from inadequate back diffusion of water and/or back permeation (hydraulic) of water—which is favored with high λ values. The literature indicates that the increase in back diffusion for membranes possessing high water contents offsets the potentially deleterious increase in EOD.

The paradox facing researchers attempting to develop novel polymer systems for PEM applications is that modification of a membrane to improve a particular physical property often deleteriously affects another. Case in point: increasing the IEC of a membrane has a positive influence on proton mobility and back diffusion of water, by virtue of the membrane absorbing more water. However, this results in an increase in EOD, may induce weaker mechanical properties and poorer adhesion, and may limit the thinness to which membranes may be prepared. Increasing the IEC may increase proton concentration or decrease it depending on the IEC regime. Increasing the water content of the membrane has a positive influence on proton mobility and back diffusion of water, but increases the n_{drag}, decreases the proton concentration, and decreases the membrane's mechanical strength and adhesive properties. For a given composition, enhancement of the connectivity of hydrophilic domains by control of the morphology is expected to exert the following positive influences: an increase in proton mobility, an increase in back diffusion of water, an increase in proton concentration (by virtue of limiting swelling), an increase in the membrane's mechanical strength and adhesion, and the ability to form thin membranes from the polymers.

Table 7 Observed trends that illustrate the effect of changing membrane variables on measurable performance[a]

	Proton conductance			Water transport		
	μ_{H^+}	$[H^+]$	Attainable thinness of membrane	EOD	D_{H_2O}	$P^i_{H_2O}$
Increase in IEC	✓[b]	✓, x[c]	x[d]	x[b]	✓[b]	✓[b]
Increase in λ	✓[b]	x[e]	x[f]	x	✓	✓
Increase in continuity of ionic regions	✓	✓[g]	✓[h]	x	✓	✓

[a] ✓ and x represent a positive and negative influence, respectively, on the property as it relates to fuel cell performance
[b] Due to an increased lambda value
[c] This depends on the IEC regime—see text
[d] Increasing the IEC lowers the mechanical strength and the ability to obtain thin membranes
[e] A dilution effect
[f] Due to a lower mechanical strength
[g] Due to a reduced lambda value, and if lambda $>\sim 30$
[h] Due to a greater mechanical strength
[i] Hydraulic permeability of water

The only deleterious factor observed upon enhancement of the connectivity of hydrophilic domains is an increase in n_{drag}—but this is expected to be well compensated by an increase in back diffusion of the membrane. These trends and inferences are tabulated in Table 7. In conclusion, the extensive body of literature on the structure–property relationships of proton-bearing polymers and PEMs indicates that enhancement of phase separation may be of great importance in the design of next-generation membranes for fuel cells.

References

1. EG & G Technical Services Inc. Science Application International Corporation (2002) Fuel Cell Handbook. US Department of Energy, p 82
2. Miyatake K, Shouji E, Yamamoto K, Tsuchida E (1997) Macromolecules 30:2941
3. Lufrano F, Squadrito G, Patti A, Passalacqua E (2000) J Appl Polym Sci 77:1250
4. Lufrano F, Gatto I, Staiti A, Antonucci V, Passalacqua E (2001) Solid State Ionics 145:47
5. Karlsson LE, Jannasch P (2004) J Membr Sci 230:61
6. Miyatake K, Oyaizu K, Tsuchida E, Hay AS (2001) Macromolecules 34:2065
7. Miyatake K, Hay AS (2001) J Polym Sci A Polym Chem 39:3211
8. Miyatake K, Hay AS (2001) J Polym Sci A Polym Chem 39:3770

9. Meng YZ, Tjong SC, Hay AS, Wang SJ (2001) J Polym Sci A Polym Chem 39:3218
10. Meng YZ, Tjong SC, Hay AS, Wang SJ (2003) Eur Polym J 39:627
11. Vetter S, Ruffmann B, Buder I, Nunes SP (2005) J Membr Sci 260:181
12. Zhang Y, Litt M, Savinell RF, Wainright JS (1999) Polym Prepr 40:480
13. Zhang Y, Litt M, Savinell RF, Wainright JS, Vendramini J (2000) Polym Prepr 41:1561
14. Yin Y, Fang J, Cui Y, Tanaka K, Kita H, Okamoto K (2003) Polymer 44:4509
15. Einsla BR, Hong YT, Kim HS, Wang F, Gunduz N, McGrath JE (2004) J Polym Sci A Polym Chem 42:862
16. Guo Q, Pintauro PN, Tang H, O'Connor S (1999) J Membr Sci 154:175
17. Tang H, Pintauro PN (2001) J Appl Polym Sci 79:49
18. Allcock HR, Hofmann MA, Ambler CM, Lvov SN, Zhou XY, Chalkova E, Weston J (2002) J Membr Sci 201:47
19. Allcock HR, Hofmann MA, Ambler CM, Morford RV (2002) Macromolecules 35:3484
20. Gupta B, Büchi FN, Staub M, Grman D, Scherer GG (1996) J Polym Sci A Polym Chem 34:1873
21. Nasef MM, Saidi H, Nor HM, Foo OM (2000) Polym Int 49:1572
22. Peled E, Duvdevani T, Aharon A, Melman A (2000) Electrochem Solid State Lett 3:525
23. Miyake N, Wainright JS, Savinell RF (2001) J Electrochem Soc 148:A905
24. Bonnet B, Jones DJ, Rozière J, Tchicaya L, Alberti G, Casciola M, Massinelli L, Bauer B, Peraio A, Ramunni E (2000) J New Mater Electrochem Syst 3:87
25. Mokrini A, Acosta JL (2001) Polymer 42:9
26. Nácher A, Escribano P, Del Río C, Rodríguez A, Acosta JL (2003) J Polym Sci A Polym Chem 41:2809
27. Edmondson CA, Fontanella JJ, Chung SH, Greenbaum SG, Wnek GE (2001) Electrochim Acta 46:1623
28. Serpico JM, Ehrenberg SG, Fontanella JJ, Jiao X, Perahia D, McGrady KA, Sanders EH, Kellogg GE, Wnek GE (2002) Macromolecules 35:5916
29. Jones DJ, Rozière J (2001) J Membr Sci 185:41
30. Ma YL, Wainright JS, Litt M, Savinell RF (2004) J Electrochem Soc 151:A8
31. Savadogo O (1998) J New Mater Electrochem Syst 1:47
32. Rikukawa M, Sanui K (2000) Prog Polym Sci 25:1463
33. Kerres JA (2001) J Membr Sci 185:3
34. Alberti G, Casciola M, Massinelli L, Bauer B (2001) J Membr Sci 185:73
35. Li Q, He R, Jensen JO, Bjerrum NJ (2003) Chem Mater 15:4896
36. Alberti G, Casciola M (2003) Annu Rev Mater Res 33:129
37. Schuster MFH, Meyer WH (2003) Annu Rev Mater Res 33:233
38. Rozière J, Jones DJ (2003) Annu Rev Mater Res 33:503
39. Jannasch P (2003) Curr Opin Colloid Interf Sci 8:96
40. Hickner MA, Ghassemi H, Kim YS, Einsla BR, McGrath JE (2004) Chem Rev 104:4587
41. Savadogo O (2004) J Power Source 127:135
42. Smitha B, Sridhar S, Khan AA (2005) J Membr Sci 259:10
43. Harrison WL, Hickner MA, Kim YS, McGrath JE (2005) Fuel Cells 5:201
44. Kerres JA (2005) Fuel Cells 5:230
45. Jannasch P (2005) Fuel Cells 5:248
46. Gubler L, Gürsel SA, Scherer GG (2005) Fuel Cells 3:317
47. Souzy R, Ameduri B (2005) Prog Polym Sci 30:644

48. Nakao M, Yoshitake M (2003) In: Vielstich W, Lamm A, Gasteiger HA (eds) Handbook of fuel cells: fundamentals, technology and applications, part 3, vol 3. Wiley, Chichester, p 412
49. Kreuer KD (2003) In: Vielstich W, Lamm A, Gasteiger HA (eds) Handbook of fuel cells: fundamentals, technology and applications, part 3, vol 3. Wiley, Chichester, p 420
50. Wainright JS, Litt MH, Savinell RF (2003) In: Vielstich W, Lamm A, Gasteiger HA (eds) Handbook of fuel cells: fundamentals, technology and applications, part 3, vol 3. Wiley, Chichester, p 436
51. Jones DJ, Rozière J (2003) In: Vielstich W, Lamm A, Gasteiger IIA (eds) Handbook of fuel cells: fundamentals, technology and applications, part 3, vol 3. Wiley, Chichester, p 447
52. Harris FW (2003) Advances in materials for proton exchange membranes fuel cell systems. Preprint Number 12, Asilomar CA
53. Genies C, Mercier R, Sillion B, Cornet N, Gebel G, Pinéri M (2001) Polymer 42:359
54. Guo X, Fang J, Watari T, Tanaka K, Kita H, Okamoto K (2002) Macromolecules 35:6707
55. Zaluski C, Xu G (1994) Macromolecules 27:6750
56. Mauritz KA, Moore RB (2004) Chem Rev 104:4535
57. Yang Y, Holdcroft S (2005) Fuel Cells 5:171
58. Hickner MA, Pivovar BS (2005) Fuel Cells 5:213
59. Büchi FN, Scherer GG (1996) J Electroanal Chem 404:37
60. Slade S, Campbell SA, Ralph TR, Walsh FC (2002) J Electrochem Soc 149:A1556
61. Wakizoe M, Velev OA, Srinivasan S (1995) Electrochim Acta 40:335
62. Healy J, Hayden C, Xie T, Olson K, Waldo R, Brundage A, Gasteiger H, Abbott J (2005) Fuel Cells 5:302
63. Scrosati B (1987) Electrode kinetics and electrochemical cells. In: MacCallum JR, Vincent CA (eds) Polymer electrolyte reviews 1. Elsevier, London, p 315
64. Hadjichristidis N, Pispas S, Floudas GA (2003) Block copolymers: synthetic strategies, physical properties, and applications. Wiley, New Jersey
65. Hamley IW (1998) The physics of block copolymers. Oxford University Press, New York
66. Morrison FA, Winter HH (1989) Macromolecules 22:3533
67. Kotaka T, Okamoto M, Kojima A, Kwon YK, Nojima S (2001) Polymer 42:3223
68. Stevens MP (1999) Polymer chemistry: an introduction, 3rd edn. Oxford University Press, New York, p 63
69. Brinkmann S, Stadler R, Thomas EL (1998) Macromolecules 31:6566
70. Funaki Y, Kumano K, Nakao T, Jinnai H, Yoshida H, Kimishima K, Tsutsumi K, Hirokawa Y, Hashimoto T (1999) Polymer 40:7147
71. Kim G, Libera M (1998) Macromolecules 31:2569
72. Heck B, Arends P, Ganter M, Kressler J, Stühn B (1997) Macromolecules 30:4559
73. Moore RB, Martin CR (1986) Anal Chem 58:2569
74. Gebel G, Aldebert P, Pinéri M (1987) Macromolecules 20:1425
75. Banerjee S, Curtin DE (2004) J Fluorine Chem 125:1211
76. Gierke TD, Hsu WY (1982) In: Eisenberg A, Yeager HL (eds) Perfluorinated ionomer membranes. ACS Symp Ser 180, chap 13, p 283
77. Pourcelly G, Gavach C (1992) In: Colomban P (ed) Proton conductors, solids, membranes, and gels—materials and devices. Cambridge University Press, Cambridge
78. Heitner-Wirguin C (1996) J Membr Sci 120:1

79. Grot WG (1994) Macromol Symp 82:161
80. Doyle M, Rajendran G (2003) Perfluorinated membranes. In: Vielstich W, Lamm A, Gasteiger HA (eds) Handbook of fuel cells: fundamentals, technology and applications, part 3, vol 3. Wiley, Chichester, p 351
81. Fernandez RE (1999) Perfluorinated ionomers. In: Polymer data handbook. Oxford University Press, Oxford, p 233
82. Eisman GA (1990) J Power Sources 29:389
83. Prater K (1990) J Power Sources 29:239
84. Ghielmi A, Vaccarono P, Troglia C, Arella V (2003) Ann NY Acad Sci 984:226
85. Rivard LM, Pierpont D, Freemeyer HT, Thaler A, Hamrock SJ (2003) Development of a new electrolyte membrane for PEM fuel cells. Fuel cell seminar, Miami Beach
86. Hoshi N, Uematsu N, Saito H, Hattori M, Aoyagi T, Ikeda M (2005) US Patent 2 005 130 006
87. Watakabe A (2005) US Patent 2 005 037 265
88. Savett SC, Atkins JR, Sides CR, Harris JL, Thomas BH, Creager SE, Pennington WT, DesMarteau DD (2002) J Electrochem Soc 149:A1527
89. Atkins JR, Sides CR, Creager SE, Harris JL, Pennington WT, Thomas BH, DesMarteau DD (2003) J New Mater Electrochem Syst 6:9
90. Zhou Z, Dominey RN, Rolland JP, Maynor BW, Pandya AA, De Simone JM (2006) J Am Chem Soc 128:12963
91. Dargaville TR, George GA, Hill DJT, Whittaker AK (2003) Prog Polym Sci 28:1355
92. Nasef MM, Hegazy ESA (2004) Prog Polym Sci 29:499
93. Kabanov VY (2004) High Energ Chem 38:57
94. Ogawa T, Marvel CS (1985) J Polym Sci Polym Chem 23:1231
95. Bishop MT, Karasz FE, Russo PS, Langley KH (1985) Macromolecules 18:86
96. Wang F, Hickner M, Ji Q, Harrison W, Mecham J, Zawodzinski TA, McGrath JE (2001) Macromol Symp 175:387
97. Wang F, Hickner M, Kim YS, Zawodzinski TA, McGrath JE (2002) J Membr Sci 197:231
98. Harrison WL, Wang F, Mecham JB, Bhanu VA, Hill M, Kim YS, McGrath JE (2003) J Polym Sci A Polym Chem 41:2264
99. Xing P, Robertson GP, Guiver MD, Mikhailenko SD, Kaliaguine S (2004) Macromolecules 37:7960
100. Gao Y, Robertson GP, Guiver MD, Mikhailenko SD, Li X, Kaliaguine S (2005) Macromolecules 38:3237
101. Xing P, Robertson GP, Guiver MD, Mikhailenko SD, Kaliaguine S (2005) Polymer 46:3257
102. Asensio JA, Borrs S, Gmez-Romero P (2002) J Polym Sci A Polym Chem 40:3703
103. Einsla BR, McGrath JE (2004) Prepr Symp Am Chem Soc Div Fuel Chem 49:616
104. Ghassemi H, McGrath JE (2004) Polymer 45:5847
105. Schuster M, Kreuer KD, Anderson HT, Maier J (2007) Macromolecules 40:598
106. Schmeisser J, Holdcroft S, Yu J, Ngo T, McLean G (2005) Chem Mater 17:387
107. Cho KY, Jung HY, Shin SS, Choi NS, Sung SJ, Park JK, Choi JH, Park KW, Sung YE (2004) Electrochim Acta 50:588
108. Gebel G, Diat O (2005) Fuel Cells 5:261
109. Hsu WY, Gierke TD (1983) J Membr Sci 13:307
110. Fujimura M, Hashimoto T, Kawai H (1982) Macromolecules 15:136
111. Dreyfus B, Gebel G, Aldebert P, Pinéri M, Escoubes M, Thomas M (1990) J Phys France 51:1341

112. Litt MH (1997) Polym Prepr 38:80
113. Haubold HG, Vad T, Jungbluth H, Jiller P (2001) Electrochim Acta 46:1559
114. Gebel G (2000) Polymer 41:5829
115. Rubatat L, Rollet AL, Gebel G, Diat O (2002) Macromolecules 35:4050
116. Gierke TD, Munn GE, Wilson FC (1981) J Polym Sci Polym Phys Ed 19:1687
117. Xue T, Trent JS, Osseo-Asare K (1989) J Membr Sci 45:261
118. Porat Z, Fryer JR, Huxham M, Rubinstein I (1995) J Phys Chem 99:4667
119. McLean RS, Doyle M, Sauer BB (2000) Macromolecules 33:6541
120. Moore RB, Martin CR (1988) Macromolecules 21:1334
121. Laporta M, Pegoraro M, Zanderighi L (2000) Macromol Mater Eng 282:22
122. Halim J, Scherer GG, Stamm M (1994) Macromol Chem Phys 195:3783
123. Grot WG, Chadds C (1982) European Pat 0 066 369
124. Martin CR, Rhoades TA, Ferguson JA (1982) Anal Chem 54:161
125. Aldebert P, Dreyfus B, Pinéri M (1986) Macromolecules 19:2651
126. Aldebert P, Dreyfus B, Gebel G, Nakamura N, Pinéri M Volino F (1988) J Phys (Paris) 49:2101
127. Loppinet B, Gebel G, Williams CE (1997) J Phys Chem B 101:1884
128. Loppinet B, Gebel G (1998) Langmuir 14:1977
129. Lee SJ, Yu TL, Lin HL, Liu WH, Lai CL (2004) Polymer 45:2853
130. pK_a database 4.0 by Advanced Chemistry Development, Inc. (Toronto), http://www.acdlabs.com/products/phys_chem_lab/pKa
131. Ma C, Zhang L, Mukerjee S, Ofer D, Nair N (2003) J Membr Sci 219:123
132. Weast RC (ed) (1983) CRC handbook of chemistry and physics, 64th edn. CRC, Boca Raton, p 167
133. Essafi W, Gebel G, Mercier R (2004) Macromolecules 37:1431
134. Kreuer KD (2001) J Membr Sci 185:29
135. Ise M (2000) Ph.D. thesis, University of Stuttgart
136. Kim YS, Dong L, Hickner MA, Pivovar BS, McGrath JE (2003) Polymer 44:5729
137. Kim YS, Wang F, Hickner M, Mccartney S, Hong YT, Harrison W, Zawodzinski TA, McGrath JE (2003) J Polym Sci B Polym Phys 41:2816
138. Beattie PD, Orfino FP, Basura VI, Zychowska K, Ding JF, Chuy C, Schmeisser J, Holdcroft S (2001) J Electroanal Chem 503:45
139. Yarusso DJ, Cooper SL (1983) Macromolecules 16:1871
140. Orifino FP, Holdcroft S (2000) J New Mater Electrochem Syst 3:287
141. Gebel G, Diat O, Stone C (2003) J New Mater Electrochem Syst 6:17
142. Kroschwitz JI et al. (eds) (1985) Encyclopedia of polymer science and technology, vol 2. Wiley, New York, p 324
143. Cornet N, Diat O, Gebel G, Jousse F, Marsacq D, Mercier R, Pinéri M (2000) J New Mater Electrochem Syst 3:33
144. Faure S, Cornet N, Gebel G, Mercier R, Pinéri M, Sillion B (1997) In: Savadogo O, Roberge PR (eds) Proceedings of the second international symposium on new materials for fuel cell and modern battery systems. Montreal, 6–10 July 1997, p 818
145. Cornet N, Beaudoing G, Gebel G (2001) Sep Purif Technol 22–23:681
146. Piroux F, Espuche E, Mercier R, Pinéri M (2003) J Membr Sci 223:127
147. Kim J, Kim B, Jung B (2002) J Membr Sci 207:129
148. Kim J, Kim B, Jung B, Kang YS, Ha HY, Oh IH, Ihn KJ (2002) Macromol Rapid Commun 23:753
149. Kim B, Kim J, Jung B (2005) J Membr Sci 250:175
150. Elabd YA, Walker CW, Beyer FL (2004) J Membr Sci 231:181

151. Elabd YA, Napadensky E, Walker CW, Winey KI (2006) Macromolecules 39:399
152. Webster OW (1991) Science 251:887
153. Wang J, Matyjaszewski K (1995) J Am Chem Soc 117:5614
154. Wang J, Matyjaszewski K (1995) Macromolecules 28:7901
155. Matyjaszewski K (ed) (1998) Controlled radical polymerization, ACS Symp Ser 685. American Chemical Society, Washington
156. Matyjaszewski K, Xia J (2001) Chem Rev 101:2921
157. Zhang Z, Ying S, Shi Z (1999) Polymer 40:1341
158. Yang YS, Shi Z, Holdcroft S (2004) Eur Polym J 40:531
159. Yang YS, Shi Z, Holdcroft S (2004) Macromolecules 37:1678
160. Kerres J, Ullrich A, Häring T, Baldauf M, Gebhardt U, Preidel W (2000) J New Mater Electrochem Syst 3:229
161. Destarac M, Matyjaszewski K, Silverman E, Ameduri B, Boutevin B (2000) Macromolecules 33:4613
162. Shi Z, Holdcroft S (2004) Macromolecules 37:2084
163. Shi Z, Holdcroft S (2005) Macromolecules 38:4193
164. Jokela K, Serimaa R, Torkkeli M, Sundholm F, Kallio T, Sundholm G (2002) J Polym Sci B Polym Phys 40:1539
165. Hietala S, Paronen M, Holmberg S, Näsman J, Juhanoja J, Karjalainen M, Serimaa R, Toivola M, Lehtinen T, Parovuori K, Sundholm G, Ericson H, Mattsson B, Torell L, Sundholm F (1999) J Polym Sci A Polym Chem 37:1741
166. Ding J, Chuy C, Holdcroft S (2002) Macromolecules 35:1348
167. Ding J, Chuy C, Holdcroft S (2001) Chem Mater 13:2231
168. Ding J, Chuy C, Holdcroft S (2002) Adv Funct Mater 12:389
169. Norsten TB, Guiver MD, Murphy J, Astill T, Navessin T, Holdcroft S, Frankamp BL, Rotello VM, Ding JF (2006) Adv Funct Mater 16:1814
170. Basura VI, Chuy C, Beattie PD, Holdcroft S (2001) J Electroanal Chem 501:77
171. Doyle M, Lewittes ME, Roelofs MG, Perusich SA (2001) J Phys Chem B 105:9387
172. Carretta N, Tricoli V, Picchioni F (2000) J Membr Sci 166:189
173. Elabd YA, Napadensky E, Sloan JM, Crawford DM, Walker CW (2003) J Membr Sci 217:227
174. Kim YS, Hickner MA, Dong L, Pivovar BS, McGrath JE (2004) J Membr Sci 243:317
175. Silva RF, Francesco MD, Pozio A (2004) Electrochim Acta 49:3211
176. Slade S, Campbell SA, Ralph TR, Walsh FC (2002) J Electrochem Soc 149:A1556
177. Dimitrova P, Friedrich KA, Vogt B, Stimming U (2002) J Electroanal Chem 532:75
178. Kreuer KD, Paddison SJ, Spohr E, Schuster M (2004) Chem Rev 104:4637
179. Eigen M, De Maeyer L (1958) Proc R Soc Lond A Math Phys Sci 247:505
180. Zundel G, Metzger H (1968) Z Phys Chem 58:225
181. Zundel G, Fritch J (1986) In: Dogonadze R, Kalman E, Kornyshev AA, Ulstrup J (eds) Spectroscopy of solvation, the chemical physics of solvation, part B. Elsevier, Amsterdam, p 21
182. Spohr E, Commer P, Kornyshev AA (2002) J Phys Chem B 106:10560
183. Agmon N (1995) Chem Phys Lett 244:456
184. Kreuer KD (2000) Solid State Ionics 136–137:149
185. Bender G (1999) Master thesis. University of Stuttgart
186. Paddison SJ, Bender G, Kreuer KD, Nicoloso N, Zawodzinski TA (2000) J New Mater Electrochem Syst 3:293
187. Tuckerman ME, Marx D, Klein ML, Parrinello M (1997) Science 275:817
188. Kreuer KD (1996) Chem Mater 8:610

189. Siu A, Schmeisser J, Holdcroft S (2006) J Phys Chem B 110:6072
190. Peckham TJ, Schmeisser J, Holdcroft S. J Phys Chem B (in press)
191. Zawodzinski TA, Derouin C, Radzinski S, Sherman RJ, Smith VT, Springer TE, Gottesfeld S (1993) J Electrochem Soc 140:1041
192. Pivovar AA, Pivovar BS (2005) J Phys Chem B 109:785
193. Sondheimer SJ, Bunce NY, Fyfe CA (1986) J Macromol Sci Rev Macromol Chem Phys C26:353
194. Hatakeyama H, Hatakeyama T (1998) Thermochim Acta 308:3
195. Kim YS, Dong L, Hickner MA, Glass TE, Webb V, McGrath JE (2003) Macromolecules 36:6281
196. Gupta B, Haas O, Scherer GG (1995) J Appl Polym Sci 57:855
197. Kanekiyo M, Kobayashi M, Ando I, Kurosu H, Ishii T, Amiya S (1998) J Mol Struct 447:49
198. Ghi PY, Hill DJT, Whittaker AK (2002) Biomacromolecules 3:991
199. Saito M, Hayamizu K, Okada T (2005) J Phys Chem B 109:3112
200. Elomaa M, Hietala S, Paronen M, Walsby N, Jokela K, Serimaa R, Torkkeli M, Lehtinen T, Sundholm G, Sundholm F (2000) J Mater Chem 10:2678
201. Lafitte B, Karlsson LE, Jannasch P (2002) Macromol Rapid Commun 23:896
202. Siu A, Pivovar B, Horsfall J, Lovell KV, Holdcroft S (2006) J Polym Sci B Polym Phys 44:2240
203. Randin JP (1982) J Electrochem Soc 129:1215
204. Paddison SJ (2003) Annu Rev Mater Res 33:289
205. Uosaki K, Okazaki K, Kita H (1990) J Electroanal Chem 287:163
206. Cappadonia M, Erning JW, Stimming U (1994) J Electroanal Chem 376:189
207. Cappadonia M, Erning JW, Niaki SMS, Stimming U (1995) Solid State Ionics 77:65
208. Zawodzinski TA, Neeman M, Sillerud LO, Gottesfeld S (1991) J Phys Chem 95:6040
209. Zawodzinski TA, Springer TE, Davey J, Jestel R, Lopez C, Valerio J, Gottesfeld S (1993) J Electrochem Soc 140:1981
210. Edmondson CA, Stallworth PE, Chapman ME, Fontanella JJ, Wintersgill MC, Chung SH, Greenbaum SG (2000) Solid State Ionics 135:419
211. Edmondson CA, Fontanella JJ (2002) Solid State Ionics 152–153:355
212. Saito M, Arimura N, Hayamizu K, Okada T (2004) J Phys Chem B 108:16064
213. Dong L, Kim YS, Wang F, Hickner M, Glass TE, McGrath JE (2003) Polym Prepr 44:355
214. Collings AF, Mills R (1970) J Chem Soc Faraday Trans 66:2761
215. Ye X, LeVan MD (2003) J Membr Sci 221:147
216. Ge S, Li X, Yi B, Hsing IM (2005) J Electrochem Soc 152:A1149
217. Rivin D, Kendrick CE, Gibson PW, Schneider NS (2001) Polymer 42:623
218. Gates CM, Newman J (2000) AIChE J 46:2076
219. Krtil P, Trojánek A, Samec Z (2001) J Phys Chem 105:7979
220. Okada T, Xie G, Gorseth O, Kjelstruo S, Nakamura N, Arimura T (1998) Electrochim Acta 43:3741
221. Yeo SC, Eisenberg A (1977) J Appl Polym Sci 21:875
222. Motupally S, Becker AJ, Weidner JW (2000) J Electrochem Soc 147:3171
223. Webber AZ, Newman J (2004) J Electrochem Soc 146:2049
224. Fuller TF, Newman J (1992) J Electrochem Soc 139:1332
225. Breslau BR, Miller IF (1971) Ind Eng Chem Fundam 10:554
226. Xie G, Okada T (1995) J Electrochem Soc 142:3057
227. Okada T, Xie G, Meeg M (1998) Electrochim Acta 43:2141

228. Zawodzinski TA, Davey J, Valerio J, Gottesfeld S (1995) Electrochim Acta 40:297
229. Ren X, Henderson W, Gottesfeld S (1997) J Electrochem Soc 144:L267
230. Ren X, Gottesfeld S (2001) J Electrochem Soc 148:A87
231. Ise M, Kreucr KD, Maier J (1999) Solid State Ionics 125:213
232. Pivovar BS, Smyrl WH, Cussler EL (2005) J Electrochem Soc 152:A53
233. Antolini E (2004) J Appl Electrochem 34:563
234. Mathias MF, Roth J, Fleming J, Lehnert W (2003) Diffusion media materials and characterization. In: Vielstich W, Lamm A, Gasteiger HA (eds) Handbook of fuel cells: fundamentals, technology and applications, part 3, vol 3. Wiley, Chichester, p 517
235. Kocha SS (2003) Principles of MEA preparation. In: Vielstich W, Lamm A, Gasteiger HA (eds) Handbook of fuel cells: fundamentals, technology and applications, part 3, vol 3. Wiley, Chichester, p 538
236. Kinoshita K, Stonehart P (1977) In: Bockris JO, Conway BE (eds) Modern aspects of electrochemistry, vol 12. Plenum, New York, chap 4
237. Lee SJ, Mukerjee S, McBreen J, Rho YW, Kho YT, Lee TH (1998) Electrochim Acta 43:3693
238. Cheng XL, Yi BL, Han M, Zhang JX, Qiao YG, Yu JR (1999) J Power Sources 79:75
239. Navessin T, Holdcroft S, Wang Q, Song D, Lin Z, Eikerling M, Horsfall J, Lovell KV (2004) J Electroanal Chem 567:111
240. Navessin T, Eikerling M, Wang Q, Song D, Lin Z, Horsfall J, Lovell KV, Holdcroft S (2005) J Electrochem Soc 152:A796
241. Chuy C, Basura VI, Simon E, Holdcroft S, Horsfall J, Lovell KV (2000) J Electrochem Soc 147:4453
242. Horsfall J, Lovell KV (2001) Fuel Cells 1:186
243. Huslage J, Rager T, Schnyder B, Tsukada A (2002) Electrochim Acta 48:247
244. Easton EB, Langsdorf BL, Hughes JA, Sultan JM, Qi Z, Kaufman A, Pickup PG (2003) J Electrochem Soc 150:C735
245. Baradie B, Dodelet JP, Guay D (2000) J Electroanal Chem 489:101
246. Kim YS, Sumner MJ, Harrison WL, Riffle JS, McGrath JE, Pivovar BS (2004) J Electrochem Soc 151:A2150
247. Easton EB, Astill TD, Holdcroft S (2005) J Electrochem Soc 152:A752
248. Horsfall JA, Lovell KV (2002) Polym Adv Technol 13:381
249. Brack HP, Büchi FN, Huslage J, Scherer GG (1998) In: Gottesfeld S, Fuller TF (eds) Proceedings of the 2nd symposium on proton conducting membrane fuel cells II, proceedings vol 98-27. The Electrochemical Society, Boston, p 52
250. Gubler L, Kuhn H, Schmidt TJ, Scherer GG, Brack HP, Simbeck K (2004) Fuel Cells 4:196
251. Steck AE, Stone C (1997) In: Savadogo O, Roberge PR Valeriu A (eds) New materials for fuel cell and modern battery systems II: proceedings of the second international symposium on new materials for fuel cell and modern battery systems. l'Ecole Polytechnique de Montreal, Montreal, p 792
252. Bauer B, Jones DJ, Rozière J, Tchicaya L, Alberti G, Casciola M, Massinelli L, Peraio A, Besse S, Ramunni E (2000) J New Mater Electrochem Syst 3:93
253. Soczka-Guth T, Baurmeister J, Frank G, Knauf R (1999) WO Patent 99/29763
254. Blazquez JA, Iruin JJ, Eceolaza S, Marestin C, Mercier R, Mecerreyes D, Miguel O, Vela A, Marcilla R (2005) J Power Sources 151:63
255. Besse S, Capron P, Diat O, Gebel G, Jousse F, Marsacq D, Pinéri M, Marestin C, Mercier R (2002) J New Mater Electrochem Syst 5:109

256. Ehrenberg SG, Serpico JM, Sheikh-Ali BM, Tangredi TN, Zador E, Wnek GE (1997) In: Savadogo O, Roberge PR, Valeriu A (eds) New materials for fuel cell and modern battery systems II: proceedings of the second international symposium on new materials for fuel cell and modern battery systems. l'Ecole Polytechnique de Montreal, Montreal, p 828
257. Kallio T, Lundstrom M, Sundholm G, Walsby N, Sundholm F (2002) J Appl Electrochem 32:11
258. Kallio T, Jokela K, Ericson H, Serimaa R, Sundholm G, Jacobsson P, Sundholm F (2003) J Appl Electrochem 33:505

Perfluorinated Ionic Polymers for PEFCs (Including Supported PFSA)

M. Yoshitake (✉) · A. Watakabe

Asahi Glass Co., Ltd., 1150 Hazawa-cho, Kanagawa-ku, Yokohama-shi,
221-8755 Kanagawa, Japan
masaru-yoshitake@agc.co.jp

1	Introduction and History of Perfluorinated Membranes	128
2	Fundamentals of PFSA Membranes	130
2.1	Conventional Perfluorinated Membranes for PEFCs	132
2.2	Membranes with Short Side Chains	134
2.3	Sulfonimide Membranes	138
3	High Temperature Membranes	138
3.1	Terpolymers	140
4	Chemical Degradation Problem	142
5	Reinforcement Technology	144
6	DMFC Application	146
6.1	Examples of PFSA Type Membranes for DMFC	146
7	Topics Related to New Analytical Results on the Structure of PFSA Dispersions and Membranes	147
8	Summary	151
	References	151

Abstract This article outlines some history of and recent progress in perfluorinated membranes for polymer electrolyte fuel cells (PEFCs). The structure, properties, synthesis, degradation problems, technology for high temperature membranes, reinforcement technology, and characterization methods of perfluorosulfonic acid (PFSA) membranes are reviewed.

Keywords Analysis · Cross leak degradation · High temperature membrane · Perfluorosulfonic acid · PFSA · Reinforcement

Abbreviations
DLS Dynamic light scattering
DMFC Direct methanol fuel cell
ETFE Ethylene–tetrafluoroethylene copolymer
EW Equivalent weight

IEC   Ion-exchange capacity
MEA   Membrane electrode assembly
MW   Molecular weight
PEFC Polymer electrolyte fuel cell
PFSA Perfluorosulfonic acid
PSVE Perfluorosulfonyl vinyl ether
PTFE Polytetrafluoroethylene
SANS Small-angle neutron scattering
SAXS Small-angle X-ray scattering
TFE   Tetrafluoroethylene
UAS   Ultrasonic attenuation spectroscopy

1
Introduction and History of Perfluorinated Membranes

Perfluorosulfonic acid (PFSA) membranes as shown in Fig. 1 were first developed for fuel cells by DuPont as "Nafion"® and installed into the Biosatellite spacecraft in 1967 [1, 2]. Various types of PFSA polymers, such as Flemion®, Aciplex®, and Dow membrane, were developed subsequently. They have excellent chemical stability, high proton conductivity, and high water diffusivity in a wide range of temperatures, brought about by the nature of fluorinated compounds and these non-cross-linked structures [3–5].

Although the PFSA membrane has not been utilized for spaceships after the Biosatellite program up to now, PFSA technology has been utilized in the chlor-alkali electrolysis industry and water electrolysis as one of the key materials [6]. Furthermore, it has also been utilized as a superacid catalyst, separator for the redox flow battery, an ion-permeating membrane for organoelectrosynthesis, and in water-permeating devices such as pervaporation membrane humidifiers and hollow-fiber dryers, etc. Perfluorinated carboxylic acid polymers have shouldered the important roles in chlor-alkali electrolysis and the technology of actuator and artificial muscle, etc.

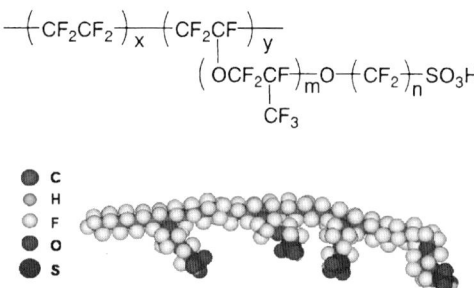

Fig. 1 Chemical structure of perfluorosulfonic acid (PFSA) membranes

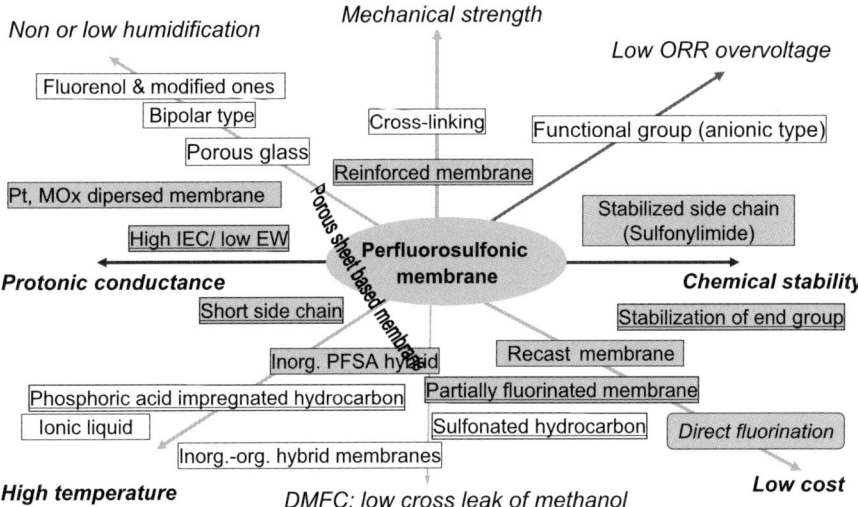

Fig. 2 Typical proposed candidate electrolytes and technologies for fuel cells. (*Underlined* electrolytes are in/after the stage of stack testing)

The application to fuel cells was reopened by Ballard stacks using a new Dow membrane that is characterized by short side chains. The extremely high power density of the polymer electrolyte fuel cell (PEFC) stacks was achieved not only by the higher proton conductance of the membrane, but also by the usage of PFSA polymer dispersed solution, serpentine flow separators, the structure of the thin catalyst layer, and the gas diffusion layer. Although PFSA membranes remain the most commonly employed electrolyte up to now, their drawbacks, such as decrease in mechanical strength at elevated temperature and necessity for humidification to keep the proton conductance, caused the development of various types of new electrolytes and technologies [7], as shown in Fig. 2.

PFSA membranes with high ion-exchange capacity (IEC) of 1.1 meq/g and long side chains applicable to PFECs [8] were examined. Thin membranes cast from PFSA solution [9, 10] and membranes in which fine particles of catalyst or metal oxides are dispersed [11–13] were proposed for low- or no-humidity operation. Thin membranes can offer not only lower cell resistance but also higher water diffusivity and hence higher power density [8]. To improve the handling characteristics of thin membranes, various reinforcement techniques were proposed [14]. Especially, polytetrafluoroethylene (PTFE) reinforcement technology is typically used for thin membranes. The sulfonimide structure was applied to strengthen the stability of the end sulfonic group [15]. Fluorination of the end groups of the PFSA polymer [16] was reported to contribute to increasing the chemical stability of PFSA under severe conditions, such as low humidity and higher temperature or open circuit voltage (OCV). As for cost reduction, the casting method [17] and various styrene-grafted membranes based on the commercial fluoro compound sheet [18] have been investigated. A newly developed direct fluorination method [19] is expected to be applied to the PEFC membranes. Surface treatment and impregnation of inorganics have been investigated to decrease the swelling in methanol solution for direct methanol fuel cell (DMFC) application. For temperatures over 100 °C, various composite materials with inorganics have been proposed to improve the proton conductivity and stability. Operation under lower humidification, which is convenient to obtain a simpler system and lower system cost, was reported to often cause abnormally rapid degradation of membrane electrode assemblies (MEAs), and the mechanism of the membrane degradation started to be investigated around 2001 by involving many researchers and organizations. As a result, various interesting findings and measures to improve the stability of PFSA have been reported and proposed. Basically, they are based on the analysis results by GE [20]. Hydrogen peroxide formed by the cross leak of hydrogen and oxygen through the membrane is estimated to cause rapid degradation of membranes, especially under low humidity and OCV conditions. Operation over 100 °C is expected to offer such merits as better cell efficiency and smaller radiators in fuel cell electric vehicle applications. Recently, it has been reported that over 5000 h operation is possible at 120 °C and 50%RH using a new perfluoro composite membrane [21].

Decrease in swelling and methanol cross leak of membranes in DMFC applications is one of the most important development items, and various efforts and trials have been performed to improve these characteristics for both perfluorinated and hydrocarbon membranes.

Various reviews on PFSA technology development have been published and detailed explanations of the individual items are available from those materials. In this chapter, the fundamentals of PFSA membranes, the requirements for advanced PEFCs, development trends for high temperature membranes, reinforcement technology, membranes for DMFC, and topics on analysis technology are reviewed.

2
Fundamentals of PFSA Membranes

Fluorine is "a small atom with a big ego". It took a very long time for it to be handled with ease since its isolation by Moissan. Carbon was found to be able to give extremely comfortable "seats" for it as PTFE by Plankett in 1938, and various fluorinated compounds were prepared and commercialized. The bonding energy of C–F is $485\,\text{kJ}\,\text{mol}^{-1}$ and larger compared with those of C–H ($350\,\text{kJ}\,\text{mol}^{-1}$ for aliphatics and $435\,\text{kJ}\,\text{mol}^{-1}$ for aromatics). This basic feature gives remarkable stability to various fluoro compounds.

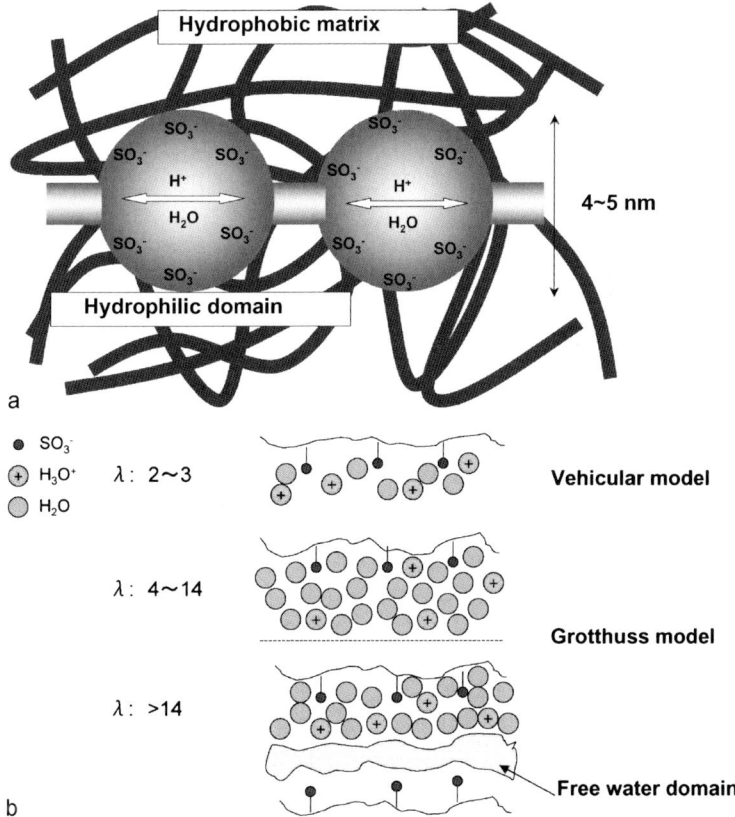

Fig. 3 Schematic illustration of the PFSA polymer structure. **a** Cluster-network model. **b** Hydration and proton conduction mechanism

PFSA membranes, whose representative polymer structure is shown in Fig. 1, are preferably used for the development of solid PEFCs because they have excellent chemical stability, high thermal stability, and high proton conductivity and, furthermore, are available in larger quantities than other developing membranes. The excellent performance, such as high proton conductivity and water mobility, has been explained by the cluster-network model for the hydrated polymer, as shown in Fig. 3a. This model was originally proposed by Gierke on the basis of analysis by small-angle X-ray scattering (SAXS) [22]. The mechanism of proton conduction is said to be dependent on the molecular ratio of water and the sulfonic acid group at the end of the side chains ($\lambda = H_2O/SO_4^-$), as shown in Fig. 3b.

As described before, the commercialization of PEFCs requires lower prices of the membranes and generation systems, compactness of the system, and higher efficiency of power generation. New types of PFSA membranes, new processes of membrane fabrication, and high temperature membranes have been investigated. This article deals with (1) conventional PFSA membranes with long side chains, (2) membranes with short side chains, (3) sulfonimide membranes, and (4) other miscellaneous types of membranes.

2.1
Conventional Perfluorinated Membranes for PEFCs

The molecular structure of a conventional polymer used for a PFSA membrane is shown in Fig. 1. Membranes registered as Nafion® (DuPont), Flemion®, (Asahi Glass), and Aciplex® (Asahi Chemical) have been commercialized for brine electrolysis and they are used in the form of alkali metal salt. Figure 4 shows a schematic illustration of a membrane for chlor-alkali electrolysis. The PFSA layer is laminated with a thin perfluorocarboxylic acid layer, and both sides of the composite membrane are hydrophilized to avoid the sticking of evolved hydrogen and chlorine. The membrane is reinforced with PTFE cloth. The technology was applied to PEFC membranes with thickness of over 50 µm [14].

For the synthesis of this type of polymer, a fluorosulfonyl monomer is frequently copolymerized with tetrafluoroethylene (TFE). The synthetic scheme of this monomer is shown in Fig. 5 [24]. The IEC is about 0.9 to 1.1 meq/g dry polymer. As the IEC increases, water absorption increases, and the crystallinity based on successive sequences of the TFE monomer unit becomes smaller, which lowers the mechanical strength. On the other hand, when the IEC decreases, water absorption becomes smaller, which brings lower proton conductivity.

The term IEC means the number of ion-exchange groups, in the unit of milliequivalents, contained in one gram of dry resin. The term equivalent weight (EW) is also used for expressing the number of ion-exchange groups.

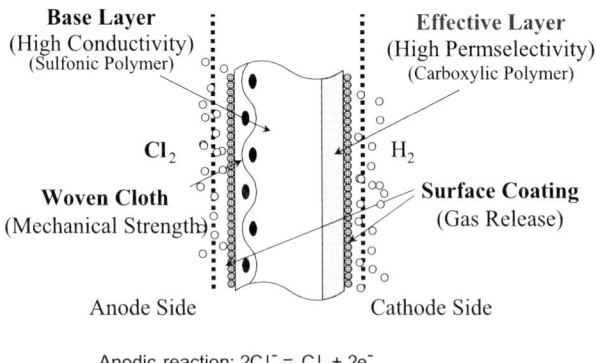

Anodic reaction: $2Cl^- = Cl_2 + 2e^-$
Cathodic reaction: $H_2O + 2e^- = 2OH^- + H_2$

Fig. 4 Design model of perfluorinated membrane for chlor-alkali electrolysis

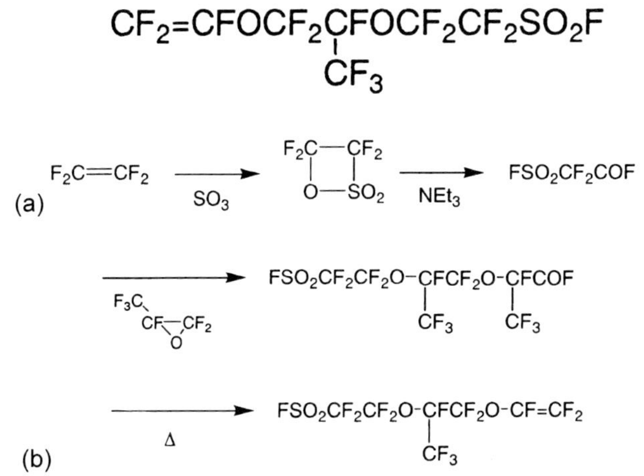

$$CF_2=CFOCF_2CFOCF_2CF_2SO_2F$$
$$|$$
$$CF_3$$

(a) $F_2C=CF_2 \xrightarrow{SO_3}$ [F₂C—CF₂ / O—SO₂ ring] $\xrightarrow{NEt_3}$ FSO_2CF_2COF

$\xrightarrow{F_3C-CF-CF_2 \text{ (epoxide)}}$ $FSO_2CF_2CF_2O-CFCF_2O-CFCOF$ with CF_3 groups

(b) $\xrightarrow{\Delta}$ $FSO_2CF_2CF_2O-CFCF_2O-CF=CF_2$ with CF_3

Fig. 5 Chemical structure and synthetic scheme of fluorosulfonyl vinyl ether (PSVE)

$+(CF_2-CF_2)_x(CF_2-CF)_y+$
$(O-CF_2-CF)_m-O-(CF_2)_n-SO_2F$
CF_3

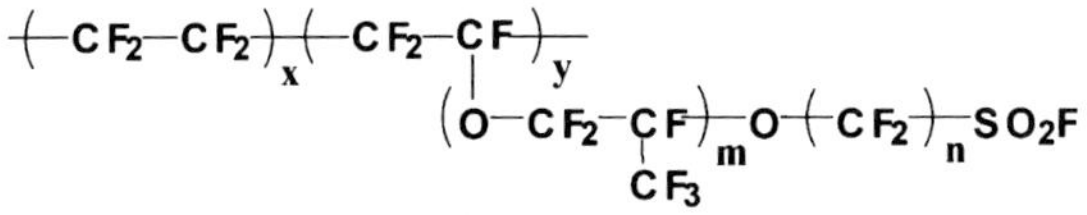

Fig. 6 Chemical structure of a copolymer of TFE and PSVE

This value corresponds to the average molecular weight per ionic group, according to the following relationship between IEC and EW:

IEC = 1000/EW .

A copolymer of TFE and perfluorosulfonyl vinyl ether (PSVE; Fig. 5) given in Fig. 6 is a thermoplastic polymer. It can be extruded to form a thin film. The films with $-SO_2F$ groups can be converted to proton-conductive ion-exchange resins having $-SO_3H$ groups by alkali hydrolysis followed by acidification. This ion-exchange resin can be dissolved or dispersed in polar solvents, such as water, lower alcohols, dimethylformamide, and dimethylacetamide. This liquid composition is used for cast film formation and for an ink to form catalyst layers. An understanding of the characteristics of ionomer dispersed solutions, such as viscosity and structure, is important to improve recast membranes and MEAs. Recent analysis results are explained in Sect. 2.2.

The membrane development for PEFCs comprises many steps, as shown in Fig. 7. The multiple steps in monomer synthesis for PFSA preparation is often said to be one of the major factors in the high cost of the PFSA membranes. In fact, PTFE and ethylene–tetrafluoroethylene copolymer (ETFE), whose synthetic scheme is very simple, are available at comparably low price. As a matter of fact, a considerably large quantity of the ETFE sheet was used

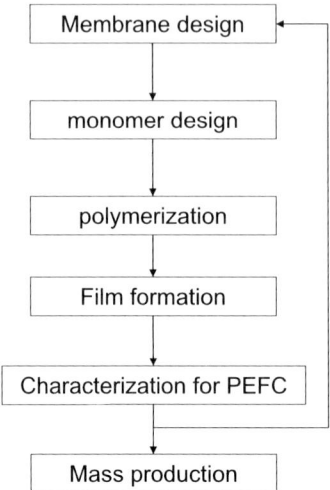

Fig. 7 General flow chart of membrane development

in the German soccer stadium Allianz-Arena (about 200 000 m^2 for the main stadium) [25], and it is to be used for main and swimming stadiums of the Olympic games in Beijing, China (about 300 000 m^2 of the film) [26].

Different from PTFE and ETFE, cost reduction of the PSVE polymer is one of the problems for prevailing perfluorinated membranes. Recently, a new direct fluorination process with elementary fluorine was reported to produce perfluorinated fluorosulfonyl polymers, as shown in Fig. 8 [19]. This new process consists of the following features and is free from explosion in the gas phase and the difficulty in finding solvents with stability to fluorine gas and solubility for hydrocarbon compounds, these problems having limited the application of the conventional direct fluorination methods:

1. Alcohol (ROH) as a raw material is reacted with $R_F COF$ into an ester ($R_F COOR$) which is not very volatile and soluble in $R_F COF$.
2. Direct fluorination with elementary fluorine gas.
3. Decomposition of the ester and regeneration of $R_F COF$.

This process can offer wide selection of the synthetic routes of precursor hydrocarbon, shorter steps, and low-cost production of the perfluorinated polymers [27].

2.2
Membranes with Short Side Chains

So-called Dow membrane with short side chains once attracted much attention, because it showed excellent performance in fuel cell operation compared with Nafion® membranes [28–30]. The polymer structure is shown in Fig. 9.

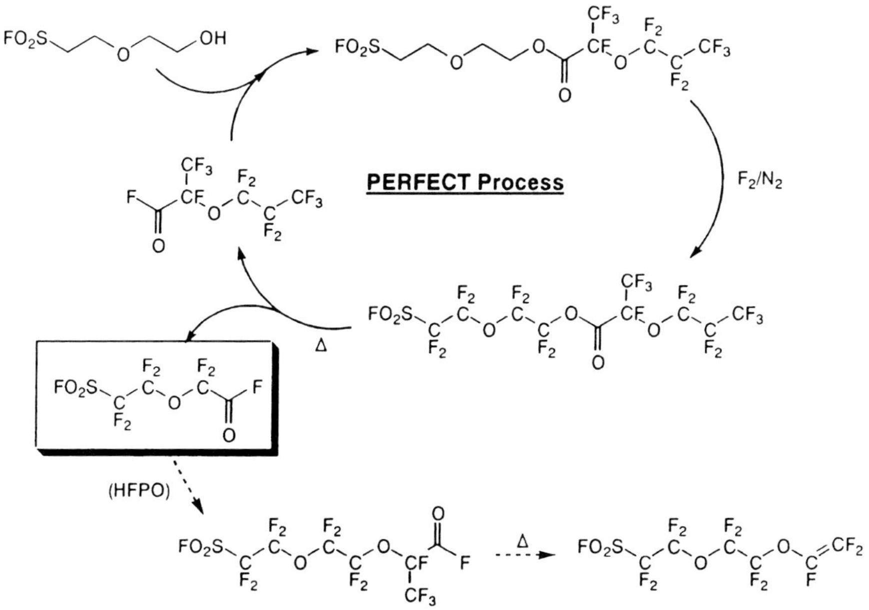

Fig. 8 Direct fluorination with elementary fluorine to synthesize perfluoro compounds. "PERFECT" means PERFluorination of Esterified Compounds then Thermal elimination

$$-(CF_2CF_2)_x-(CF_2CF)_y-$$
$$\qquad\qquad\qquad |$$
$$\qquad\qquad\qquad OCF_2CF_2SO_3H$$

Fig. 9 Chemical structure of the Dow membrane

$$CF_2=CFOCF_2CF_2SO_2F$$

Fig. 10 Structure of the perfluorosulfonyl monomer of the short side chain membrane

The Dow membranes are prepared by the copolymerization of TFE with PSVE, which is shown in Fig. 10.

An important difference between the Dow and Nafion® membranes was considered to be membrane resistivity, which depends on membrane thickness and IEC [30]. The mechanical strength of the perfluorinated membranes largely depends on the molar content of the TFE unit in the copolymers. Since the molecular weight of the Dow monomer is smaller than that of the Nafion® monomer, a higher IEC can be obtained with the Dow monomer compared with the Nafion® monomer at a similar molar content of TFE unit in the copolymers, while keeping the membrane mechanical strength. Recently, thinner PSVE membranes with long side chains have been available for reduction of membrane resistance [31]; nevertheless, the short side chain membrane is actively being developed. The reason is that the short side chain

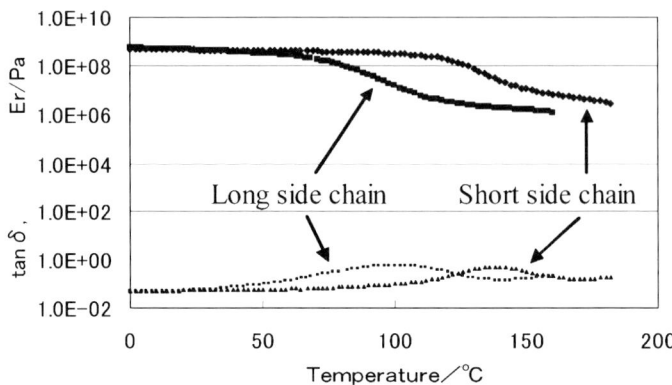

Fig. 11 Mechanical characteristics of a short side chain membrane

membrane has a higher glass transition temperature than conventional perfluorinated polymers, and the former has an advantage for high temperature operation of a fuel cell. The T_g of a short side chain film in an acid form is much higher than that of the conventional PFSA membrane with long side chains [32]. Although the glass transition temperature of the latter is about 100–130 °C, where the $\tan\delta$ peak appears in the dynamic elastic modulus measurement, the modulus of the film suddenly decreases at 70 to 80 °C. As for the short side chain membrane, its softening temperature is over 100 °C, and it is considered to be a preferred material for higher temperature operation. The dynamic mechanical properties of a short side chain polymer are given in Fig. 11 together with those of a conventional perfluorinated polymer. The functional groups of the polymers are sulfonic acid (–SO$_3$H) groups.

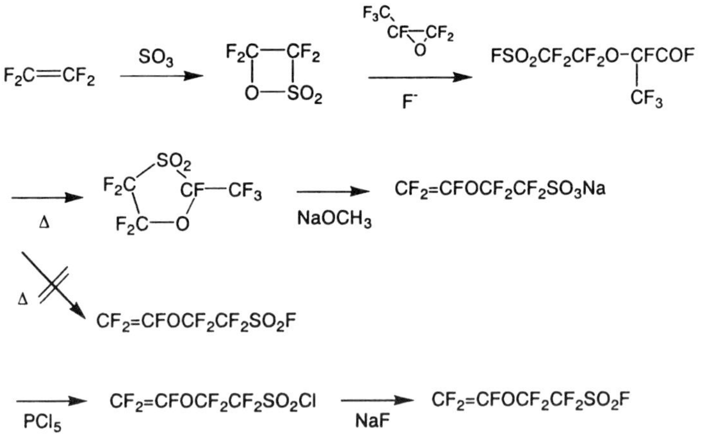

Fig. 12 Synthetic scheme of the Dow membrane PSVE monomer

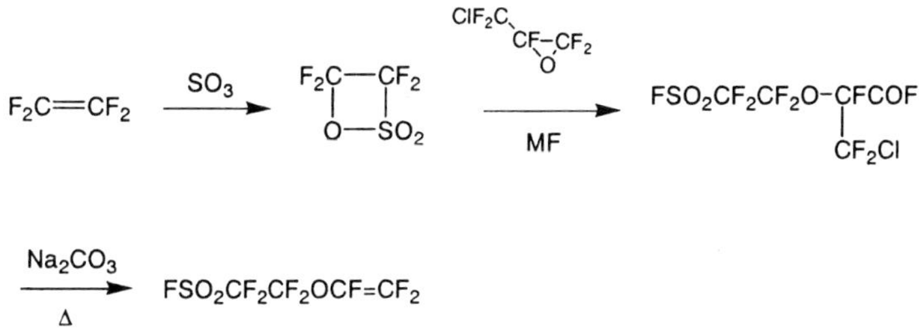

$F_2C=CF_2 \xrightarrow{SO_3} \underset{O-SO_2}{F_2C-CF_2} \xrightarrow{MF} \underset{ClF_2C}{\overset{CF-CF_2}{\underset{O}{\diagup}}} \quad FSO_2CF_2CF_2O-\underset{CF_2Cl}{CFCOF}$

$\xrightarrow[\Delta]{Na_2CO_3} FSO_2CF_2CF_2OCF=CF_2$

Fig. 13 New synthetic route to the Dow membrane PSVE monomer

Originally the monomer in Fig. 9 was prepared by DuPont by the synthetic scheme shown in Fig. 12 [33]. Thermolysis of the acyl fluoride in Fig. 12 did not give a desired monomer but gave a cyclo compound. In order to prevent the cyclization, a new synthetic route was developed as shown in Fig. 13, which was applied to the synthesis of Dow membranes [34]. A chlorine atom was introduced to the acyl fluoride to improve the selectivity of vinyl ether formation. The Dow membrane was also developed for brine electrolysis, but was not commercialized probably because of its high cost. Difficulty in the preparation of the acyl fluoride in Fig. 13 is one of the causes. Recently, new synthetic processes for the short side chain monomer were developed, as represented in Fig. 14.

Recently a similar monomer and copolymer to those of the above short side chain type were reported [35–38]. The monomer has four CF_2 units (Fig. 15). Its copolymer with TFE exhibited a similar softening point to that

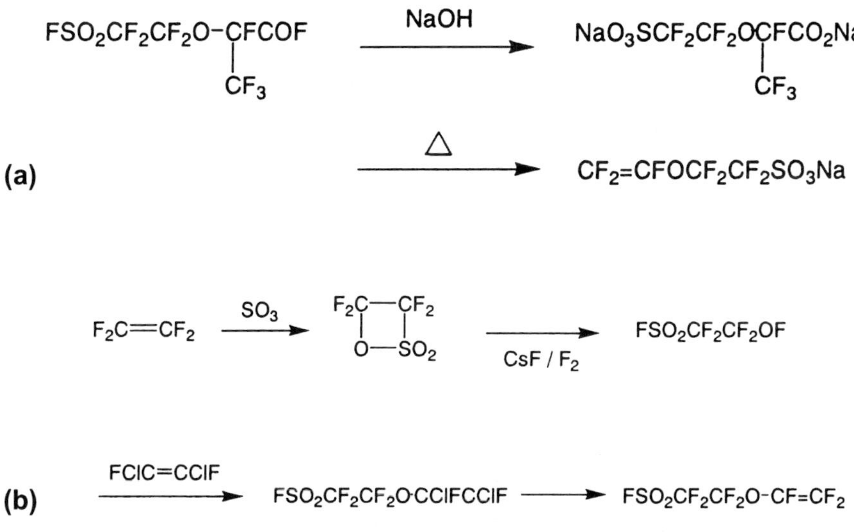

(a) $FSO_2CF_2CF_2O-\underset{CF_3}{CFCOF} \xrightarrow{NaOH} NaO_3SCF_2CF_2O\underset{CF_3}{CFCO_2Na}$

$\xrightarrow{\Delta} CF_2=CFOCF_2CF_2SO_3Na$

(b) $F_2C=CF_2 \xrightarrow{SO_3} \underset{O-SO_2}{F_2C-CF_2} \xrightarrow{CsF/F_2} FSO_2CF_2CF_2OF$

$\xrightarrow{FClC=CClF} FSO_2CF_2CF_2O\text{-}CClFCClF \longrightarrow FSO_2CF_2CF_2O\text{-}CF=CF_2$

Fig. 14 New synthetic routes to short side chain PSVE monomer

of the above short side chain monomer which has two CF_2 units. The thermal decomposition temperature of the former copolymer was much higher than that of the latter copolymer [38].

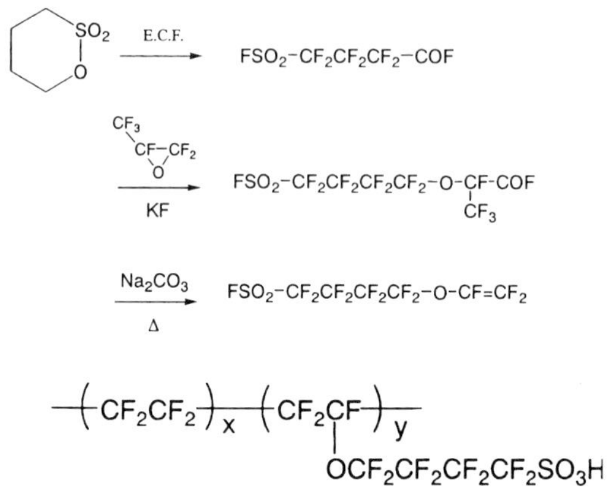
SO2
O
E.C.F.
FSO2–CF2CF2CF2–COF
CF3
CF–CF2
O
KF
FSO2–CF2CF2CF2CF2–O–CF–COF
CF3
Na2CO3
Δ
FSO2–CF2CF2CF2CF2–O–CF=CF2
CF2CF2 x CF2CF y
OCF2CF2CF2CF2SO3H

Fig. 15 New type of PFSA with straight side chains

2.3 Sulfonimide Membranes

Perfluorinated sulfonamide monomers were prepared by DesMarteau. A typical synthetic scheme is given in Fig. 16 [39, 40]. The temperature dependency and humidity dependency of proton conductivity of the sulfonamide copolymer with TFE were examined, and the properties were proved to be similar to those of a sulfonic acid type membrane [41–43]. Fuel cell performance was dependent upon membrane thickness and/or IEC, and there do not seem to be large differences depending on the species of the ion-exchange groups. Synthesis of a short side chain type sulfonimide monomer is also reported (see Fig. 17 [44]).

3 High Temperature Membranes

The development of high temperature membranes applicable to operation at up to 120 °C or higher, no or low humidity, and low pressure conditions has been expected to offer better efficiency and more compact stacks and radiators for automobiles. These expectations for membranes are typically reflected by the DOE's 2010 targets shown in Table 1. Operation at higher temperature (120 °C) and lower humidity (1.5 kPa), lower cross leak, lower

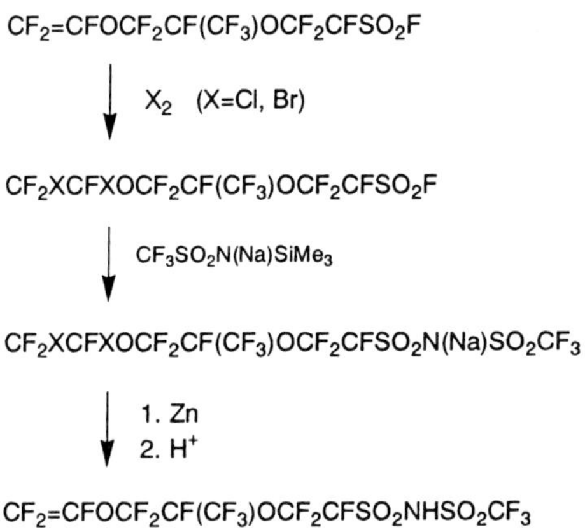

$CF_2=CFOCF_2CF(CF_3)OCF_2CFSO_2F$

$\downarrow X_2$ (X=Cl, Br)

$CF_2XCFXOCF_2CF(CF_3)OCF_2CFSO_2F$

$\downarrow CF_3SO_2N(Na)SiMe_3$

$CF_2XCFXOCF_2CF(CF_3)OCF_2CFSO_2N(Na)SO_2CF_3$

$\downarrow$ 1. Zn
2. H$^+$

$CF_2=CFOCF_2CF(CF_3)OCF_2CFSO_2NHSO_2CF_3$

Fig. 16 Typical synthetic scheme for the perfluorinated sulfonamide monomers

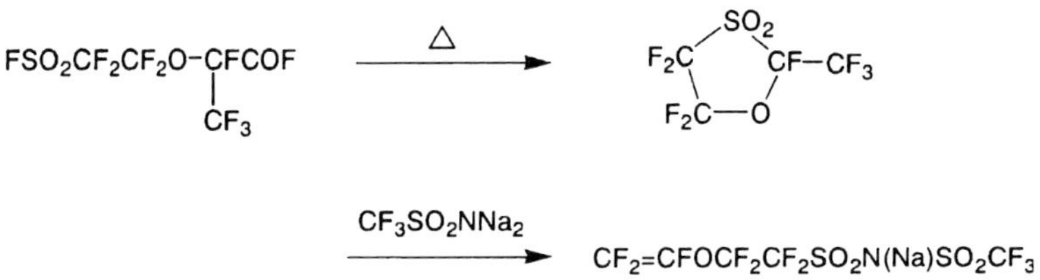

$FSO_2CF_2CF_2O-\underset{CF_3}{CFCOF} \xrightarrow{\Delta} \underset{F_2C-O}{F_2C}\overset{SO_2}{\diagdown}CF-CF_3$

$\xrightarrow{CF_3SO_2NNa_2} CF_2=CFOCF_2CF_2SO_2N(Na)SO_2CF_3$

Fig. 17 Preparation scheme of the short side chain sulfonimide monomer

area specific resistance (0.02 mΩ cm^2), lower cost, and higher durability are required by 2010.

To realize the high temperature membrane in general, the drop of mechanical strength and conductivity at elevated working temperature and the chemical degradation problem have to be overcome [46]. Typical measures

Table 1 The DOE's 2010 targets for membranes [45]

Item	Target
Proton conductivity at < 120 °C, < 1.5 kPa	0.1 S/cm
Proton conductivity at – 20 °C	0.01 S/cm
Oxygen and hydrogen cross leak	2 mA/cm^2
Durability with cycling at > 80 °C	2000 operating hours
< 80 °C	5000 operating hours
Survivability	– 40 °C
Cost	$40/m^2

to improve the high temperature performance of PFSA membranes aimed at high temperature operation are as follows:

1. Composites with inorganics to improve proton conductance at over 100 °C.
2. New molecular design which can offer lower water uptake at higher temperature and better dimensional stability.

The decrease in mechanical strength of the PFSA membranes under humidified conditions at elevated temperature is the most important issue for high temperature operation. The flexibility of the side chain and conformational changes in the backbone on hydration and proton transfer in the PFSA membranes were investigated using first principles based molecular modeling studies [47, 48]. This electronic structure calculation analysis showed that short side chain PFSA membranes gave lower water uptake and higher proton conductivity with fewer water molecules. The flexibility in both the backbone and side chains of PFSA membranes is important to the effective transport of protons under low humidity. Composite membranes comprising PFSA polymer and inorganic particles, such as silica and zirconia, as a hydrophilic material have been proposed [49]. In the case of the hydrocarbon membrane, the nanosized space in polyimide electrolytes containing trifluoromethyl groups is estimated to offer water adsorbing sites [50]. The stability of the sulfonic group of the side chains in PFSA polymer in air and water was found to offer excellent stability at 120 °C [51]. The side-chain-sulfonated hydrocarbon membranes are thought to be more resistant to hydration than backbone-sulfonated types [52, 53]. The former types, whose structure is similar to that of PFSA in the point of phase segregation in the molecule, can offer higher proton conductivity than the latter ones at higher temperature. Needless to say, a deep understanding of the PFSA structure is required for the development of high temperature PFSA membranes [54, 55]. Here, new monomer structures for PFSA except for short side chain types are reviewed.

3.1
Terpolymers

2,2-Bis(trifluoromethyl)-4,5-difluoro-1,3-dioxole (Fig. 18) was copolymerized with TFE and a PSVE monomer shown in Fig. 5. A homopolymer of this third monomer exhibits a glass transition temperature of 330 °C [56]. The terpolymer exhibits a high softening temperature like the above short side chain copolymers [57]. This is one of the another approaches to obtain high temperature membranes. The temperature dependency of the modulus of the terpolymer is compared with that of a conventional copolymer in Fig. 19 [58].

A perfluorophosphonic monomer (Fig. 20) was synthesized and copolymerized [59] with TFE. Terpolymers with TFE and PSVE were also prepared. Base hydrolysis of the copolymers caused C–P bond cleavage (Fig. 21, above).

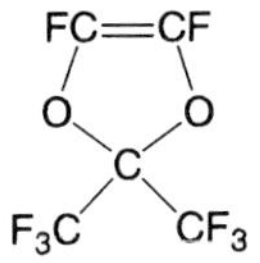

Fig. 18 2,2-Bis(trifluoromethyl)-4,5-difluoro-1,3-dioxole

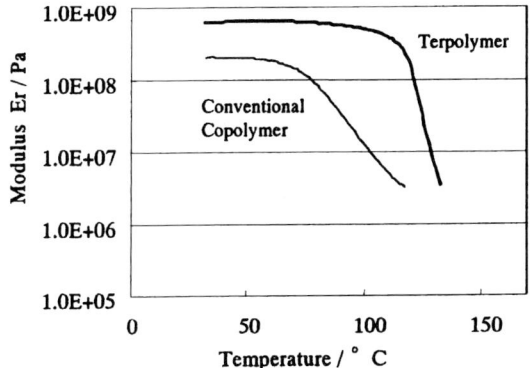

Fig. 19 Temperature dependency of the modulus of the terpolymer

The quantitative transformation of phosphinic ester polymer was carried out by acid hydrolysis (Fig. 21, below). The IECs of the copolymers were 1.4 to 2.1 meq/g.

The conductivity of the polymers at 25 °C in 1N HCl was a little smaller than that of Nafion®. Similar copolymers with an IEC of 2.5–3.5 meq/g were also prepared [60]. Their films exhibited similar proton conductivity to that of a Nafion® film at 80 °C under saturated water vapor. Preparation of phosphonic monomers requires many steps [59, 61], which makes it difficult to apply phosphonic membranes to fuel cells as well as the estimated high cathodic overpotential.

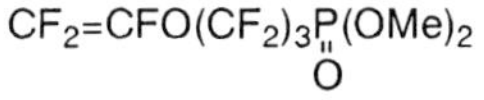

$$CF_2=CFO(CF_2)_3\overset{O}{\underset{\|}{P}}(OMe)_2$$

Fig. 20 Molecular structure of the perfluorophosphonic monomer

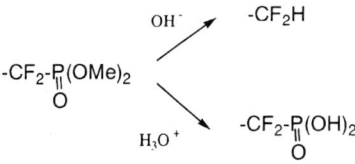

Fig. 21 Reactions of the perfluorophosphonic monomer with acid and alkali

4
Chemical Degradation Problem

Commercialization of PEFC systems requires high performance and low cost at the same time. Thinner membranes are convenient to get lower cell resistance and higher cell voltage [8] and, furthermore, can render the operation under low- or non-humidified conditions [9]. The membrane or catalyst-coated membrane (CCM) is sandwiched by the gas diffusion layer and separators, as schematically shown in Fig. 22. The usage of thinner membranes often causes not just mechanical damage, and various reinforcement methods have been developed as described in the next section. The nonuniformity of the circumstances surrounding MEAs and membranes in real cells, especially the difference between the edge region and the center, should be considered in the analysis of the degradation phenomena.

In the early stages of PEFC development after Ballard's demonstration, the chemical degradation of PFSA membranes was not considered by most fuel cell researchers except for a few analysts, because the stability of the perfluorinated membranes was believed to be proved by the analysis of membrane degradation in water electrolysis and the long operation in the chlor-alkali electrolysis industry. But the chemical degradation of perfluorinated membranes broke to the surface soon after the research on low-humidity operation began, which aimed to lower the system cost. The degradation of perfluorinated membranes has been recognized as a significant problem for the achievement of long-term durability for practical use. Evidence of membrane thinning and fluoride ion detection in the product water indicates that the polymer is undergoing chemical attack. Various research projects and programs to analyze and

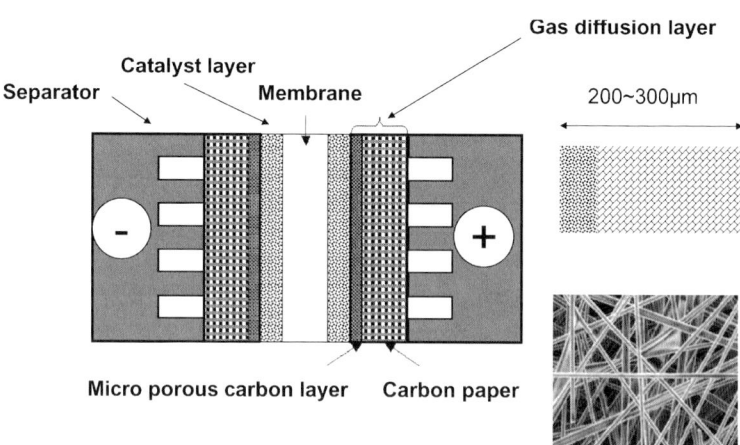

Fig. 22 Schematic illustration of a MEA and gas diffusion layer and a photo of carbon paper

solve the degradation phenomena were organized after that, and the number of reports on membrane degradation increased dramatically. Basically, the degradation problem has been investigated based on the mechanism in water electrolysis proposed by LaConti, et al. [62–66], where hydrogen peroxide or radicals caused by the cross leak of hydrogen and oxygen through the membrane play an important role. The old report [2] also proposed a typical design for mitigation applicable even to the present situation.

A mechanism of the degradation was postulated and examined. Hydrogen peroxide can be formed during the oxygen reduction reaction:

$$O_2 + 2H^+ + 2e^- \rightarrow H_2O_2 .$$

Hydrogen peroxide can decompose to give OH or OOH radicals. These materials can then attack H-containing or unsaturated groups present in the polymer. Such groups can include $-CF_2COOH$, $-CF_2H$ or $CF=CF_2$. Peroxide radical attack on such end groups is believed to be the main degradation mechanism. For a carboxylic end group the following reaction scheme has been proposed, as in Fig. 23 [16].

$$Rf-CF_2COOH + \bullet OH \longrightarrow Rf-CF_2\bullet + CO_2 + H_2O$$

$$Rf-CF_2 + \bullet OH \longrightarrow Rf-CF_2OH \longrightarrow Rf-COF + HF$$

$$Rf-COF + H_2O \longrightarrow Rf-COOH + HF$$

Fig. 23 Reaction of a carboxylic end group with an OH radical

The chemical degradation was studied both in situ (during fuel cell operation) and ex situ (by Fenton's reagent test) [68]. The structure of the examined polymer is given in Fig. 1. Except for fluoride release, the same product shown below was identified using NMR and mass spectroscopy in both the Fenton's test water and a residue extracted from MEAs that were heavily degraded during fuel cell operation.

$$CF_2=CFOCF_2\underset{\underset{CF_3}{|}}{C}FOCF_2CF_2SO_2F$$

These features demonstrate similarities between the in situ and ex situ degradation mechanism that involves degradation along a perfluorinated ionomer backbone. Similar results were also obtained after accelerated MEA durability tests under open circuit conditions [69]. For reduction of unstable polymer ends, fluorination of polymer ends by fluorine gas is effective. Membranes were exposed to gaseous hydrogen peroxide in order to examine its influence directly, which resulted in a rapid decrease in molecular weight of the polymer. This means that degradation takes place not only from polymer ends, but also main chain scission happens by chemical attack. The latter explains the decrease in molecular weight [68]. As for the attack to the main chains, several reports on the degradation of fluoropolymers, such as PTFE and PFSA, have been reported previously in alkaline fuel cells (AFCs), DMFCs, water electrolysis, and oxygen reduction gas electrodes for brine electrolysis [70–72].

In the case of the copolymer of $CF_2=CFO(CF_2)_4SO_2F$ and TFE, a fragment compound shown below was identified during peroxide testing.

$$HOOC-CF_2CF_2CF_2-SO_3H$$

Many studies on the effect of the catalyst layer and operation conditions, such as humidification and temperature, on values such as proton conductance, hydrogen leakage through the membrane, fluoride ion (F^-) release, molecular weight, and mechanical characteristics, have deepened the understanding of the mechanism of formation of H_2O_2, starting sites of decomposition in the membrane molecules, catalyst rearrangement in catalyst layers and membrane, etc. [73–77]. Various types of measures to mitigate chemical degradation have been proposed. In fact, several thousand hours of operation over 90 °C under lower humidity conditions have been reported using perfluorinated sulfonic membranes. Although polymer degradation is a serious problem, it was reported that additives can improve oxidative stability [67]. In addition, recently it was reported that a newly developed MEA with a novel fluorine-based proton-conductive polymer composite reduces the deterioration rate by 100–1000-fold compared with a conventional MEA. It can be operated continuously for more than 2000 h [79], and recently 4000 h, at 120 °C and 50% relative humidity [21]. According to the cross-sectional SEM image of the membrane in the report, the thickness of the cathode decreased by over a half, while those of the membrane and the anode remained at the initial values. The waste of the carbon support was found to be the major factor in the thinning of the cathode. It was clarified that the improvement of the oxidative stability of the carbon support is essential not only in the operation mode of automobiles, but also in the stationary mode at higher temperature. The company reported that a new catalyst with stabilized carbon support gave a more stable cell voltage of ~ 3 µV/h over 4000 h at 120 °C and 50%RH under relatively low pressure conditions (200 kPa).

5
Reinforcement Technology

The PFSA membranes have a tendency to swell considerably when they are soaked in water at higher temperature with a consequent decrease in mechanical strength. The lowering in mechanical strength is detrimental to the use of thinner membranes. On the other hand, thinner membranes have many advantages, such as low internal cell resistance or easy water management. In the early 1980s, reinforcement technologies for perfluorinated membranes were developed aimed at commercialization of chlor-alkali electrolysis, in which the robust membrane was firstly required to have a sufficient mechanical strength to insure durability for long-term operation. In the early stages of development, PTFE woven fabrics and PTFE microfibrils were proposed for reinforcement technology. Finally, PTFE woven fabrics have been recognized as a standard reinforcement method because of their good mechanical strength and chemical stability. The typical membrane for chlor-alkali electrolysis is a bilayer membrane and its major portion is a perfluorosulfonic polymer layer located on the anodic side. PTFE fabrics are embedded in the perfluorosulfonic layer. The cross section of the membrane is shown in Fig. 1. The total thickness of the membrane is approximately 200–300 µm.

On the contrary, in PEFC applications the size of the membrane is small and a higher mechanical strength is not necessarily required. The higher priority is set on the performance of the membrane, and a thin and flat membrane is required to obtain good performance. The performance is also determined by the precise structure of the porous electrodes that are bonded on both sides of the membrane. The flatness of the membrane helps to form proper electrode layers that afford good diffusivity of hydrogen, oxygen, and water vapor to the MEA. Therefore, in PEFC a specified reinforcement technology is required to prepare a thin and flat membrane that provides an appropriate mechanical strength, good chemical stability, higher proton conductivity, and water permeability. PTFE is most often selected as a reinforcing material for its excellent chemical stability and good mechanical strength. A composite comprising an expanded PTFE porous sheet and a perfluorinated ionomer was developed by W.L. Gore & Associates in the 1990s and has been commercialized as "Gore-Select®". Gore-Select® is characterized by its small thickness (20–40 µm) and excellent mechanical and electrochemical properties and has been widely used for PEFC systems. Gore-Select® shows relatively high specific resistance compared with nonreinforced membranes; however, the membrane resistance is sufficiently low because of its smaller thickness [79]. The PTFE yarn embedded type, which originated from chlor-alkali electrolysis, gives the good mechanical strength. The characteristics of this membrane were studied in NEDO's PEFC program, Japan. The PTFE fibril type was developed by Asahi Glass Co., Ltd. Even a small quantity of PTFE fibrils (a few wt.%) dispersed in a perfluorosulfonic polymer give the membrane both good mechanical strength and flatness suitable for PEFC application. Continuous film formation technology was also developed [80, 81]. Characterized values of the membrane, such as tensile strength, tear strength, creep property, and compressive property, and the cell performance were reported. Detailed explanation of the reinforcement technology of PTFE porous sheets and PTFE fibrils is available in reference [14].

Nafion® XL, one of the reinforced PFSA type membranes, was announced to give a 1.5 times increase in tensile strength and over 50% reduction in swell over Nafion®NRE211 [82].

6
DMFC Application

Experiments on DMFCs using PFSA membranes at higher temperature in the gas phase, which could give over $0.2\ \text{W/cm}^2$, ignited the development of DMFCs for electric vehicle (EV) and portable applications [83]. The PFSA membranes have a tendency to swell extremely when they are soaked in methanol aqueous solution. Reduction of the influence of methanol permeation and deformation or swelling of the membrane have been important issues in DMFC development, as shown in Table 2.

Table 2 Main research work in DMFC applications

- Operation at high temperature in the gas phase gave $0.2\ \text{W/cm}^2$
- Decrease in DMFC permeation using diluted methanol solution
- Modification of PFSA membrane with inorganic/organic fillers
- Development of new polymers: partially fluorinated types, such as styrene-grafted membranes, various hydrocarbon polymers, fullerenol-polymer composite
- Utilization of anionic membranes
- Studies on fuels other than methanol: ethanol, DME, etc.
- Gas phase DMFC using methanol clathrates

6.1
Examples of PFSA Type Membranes for DMFC

The addition of inorganic and organic fillers, introduction of the crosslinking structure, impregnation of ionomer into the porous sheet, surface modification of the PFSA membrane, and various grafted membranes have been proposed as membranes for DMFCs. These efforts decreased the methanol cross leak from several times to a tenth of that for PFSA membranes at best. Typical examples of the investigated methods are explained as follows. The membranes were recast from a PFSA dispersion in which small inorganic particles, such as silica, alumina, and titania, were able to reduce the methanol permeation by several times compared to the conventional PFSA membranes. Generally, the addition rates have optimum values. In recent research, detailed analyses have been available. For example, the interfacial properties such as zeta potential of the particles and positioning the existing sites of the particle in the membrane structure should be of more interest. The analyses with NMR and SAXS are expected to accelerate the deeper understanding of the transport mechanism of protons, water, and methanol in PFSA membranes. Discussions on the effect of size, aspect ratio, and surface properties of the organic fillers on membrane characteristics for DMFCs are expected to be developed.

7
Topics Related to New Analytical Results on the Structure of PFSA Dispersions and Membranes

Some of the properties of electrolyte membranes related to PEFCs, such as ion-exchange capacity (IEC) or equivalent weight (EW), molecular weight (MW) or corresponding parameter, water uptake after immersion in water, gas permeability, and mechanical properties in the dry state, etc., can be evaluated based on conventional procedures [84] and some protocols for measurement were made regionally [85, 86]. Progress in the development of PEFCs requires the introduction of new techniques or the creation of new measurement protocols suitable for a working fuel cell with various expected applications. As a matter of fact, it has been claimed and requested by many concerned to standardize common measuring procedures and some protocols were proposed in NEDO's PEFC programs. IEC or EW, MW, and solvent uptake in water-containing solution are the basic properties to be measured just before advanced evaluation. Properties such as water uptake, proton conductivity, water transport, gas permeation, dimensional stability, mechanical strength, compressive properties, creep properties, and chemical stability are usually evaluated in an atmosphere of controlled temperature and humidity. Pretreatment of membranes before testing is important for reproducible experimental results. More advanced analytical methods have come to be required to understand more detailed behaviors. For example, the conventional method of water uptake measurement is very delicate and requires sufficient skill [84, 87, 88]. A new method, where the amount of tritium is measured using a liquid scintillation counter after exposure to tritiated water vapor and subsequent immersion in distilled water, does not require such skill [89]. This tritium trapping method is also applicable to the measurement of the diffusion coefficients of hydrogen and oxygen in the membrane [90].

The structures of PFSA membranes have been analyzed and discussed by many researchers, and the cluster-network model for hydrated membranes proposed by Gierke [22] has been a basic model symbolic of the PFSA characteristics up to now. As for the structure of the diluted aqueous solution of PFSA, it is important to understand the structure of ionomer dispersion and catalyst ink, comprising catalyst particles, ionomer, and solvent, for the preparation of cast membrane and catalyst layer, respectively. Aldebert et al.

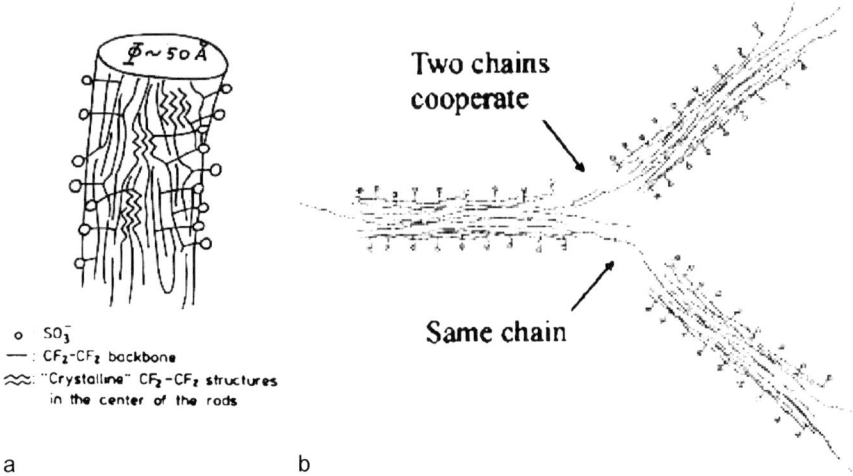

Fig. 24 Estimated models of PFSA dispersed in diluted aqueous solutions

proposed a rodlike structure by using SAXS and small-angle neutron scattering (SANS) in diluted aqueous solutions, as shown in Fig. 24a [91]. The self-assembling behavior of PFSA in more concentrated aqueous solutions was studied using dynamic light scattering (DLS) and a fringed-rod-based microgel model, as shown in Fig. 24b [93], was proposed based on a fringed-rod model [92].

Gebel proposed a schematic model of the structural evolution from dry membrane to colloidal dispersion of rodlike particles based on the results of the scattering analysis of PFSA over a wide range of water contents, combined with energetic considerations [94].

In the real fabrication process of membrane and MEA, more concentrated solutions which often contain organic solvents are used. Ultrasonic attenuation spectroscopy (UAS), whose principle is illustrated in Fig. 25, can offer detailed information on dispersed particles for a wide particle size range from 10 to 1 mm with subtle radiation energy (~ 10 mW) compared with other methods such as DLS. The application of the UAS technique to solutions of PFSA (over 1 wt. %) was tried and the structure was based on the rodlike micellar model shown in Fig. 26 [95]. The effect of molecular weight, solvent, concentration, and temperature on the size and shape of the dispersed PFSA particles was analyzed as shown in Figs. 27 and 28. Higher concentration and temperature and lower molecular weight gave smaller aspect ratios R; alcoholic solutions gave lower aspect ratios than aqueous solutions. The rigidity and hydrophobicity of perfluorinated ionomers are estimated to be related the phenomena.

The requirement to characterize the detailed structure of catalyst layers in MEAs has become more and more clamorous to improve MEA perform-

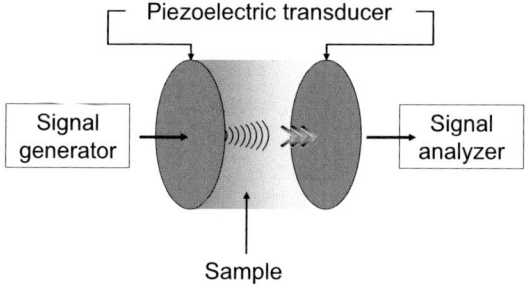

Fig. 25 Principle of ultrasonic attenuation spectroscopy

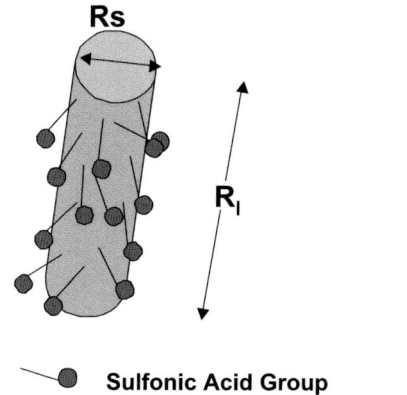

Fig. 26 Basic model assumed in the analysis of PFSA dispersion by UAS

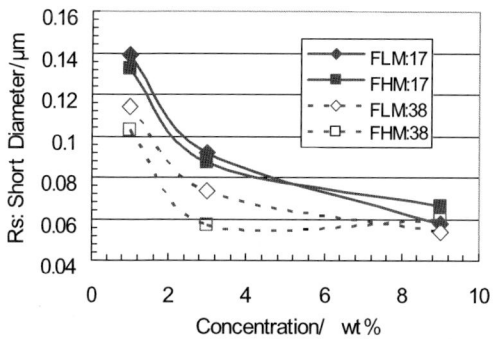

Fig. 27 Influence of concentration of Flemion® with lower molecular weight (FLM) and higher molecular weight (FHM) on R values at 17 and 38 °C in ethanol

ance. For example, many researchers have reported values around 20 to 30% [96, 97] as Pt utilization in MEA from CV measurements in N_2 atmosphere [65, 98] as Pt utilization in MEAs is strongly dependent on the preparation procedure using catalyst ink comprising catalyst and ionomer dispersion. Even though relatively higher utilization values have been reported

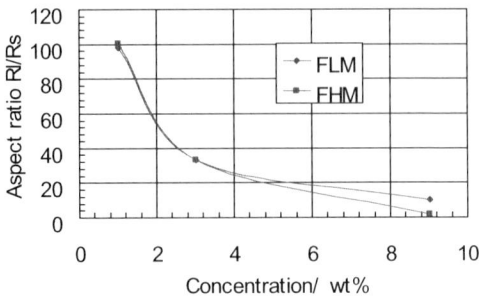

Fig. 28 Influence of concentration of Flemion® ionomers on the aspect ratio (4Rl/Rs) in aqueous solution at 17 °C

recently, a precise understanding is not sufficient at present. As a matter of fact, it is not easy to estimate the detailed structure of the catalyst layers consisting of catalyst and ionomer with microscopic methods because of the difference in contrast between catalyst metal and ionomer at high magnification. Trial measurements, such as observation of ionomer on catalyst particles in MEAs by field emission scanning electron microscopy (FE-SEM), or HR-SEM, with retarding function or TEM with elemental analysis or TEM with a cryo-system to decrease the damage of organic ionomer on catalysts by the electron beam, are reported. Figure 29 shows the existence of nonuniform coverage of ionomer [95].

Water distribution in membranes or single cells has been analyzed by various optical methods, neutron radiography (NRG) [99–103], and magnetic resonance imaging (MRI) [104–107]. Improvement of space resolution under 10 μm is expected.

As for the chemical stability, the Fenton's reagent prepared from H_2O_2 aqueous solution and $FeSO_4$ is often used. But it is well known that chemical degradation caused by H_2O_2 is most vigorous at the state of open circuit in dry conditions. Fenton tests are conducted in aqueous solution. A new

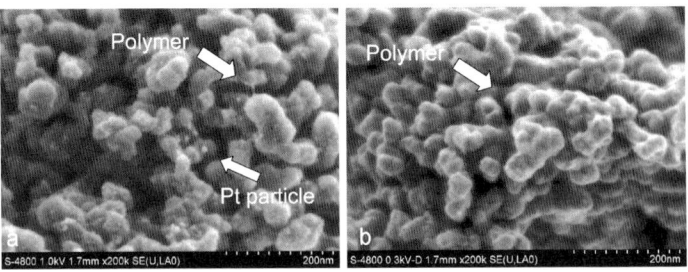

Fig. 29 **a** FE-SEM image of dried ink of catalyst and ionomer (ACV = 1 kV). **b** FE-SEM image of dried ink of catalyst and ionomer (ACV = 0.3 kV)

method in which humidified H_2O_2 gas is introduced into a circumstance-controlled chamber has been proposed to simulate the real operating state of PEFCs. A degradation mechanism consisting of unzipping of polymer ends and scission of main chains is estimated based on the changes of polymer weight, F release, and molecular weight.

Advancement in computer simulation founded on the molecular orbital method is expected to contribute to deeper understanding of the formation of ionic clusters, proton conduction, chemical deterioration of membranes, and water uptake over 100 °C, etc.

8
Summary

Over 160 years and 40 years have passed since the first exhibition of fuel cells by Grove and the first mission of the Gemini spaceship, respectively. In the meantime, "energy conversion took the wrong path in 1894" in spite of Ostwald's farsighted proposal according to the text by Bockris [108]. The PFSA membranes have been the most important components constituting the PEFC systems since the Biosatellite mission, although various new membranes have been studied and developed.

The performance of PEFCs and DMFCs has shown remarkable progress owing to the continuous development and improvement in materials such as membranes, electrocatalysts, and separators, as well as technical methods for MEA fabrication. Especially, PFSA membranes whose technologies have been raised by the chlor-alkali electrolysis industry, and may offer more massive quantity compared with other types of membranes for PEFCs, gave the greatest contribution to the development of PEFC systems. The long-term operation of over 4000 h at 120 °C under lower humidity conditions using a newly developed polymer composite membrane has expanded remarkably the possibility of polymer electrolyte and PEFC systems. The composite technology contributed to the improvement of mechanical and chemical performance.

It goes without saying that the commercialization of PEFC systems requires more progress in membrane performance and production technology. Various technologies in related fields and academic viewpoints and methods are expected to support and accelerate its attainment.

References

1. Warshay M, Prokopius PR (1989) The fuel cell in space: yesterday, today and tomorrow. Grove anniversary symposium, London, 18–21 Sept 1989
2. LaConti AB (1977) Proc Symp Electrode Mater Process Energ Convers Storage 77(6):354

3. Yeo RS, Yeager HL (1980) Structural and transport properties of perfluorinated ion-exchange membranes. In: Eisenberg A, Yeager HL (eds) Perfluorinated ionomer membranes. American Chemical Society, Washington, pp 437–505
4. Eisenberg A, Yeager HL (1982) Perfluorinated ionomer membranes, vol 283. American Chemical Society, Washington
5. Doyle M, Rajendran G (2003) Perfluorinated membranes. In: Vielstich W, Gasteiger HA, Lamm A (eds) Handbook of fuel cells: fundamentals, technology and applications. Wiley, Chichester, pp 351–395
6. Ukihashi H, Yamabe M, Miyake H (1986) Polymeric fluorocarbon acids and their applications. Prog Polym Sci 12(4):229–270
7. Takasu Y, Sugimoto W, Yoshitake M (2007) Electrochemistry 75:105
8. Yoshitake M, Yanagiawa E, Naganuma T, Kunisa Y (2000) In: Doughty DH, Nazar LF, Arakawa M, Brack H-P, Naoi K (eds) New materials for batteries and fuel cells. MRS Symp Proc 573:213–227
9. Dhar H (1993) US Patent 5242764
10. Dhar H (1994) Programs and abstracts of fuel cell seminar. San Diego, 28 Nov–1 Dec 1994, p 85
11. Watanabe M, Uchida H, Seki Y, Emori M, Stonehart P (1996) J Electrochem Soc 143:3847
12. Watanabe M, Uchida H, Emori M (1998) J Electrochem Soc 145:1137
13. Watanabe M, Uchida H, Emori M (1998) J Phys Chem B 102(17):3129
14. Nakao M, Yoshitake M (2003) Composite perfluorinated membranes. In: Vielstich W et al (eds) Handbook of fuel cells: fundamentals, technology and applications. Wiley, Chichester, pp 412–419
15. DesMarteau DD (1995) J Fluorine Chem 72:203
16. Curtin DE, Louserberg RD, Henry TJ, Tangeman PC, Tisack ME (2004) J Power Sources 131:41
17. Banerjee S, Curtin DE (2004) J Fluorine Chem 125:1211
18. Brack HP, Büchi FN, Rota M, Scherer GG (1997) ACS Polym Mater Sci Eng 77:368
19. Okazoe T, Watanabe K, Itoh M, Shirakawa D, Tatematsu S (2001) J Fluorine Chem 112:109
20. LaConti AB (1977) Proc Symp Electrode Mater Process Energ Convers Storage 77:354
21. Endoh E, Hommura S, Kawazoe H (2006) Abstract #95, 2006 fuel cell seminar, Honolulu. http://www.fuelcellseminar.com/pdf/2006/Wednesday/3B/Endoh_Eiji_0945_3B_95(rv3).pdf
22. Gierke TD, Munn GE, Wilson FC (1981) J Polym Sci 19:1687
23. Zawodzinski TA Jr, Milliken J (2002) Abstracts of 2002 fuel cell seminar, Palm Springs, p 870
24. Vaughan DJ (1973) Du Pont Innovation 43:10
25. Asahi Glass Co Ltd (2004) http://www.agc.co.jp/english/products/jirei_arena.html, accessed: January 22, 2004
26. Asahi Glass Co Ltd (2006) http://www.agc.co.jp/english/news/2006/1004e.html, accessed: October 4, 2006
27. Okazoe T, Murotani E, Watanabe K, Itoh M, Shirakawa D, Kawahara K, Kaneko I, Tatematsu S (2004) J Fluorine Chem 125:1695
28. Eisman GA (1990) J Power Sources 29:389
29. Keith P (1990) J Power Sources 29:239
30. Huff JR (1989) Prog Batteries Solar Cells 8:302
31. Banerjee S, Curtin DE (2004) J Fluorine Chem 125:1211

32. Tant MR, Darst KP, Lee KD, Martin CW (1989) Structures and properties of short-side-chain perfluorosulfonate ionomers. ACS Symp Ser 395:370–400
33. Resnick PE (1971) US Patent 3 560 568
34. Ezzell BR, Carl WP, Mod WA (1971) US Patent 4 358 412
35. Guerra MA (2004) US Patent 6 624 328
36. Hamrock SJ, Rivard LM, Moore GGI, Freemyer HT (2004) US Patent Application 20 041 212 A1
37. Hoshi N, Uematsu N, Saito H, Hattori M, Aoyagi T, Ikeda M (2005) PCT Int Appl WO 2005-29 624 A1
38. Ikeda M, Uematsu N, Saito H, Hoshii N, Hattori M, Iijima H (2005) Polym Prepr Japan 54:1R13
39. DesMarteau DD (1995) US Patent 5 463 005
40. Thomas BH, Shafer G, Ma JJ, Tu M-H, DesMarteau DD (2004) J Fluorine Chem 125:1231
41. Summer JJ, Creager SE, Ma JJ, DesMarteau DD (1998) J Electrochem Soc 145:107
42. Savet SC, Atkins JR, Sides CR, Harris JL, Thomas BH, Creager SE, Pennington WT, DesMarteau DD (2002) J Electrochem Soc 149:A1527
43. DesMarteau DD (2004) J Fluorine Chem 125:1231
44. Blau HAK (2001) PCT Int Appl WO 2001-47 872 A1
45. US Department of Energy (2006) Office of energy efficiency and renewable energy, program announcement to national laboratories for submission of proposals for research and development projects. US Department of Energy, Washington, p 10
46. Hogarth M, Glipa X (2001) High temperature membranes for solid polymer fuel cells. ETSU F/0200189/REP
47. Paddison SJ, Elliott JA (2005) J Phys Chem A 109:7583
48. Paddison SJ, Elliott JA (2006) Phys Chem Chem Phys 8:2193
49. Appleby AJ, Velev OA, LeHelloco J, Parthasarthy A, Srinivasan S, DesMarteau DD, Gillette MS, Ghosh JK (1993) J Electrochem Soc 140:109
50. Miyatake K, Zhou H, Matsuo T, Uchida H, Watanabe M (2004) Macromolecules 37:4961
51. Yoshitake M, Yanagisawa E, Terada I, Yoshida N, Naganuma T, Ishisaki T, Kunisa Y (2000) Development and characterization of perfluorinated sulfonic membranes for polymer electrolyte fuel cells. Rep Res Lab, Asahi Glass Co Ltd, Kanagawa, p 50
52. Rikukawa M, Sanui K (2000) Prog Polym Sci 25:1463
53. Rikukawa M (2007) Relationship between molecular design and functional expression for hydrocarbon polymer electrolytes, 3rd international hydrogen and fuel cell expo, FC EXPO 2007, special invitation session, 7–9 Feb 2007, Tokyo, pp 56–71
54. Mauritz KA, Moore RB (2004) Chem Rev 104:4535
55. Kreuer KD (2005) Choosing the protogenic groups in PEMS for high temperature, low humidity operation. Conference on advances in materials for proton exchange membrane fuel cell systems, Pacific Grove, 20–23 Feb 2005
56. Huang MH (1993) Macromolecules 26:5829
57. Watakabe A, Eriguchi T, Tanuma T, Kunisa Y (2002) US Patent Application 2002 142 207 A1
58. Yoshitake M, Terada I, Shimoda H, Watakabe A, Yamada K, Min K, Kunisa Y (2002) Abstracts of 2002 fuel cell seminar, Palm Springs
59. Kato M, Akiyama K, Akatsuka Y, Yamabe M (1982) Rep Res Lab Asahi Glass Co Ltd 32:117
60. Kotov SV, Pedersen SD, Qiu W, Qiu Z-M, Burton DJ (1997) J Fluorine Chem 82:13
61. Pedersen SD, Qiu W, Qiu Z-M, Kotov SV, Burton DJ (1996) J Org Chem 51:8024

62. Steck AE, Stone C (1997) Proceedings of the 2nd international symposium on new materials for fuel cell and modern battery systems, Montreal, 6–10 July 1997, pp 792–807
63. Wilkinson DP, St-Pierre J (2003) In: Vielstich W et al (eds) Handbook of fuel cells: fundamentals, technology and applications. Wiley, Chichester, pp 612–626
64. LaConti AB, Hamdan M, McDonald RC (2003) Mechanism of membrane degradation. In: Vielstich W, et al (eds) Handbook of fuel cells: fundamentals, technology and applications. Wiley, Chichester, pp 647–662
65. Gasteiger HA, Gu W, Makharia R, Mathias MF, Sompalli B (2003) Beginning-of-life MEA performance—efficiency loss contributions. In: Vielstich W et al (eds) Handbook of fuel cells: fundamentals, technology and applications. Wiley, Chichester, pp 593–610
66. Healy J, Hayden C, Xie T, Olson K, Waldo R, Brundage M, Gasteiger H, Abbott J (2005) Fuel Cells 5:302
67. Hommura S, Kawazoe K, Shimohira T (2005) 207th Meeting of the Electrochemical Society, Quebec City, Abstract #803
68. Hommura S, Kawahara K, Shimohira T (2005) Polym Prepr Japan 54:1R11
69. Hamrock S (2005) The development of new membranes for PEM fuel cells. Conference on advances in materials for proton exchange membrane fuel cell systems, 20–23 Feb 2005, Pacific Grove
70. Scherer GG (1990) Ber Bunsenges Phys Chem 94:1008
71. Guelzow E, Fischer M, Helmhold A, Reissner R, Schulze M, Wagner N, Lorenz M, Mueller B, Kaz T (1998) Fuel Cell Semin Abstr, p 469
72. Schulze M, Reissner R, Guelzow E (2006) Fuel Cell Semin Abstr Oral Present, p 370
73. Cipollini NE (2006) Mater Res Soc Symp Proc 885E:0885-A01-06
74. Kodani H, Wakizoe M, Miyake N (2005) High temperature membrane with durability for PEFCs. Proceedings of the international fuel cell workshop 2005, Kofu, p 55
75. Kato H (2007) Current status and future perspective of MEA development for PEFC, 3rd international hydrogen and fuel cell expo, FC EXPO 2007, FC-8, the most-developed element technology of polymer electrolyte fuel cells, Tokyo, pp 45–72
76. Pozio A, Silva RF, De Francesco M, Giorgi L (2003) Electrochim Acta 00:1
77. Schiraldi D, Zhou C (2005) DOE Briefing, 26 May 2005
78. Asahi Glass Co Ltd (2004) http://www.agc.co.jp/english/news/2004/0928_e.pdf, accessed: September 28, 2004
79. Cleghorn SJC (2003) http://www.gore.com/MungoBlobs/fuel_cells_presentation_SAE_auto.pd, accessed: April 9, 2003
80. Yanagisawa E, Kunisa Y, Ishisaki T, Terada I, Yoshitake M (2000) In: Yamamoto O, Takeda Y, Noda S, Kawatsu S, Imanishi N (eds) Proceedings of the international symposium on fuel cells for vehicles, pp 172–177
81. Hommura S, Kunisa Y, Terada I, Yoshitake M (2003) J Fluorine Chem 120:151
82. DuPont (2006) http://www.dupont.com/fuelcells/pdf/pressrel_11162006.pdf, accessed: November 16, 2006
83. Ren X, Wilson MS, Gottesfeld S (1996) J Electrochem Soc 143:L12
84. Yoshitake M, Tamura M, Yoshida N, Ishisaki T (1996) Denki Kagaku 64:727
85. Yoshida N, Ishisaki T, Watakabe A, Yoshitake M (1998) Electrochim Acta 43:3749
86. http://www.usfcc.com/usfcc/wgroup_codes.html
87. Zawodzinski TA, Davey J, Valerio J (1992) J Electrochem Soc 139:1332
88. Hinatsu JT, Mizuhata M, Takenaka H (1994) J Electrochem Soc 141:1493

89. Takata H, Nishikawa M, Arimura Y, Egawa T, Fukada S, Yoshitake M (2005) Int J Hydrogen Energy 30:1017
90. Takata H, Mizuno N, Nishikawa M, Fukada S, Yoshitake M (2007) Int J Hydrogen Energy 32:371
91. Aldebert P, Dreyfus B, Gebel G, Nakamura N, Pineri M, Volino F (1988) J Phys France 49:2101
92. Szajdzinska-Pietek E, Schlick S, Plonka A (1994) Langmuir 10:2188
93. Jiang S, Xia K-Q, Xu G (2001) Macromolecules 34:7783
94. Gebel G (2000) Polymer 41:5829
95. Yoshitake M (2006) Abstracts of pre-symposium of TOCAT5: catalysis in relation to fuel cell technology, Fukuoka
96. Ralph TR, Hards GA, Keating JE, Campbell SA, Wilkinson DP, Davis M, St-Pierre J, Johnson MC (1997) J Electrochem Soc 144:3845
97. Kocha SS (2003) Principles of MEA preparation. In: Vielstich W, Gasteiger HA, Lamm A (eds) Handbook of fuel cells: fundamentals, technology and applications, chap 43. Wiley, Chichester
98. Thompsett D (2003) Catalysts for the proton exchange membrane fuel cell. In: Hoogers G (ed) Fuel cell technology handbook, chap 6. CRC, Boca Raton
99. Bellows RJ, Lin MY, Arif M, Thompson AK, Jacobson D (1999) J Electrochem Soc 146:1099
100. Satija R, Jacobson DL, Arif M, Werner SA (2004) J Power Sources 129:238
101. Kramer D, Zhang J, Shimoi R, Lehman E, Wokaun A, Shinohara K, Scherer GG (2005) Electrochim Acta 50:2603
102. Schneider IA, Kramer D, Wokaun A, Scherer GG (2005) Electrochem Commun 7:1393
103. Zhang J, Kramer D, Shimoi R, Lehmann E, Wokaun A, Shinohara K, Scherer GG (2006) Electrochim Acta 51:2715
104. Tsushima S, Teranishi K, Hirai S (2004) Electrochem Solid-State Lett 7(9):A269
105. Teranishi K, Tsushima S, Hirai S (2005) Electrochem Solid-State Lett 8(6):A281–A284
106. Tsushima S, Nanjo T, Nishida K, Hirai S (2005) ECS Transact 1(6):199–205
107. Tsushima S, Teranishi K, Hirai S (2006) ECS Transact 3(1):91–96
108. Bockris JOM, Reddy AKN (1970) Modern Electrochemistry. Plenum, New York, pp 1350–1400

Radiation Grafted Membranes

Selmiye Alkan Gürsel[1,3] · Lorenz Gubler[1] (✉) · Bhuvanesh Gupta[2] · Günther G. Scherer[1]

[1]Electrochemistry Laboratory, Paul Scherrer Institut, 5232 Villigen PSI, Switzerland
lorenz.gubler@psi.ch

[2]Department of Textile Technology, Indian Institute of Technology, 110016 New Delhi, India

[3]*Present address:*
Faculty of Engineering and Natural Sciences, Sabanci University, 34956 Tuzla/Istanbul, Turkey

1	Introduction	159
2	Preparation of Radiation Grafted Membranes	160
2.1	Nature of Radiation	161
2.2	Graft Polymerization	162
2.3	Radiation Effects on Polymers	164
2.4	Grafting Parameters	167
2.4.1	Nature of Base Polymer	169
2.4.2	Irradiation Dose and Dose Rate	171
2.4.3	Monomer Concentration	174
2.4.4	Grafting Temperature	176
2.4.5	Grafting Medium	177
2.4.6	Additives	180
2.5	Crosslinking	180
2.6	Sulfonation	184
3	Characterization and Structure of Grafted Films and Membranes	185
3.1	Graft Mapping	185
3.2	Surface Chemistry and Surface Morphology	188
3.3	Thermal Characterization	192
3.4	Mechanical Properties	195
4	Fuel Cell Application	196
4.1	Membrane Properties Relevant to Fuel Cell Application	196
4.1.1	Ion Exchange Capacity	197
4.1.2	Water Uptake	199
4.1.3	Conductivity	201
4.2	Performance in Fuel Cells	203
4.2.1	MEA Fabrication	204
4.2.2	Fuel Cell Testing	204
4.2.3	Water States and Water Management	204
4.2.4	Reactant Permeability	206
4.2.5	Chemical Stability	206
4.2.6	Mechanical Integrity	207

4.2.7 Fuel Cell Performance 207
4.2.8 Performance in Direct Methanol Fuel Cells 209

5 Conclusions 210

References 211

Abstract The development of proton-exchange membranes for fuel cells has generated global interest in order to have a potential source of power for stationary and portable applications. The membrane is the heart of a fuel cell and the performance of a fuel cell depends largely on the physico-chemical nature of the membrane and its stability in the hostile environment of hydrogen and oxygen at elevated temperatures. Efforts are being made to develop membranes that are similar to commercial Nafion® membranes in performance and are available at an affordable price. The radiation grafting of styrene and its derivatives onto existing polymer films and subsequent sulfonation of the grafted films has been an attractive route for developing these membranes with required chemistry and properties. The process of radiation grafting offers enormous possibilities for design of the polymer architecture by careful variation of the irradiation and the grafting conditions. A wide range of crosslinkers are available, which introduce stability to the membrane during its operation in fuel cells. Crosslinking of the base polymer prior to grafting has also been an attractive means of obtaining membranes with better performance. A systematic presentation is made of the grafting process into different polymers, the physical properties of the resultant membranes, and the fuel cell application of these membranes.

Keywords Polymer electrolyte fuel cell · Proton exchange membrane · Radiation grafting

Abbreviations
ATR   Attenuated total reflection spectroscopy
c_{H^+} Volumetric density of protons
DG   Degree of grafting
D_{H^+} Proton diffusion coefficient
DSC   Differential scanning calorimetry
DVB   Divinylbenzene
ESR   Electron spin resonance
ETFE   Poly(ethylene-*alt*-tetrafluoroethylene)
FEP   Poly(tetrafluoroethylene-*co*-hexafluoropropylene)
FTIR   Fourier transform infrared spectroscopy
G value   Radiation chemical yield
Gy   Gray
IEC   Ion exchange capacity
MEA   Membrane electrode assembly
MFI   Melt flow index
m Mass
$n(H_2O)$ Number of water molecules
$n(SO_3H)$ Number of exchange sites
PEFC   Polymer electrolyte fuel cell
PFA   Poly(tetrafluoroethylene-*co*-perfluorovinyl ether)
pK_a Acid dissociation constant

PSSA	Polystyrene sulfonic acid
PTFE	Poly(tetrafluoroethylene)
PVDF	Poly(vinylidene fluoride)
SANS	Small angle neutron scattering
SAXS	Small angle X-ray scattering
SEM	Scanning electron microscopy
$-SO_3H$	Sulfonic acid
TAC	Triallylcyanurate
TFS	α,β,β-Trifluorostyrene
T_g	Glass transition temperature
TGA	Thermogravimetric analysis
T_m	Melting temperature
XMA	X-ray microprobe analysis
XPS	X-ray photoelectron spectroscopy
ϕ	Water uptake
λ	Hydration number
σ_{H^+}	Proton conductivity

1
Introduction

Membrane science and technology is the fascinating world of polymers, which extends from separation science and bioreactors to environmental care and electrochemistry [1]. The attraction of membranes lies in their energy-efficient processes combined with their low cost separation, as compared to conventional techniques. The versatile nature of membranes has made their application areas grow enormously. Membranes with different shapes and chemical designs are available, which makes them suitable for processes such as nanofiltration, reverse osmosis, pervaporation, bioreactors, dialysis, electrodialysis, electrolysis, and fuel cells. Membranes have generated considerable interest as solid polymer electrolytes in fuel cells, which have been identified as a promising source of power for stationary and portable applications [2]. The fuel cell offers several advantages in terms of the high power densities and having water as a by-product, which makes it an eco-friendly alternative for energy production. The membrane in a fuel cell offers support structure for the electrodes and allows proton transport across its matrix from anode to cathode. The fuel cell requires a proton exchange membrane that shows good mechanical strength, high chemical stability, and appropriate ionic conductivity (e.g., $> 10^{-2}$ S cm^{-1}). In the current state of technology, perfluorinated membrane materials such as Nafion® (DuPont, USA), Flemion (Asahi Glass, Japan), and Aciplex (Asahi Kasei, Japan) are used predominantly in polymer electrolyte fuel cells, due to their attractive conductivity and chemical stability. However, for market introduction of fuel cell products, cost-competitive membrane technology has to be developed.

The Nafion® membrane, for instance, has shown good performance in fuel cells but has certain limitations, i.e., it has poor ionic conductivity at low humidity and is available at an expensive rate of $\sim 500\ \$/m^2$. The costs for Nafion®, for example, become attractive only at high production volumes [3]. Consequently, the search for new membrane materials with low cost and the required electrochemical characteristics, along with performances matching those of Nafion®, is continuing and has become the most focused research area in the design of polymer electrolyte fuel cells.

Both the physical and chemical factors are essential for the establishment of a critical relationship between the structure and performance of a membrane in operation. Therefore, designing a membrane needs proper understanding of both the polymeric material and the fuel cell requirements. With no other membrane in sight and under the complexity of inventing new materials, it becomes necessary to modify existing materials into required membrane structures. A great deal of research effort has been directed to the development of membranes by introducing ionic functionality into different polymers. The sulfonation of polymer films such as in polyetheretherketone and polysulfone is one such approach being used to develop ionic membranes [4–6]. However, the ionic character of membranes needs to be accompanied by their good performance in fuel cell application. That is why the current efforts have been directed to the modification of existing polymer films in such a way that the modified material acquires desired functionality and performs well. Although the base matrix may be any type of polymer, the selection of the fluorinated or perfluorinated polymer matrix has been a prime consideration due to the better chemical and thermal resistance that these polymers provide. Consequently, the functionalization of these polymers by radiation grafting of appropriate monomers has become an attractive way to develop such membranes. It is quite spectacular to envisage that polymers can be altered into materials that display a unique combination of characteristics such as ionic nature, water absorption, and high conductivity. Enormous work has been carried out on the graft modification of polymers and several reviews have been published in this domain [7–13]. Recent reviews related to radiation grafting on fluoropolymers provide thorough knowledge in this area [14–20].

We have confined our goal to reviewing the state-of-the-art in the development of radiation grafted proton-exchange membranes. This review provides an up-to-date summary of the synthesis, properties, and applications of radiation grafted membranes as solid polymer electrolytes in fuel cells.

2
Preparation of Radiation Grafted Membranes

A graft copolymer, in general, can be defined as consisting of one or more types of molecules, as block, connected as side chains to a main chain. These side chains should have constitutional or configurational features that differ from those of the main chain. The modification of polymers through graft polymerization offers an interesting route for achieving membranes with desirable characteristics. Depending on the chemical nature of the monomer, membranes with desired physico-chemical properties may be fabricated. Therefore, if the monomer is ionic in nature, the grafted membrane acquires ionic character with little influence on most of its inherent characteristics. In this section we describe vital aspects that influence membrane fabrication and performance.

2.1
Nature of Radiation

Membrane development requires activation of the entire bulk of the film so that modification across the film may be achieved. This makes it necessary to use high energy radiation, which may penetrate and produce ionization of the polymer matrix. The nature of the radiation has significant impact on the physical and chemical properties of the resultant membrane. A wide range of types of high energy radiation are available to be used for the grafting process. The radiation may be either electromagnetic in nature, such as X-rays and gamma rays, or charged particles, such as beta particles and electrons. The basic difference between the two types of radiation lies in the higher penetration of the electromagnetic radiation. Charged particles lose energy almost continuously through a large number of small energy transfers while passing through matter. However, photons tend to lose a relatively large amount of intensity by interaction with matter. The advantage of electromagnetic radiation, such as gamma rays, is that the fractions of photons that do not interact with a finite thickness of the material are transmitted with their original energies and directions (exponential attenuation law). Hence, the dose rate of radiation may be easily controlled by the use of a suitable attenuator without influencing the photon energy, which is a very important aspect in radiation-initiated polymerization of monomers.

Although different gamma sources are available today, the most versatile gamma radiation source is Co^{60}, which has a long half-life of 5.3 years and emits radiation of 1.17 and 1.33 MeV (mean value of 1.25 MeV). Two different types of gamma radiation source are available for irradiation. One of the sources is a cavity-type unit where a hollow source in the form of a cylinder remains stationary. The Co^{60} remains in this cylindrical structure as the pins. The sample is introduced into this cylinder cavity by means of a moving drawer. The sample moves down inside the cavity during the exposure stage. Once the irradiation is over, the sample is drawn out and may be subsequently removed. The second type of source is a cave-type where Co^{60} is kept in a shielded container. The whole unit is kept underground and the source moves out with the help of a moving belt for irradiation of a stationary sample. The latter type is usually used for the irradiation of samples at an industrial scale.

Exposure of the polymer to radiation is expressed as the absorbed dose. The absorbed radiation dose is defined as the amount of energy imparted to the matter. The units initially used for the radiation dose were rad and Mrad. The most recent unit of radiation is Gray (Gy), which corresponds to 10^4 erg g^{-1}. For higher doses, another unit, kilogray (kGy), is used. The dose rate, therefore, is defined as the adsorbed dose per unit time (Gy min^{-1}). Since radiation grafting proceeds by the generation of free radicals on the polymer as well as on the monomer, the G value (i.e., radiation chemical yield, expressed as the number of free radicals generated for 100 eV energy absorbed per gram) plays an important role in the grafting process. For most polymers the G value remains in the range 2–3.

2.2
Graft Polymerization

Radiation-induced grafting is a process where, in a first step, an active site is created in the preexisting polymer. This site is usually a free radical, where the polymer chain behaves like a macroradical. This may subsequently initiate the polymerization of a monomer, leading to the formation of a graft copolymer structure where the backbone is represented by the polymer being modified, and the side chains are formed from the monomer (Fig. 1). This method offers the promise of polymerization of monomers that are difficult to polymerize by conventional methods without residues of initiators and catalysts. Moreover, polymerization can be carried out even at low temperatures, unlike polymerization with catalysts and initiators. Another interesting as-

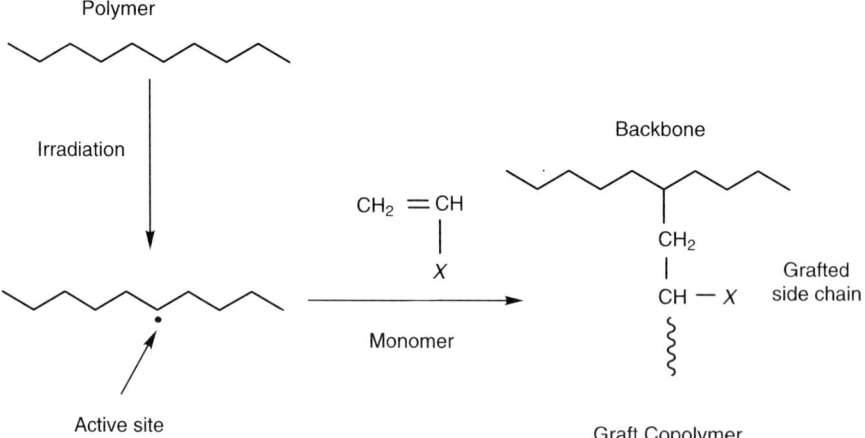

Fig. 1 Radiation-induced grafting

pect of the radiation grafting process is that the grafting may be carried out onto a polymer irrespective of its shape or form. Still, membrane development requires that the grafting is carried out on polymers already existing in the form of a film so that the resultant material remains in sheet form. This overcomes the problem of shaping a grafted polymer bulk into a thin foil. Graft polymerization using high energy radiation is one of the most convenient and the most effective way to develop membranes. By virtue of the high energy of radiation, the photon penetrates effectively into the polymer bulk and activates the matrix thoroughly. This process, therefore, offers a unique way to combine the properties of two highly incompatible polymers. Another attractive feature of radiation grafting is that the degree of grafting may be easily controlled by proper monitoring of the radiation dose, dose rate, and the reaction conditions.

Radiation grafting may be carried out by using three different options [21, 22]:

1. Simultaneous radiation grafting is where both the polymer and the monomer are exposed to radiation. In situ free radical sites are generated and the polymerization of the monomer is initiated. The limitation of this method is that the monomer is continuously exposed to radiation during the grafting reaction and hence extensive homopolymerization proceeds parallel to the grafting reaction, which leads to monomer wastage and a low level of grafting efficiency in a system.
2. Preirradiation grafting (hydroperoxide method) involves activation of the polymer by exposure to radiation under air, which results in the creation of radicals along the macromolecular backbone. These radicals subsequently interact with the oxygen and form peroxides. The graft polymerization is initiated by the decomposition of these peroxides at an elevated temperature. The drawback of this process is that significantly high irradiation doses are needed to achieve a sufficient number of hydroperoxides to accomplish reasonable graft levels, which leads to drastic changes in the physical structure of the polymer and oxidative degradation, even before any grafting is initiated and this is subsequently reflected in the membrane characteristics.
3. Preirradiation grafting (trapped radicals method) involves irradiation of the polymer under inert atmosphere or under vacuum. As a result, the radicals are formed and remain trapped within the polymer matrix. These radicals subsequently initiate the grafting of a monomer.

It is important to mention that because of the inherent differences in the irradiation approaches, the physical characteristics of the membranes will be dependent on the adopted grafting process. The extent of polymerization is expressed as the degree of grafting (DG), which is defined as the percentage mass of the grafted component within the copolymer matrix. On the other hand, grafting efficiency refers to the percentage conversion of the monomer into the grafted component with respect to the total monomer conversion.

2.3
Radiation Effects on Polymers

Knowledge of the influence of irradiation on polymers is extremely important because even a low irradiation dose may introduce significant alteration in the physical structure of the polymer prior to any grafting being accomplished. The outstanding properties of fluoropolymers, such as excellent chemical resistance, mechanical strength, high temperature stability, and good weathering make them strong candidates as membranes for a highly oxidizing environment such as in fuel cells. However, interaction of the high energy radiation with such polymers may induce significant physical and chemical changes. The irradiation causes ionization of the matrix leading to the formation of ions, radicals, and excited species. The ultimate result is reflected in the chain scission and crosslinking, along with the formation of volatiles, leading to significant variation in the molecular weight of the polymer. The magnitude of these processes will be dependent not only on the chemical nature of the polymer matrix, but also on the nature of the radiation, temperature of the irradiation, and irradiation doses. The irradiation medium may further induce chemical changes depending on the nature of the medium.

Among the fluoropolymers, poly(tetrafluoroethylene) (PTFE) undergoes severe degradation even under mild irradiation conditions both under air and in vacuum [21]. The radiation sensitivity of PTFE is so high that it is readily converted into a low molecular weight fine powder under ionizing radiation. The irradiation leads to the formation of acid fluoride ($-COF$) groups within the polymer matrix, which easily hydrolyze into carboxylic groups ($-COOH$) in contact with atmospheric humid air [23, 24]. This is the reason that surface concentration of $-COOH$ increases with increasing irradiation doses and enhances its surface energy [25]. The polymer degradation is associated with the formation of chain end free radicals, ($-CF_2-\dot{}CF_2$) or chain alkyl radicals, ($-CF_2-\dot{}CF-CF_2-$), where chain end radicals originate as a result of the main chain scission as observed by electron spin resonance (ESR) [26]. This contributes to the considerable loss in thermal stability of the irradiated polymer and becomes so pronounced that the initial decomposition temperature, as observed in thermogravimetric analysis, is brought down from 530 to 240 °C for an irradiation dose of 100 kGy [27].

The radiation chemistry of copolymers of tetrafluoroethylene with other perfluorinated moieties, such as hexafluoropropylene, is almost identical to that of PTFE with the difference that the relative magnitude of crosslinking and scission varies significantly. The various chemical moieties that have been identified under irradiation are presented in Fig. 2. Although these stud-

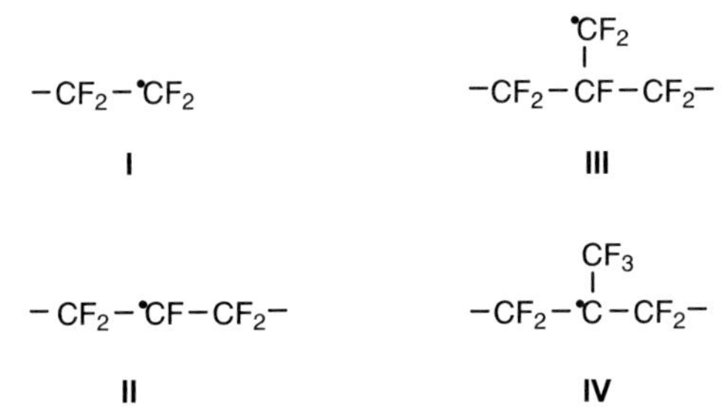

Radiation Grafted Membranes 165

$$-CF_2-\overset{\bullet}{C}F_2 \qquad -CF_2-CF-CF_2-$$
$$\overset{|}{\overset{\bullet}{C}F_2}$$

I   III

$$-CF_2-\overset{\bullet}{C}F-CF_2- \qquad -CF_2-\overset{\overset{CF_3}{|}}{\overset{\bullet}{C}}-CF_2-$$

II   IV

Fig. 2 Possible radicals formed on radiolysis of FEP (redrawn from [30])

ies on radiolysis of poly(tetrafluoroethylene-*co*-hexafluoropropylene) (FEP) are well supported by the studies of Iwasaki et al. [28], there is less agreement on the nature of the radicals and their quantification at different doses [29, 30].

The irradiation temperature of the polymer has distinct influence on the relative proportions of the radical moieties. The irradiation of FEP at a temperature as low as 77 K involves the radicals I and II as the major contributors, while very little originates in the form of III and IV. However, the irradiation at room temperature (300 K) shows a much higher contribution of chain end radicals, with the G values being 0.22 and 2.0 at 77 and 300 K, respectively. As far as the radical concentration in FEP as a function of the irradiation dose at 77 and 300 K is concerned, the radical concentration at 300 K is much higher than at the lower temperature, probably due to the enhanced molecular mobility and resultant chain scission at higher temperature [30]. Identification of the radical I as one of the principal radicals on radiolysis at 77 and 300 K is consistent with the main chain scission being the major bond-breaking step during gamma irradiation of FEP at both these temperatures. These observations are supported by the investigations on poly(tetrafluoroethylene-*co*-perfluorovinyl ether) (PFA). The nature of the radicals in PFA, as determined by ESR, was identified to be I and II. However, G values for radical formation at room temperature and 77 K were found to be 0.93 and 0.16, respectively [31], which is higher than the values for PTFE of 0.4 and 0.14 [32].

There is a systematic difference in the degradation behavior of PTFE from FEP and PFA under ionizing radiation. Both the FEP and PFA contain a pendent group in the form of $-CF_3$ and $-OC_3F_7$, respectively. This has direct bearing on the crystalline structure of the polymer due to impedance in the chain packing by these substituting groups. The higher amorphous region in these two polymers would therefore lead to greater radical mobility and subsequent chain scission as compared to PTFE. The high sensitivity of PTFE to irradiation is because the radicals have restricted movements in a highly crystalline matrix and therefore inhibit radical–radical recombination. Both PFA and FEP undergo side-chain cleavage and therefore have more chain end radicals. Recombination of the radicals is restricted and the chain scission proceeds smoothly, resulting in the formation of a higher number of radicals. This further reflects into the greater number of carboxyl groups (transformation of –COF to –COOH), which proceeds in the order FEP > PFA > PTFE [33].

The irradiation of poly(vinylidene fluoride) (PVDF) brings about little enhancement in the crystallinity for irradiation doses of about 100 kGy similar to poly(ethylene-*alt*-tetrafluoroethylene) (ETFE). However, beyond 100 kGy, ETFE shows significant loss in the crystallinity but PVDF remains almost unchanged [34].

The irradiation of fluoropolymers at elevated temperatures has been explored for the development of materials with better mechanical properties [35]. This arises because of the radiation-induced crosslinking of chains and subsequent higher network density in the resultant polymer [36]. Here, the irradiation is accomplished at a temperature higher than the melting point of the polymer. In the molten state, the polymer behaves as an amorphous matrix and the mobility of molecular chains is considerably enhanced. This promotes the mutual recombination of radicals, i.e., crosslinking involving chain end radicals and chain alkyl radicals [37].

Irradiation even at a dose as low as 5 kGy brings about a drastic improvement in the tensile strength of PTFE. As the irradiation temperature increases from room temperature towards below melting, the mechanical strength decreases quite rapidly. This is an indication that the chain scission is accelerated with increasing temperature. However, once the irradiation temperature crosses the melting temperature and reaches beyond 340 °C, both modulus and tensile strength tend to increase considerably, because the polymer enters into a molten state where the network formation is facilitated. Such behavior has been observed by other workers under different irradiation doses [38]. It is interesting to note that the crystallinity of the polymer undergoes drastic reduction with the increasing dose. This is an obvious outcome of the crosslinking of chains, which lowers the molecular mobility and prevents the chains from undergoing crystallization upon cooling. The crosslinking is so pronounced that an irradiation dose of 2 MGy leads to complete inhibition of crystallization in PTFE [32].

The radiation processing of FEP has shown that crosslinking proceeds favorably at temperatures above its glass transition temperature (70–90 °C) under vacuum. The crosslink density, as measured by the gel content, tends to increase sharply upon gamma irradiation at around 90 °C and reaches values as high as 35% at 160 °C [39]. Based on X-ray photoelectron spectroscopy (XPS), it has been found that the radical **IV** (Fig. 2) dominates over other species under gamma irradiation [40]. This structure originates from the hexafluoropropylene units in the copolymer. The combination of structure **IV** with **I** has been proposed to be the most probable route to the crosslinking reaction. This is further supported by the investigations of Sun et al. [41], where structure **IV** was proposed to be the one involved in the crosslinking reaction with other radicals. The tetrafluoroethylene component along the polymer chain still undergoes the crosslinking reaction. Forsythe et al. [42] have made comprehensive studies on the gamma irradiation-induced changes in the chemical and mechanical behavior of poly(tetrafluoroethylene-*co*-perfluoromethylvinylether). Irradiation at the temperature range 77–195 K did not result in any gel formation, indicating that the crosslinking is almost suppressed at these temperatures. Tensile strength diminished and elongation increased, suggesting that chain scission is the most appropriate change taking place. The strong evidence in favor of this degradation comes from the diminishing glass transition temperature in this temperature range. Crosslinking dominated over chain scission at 263 °C and above, where gelation also approached 80–90% and tensile strength also showed a sharp increase.

2.4
Grafting Parameters

The design of membranes by radiation grafting covers not only the covalently linked incorporation of an ionic component but also requires perfect tailor-making to govern how well the molecular architecture, physical properties, and morphology of the membranes may be controlled. A wide range of polymers have been grafted, predominantly with styrene or its derivatives, using different crosslinkers. Tables 1–3 illustrate the common base films, monomers, and crosslinkers used in radiation-induced grafting [43–46].

Graft polymerization is strongly influenced by irradiation and synthesis conditions, such as radiation dose, dose rate, monomer concentration, reaction temperature, pregrafting storage, solvents, and additives (irrespective of the base matrix). Most of the work on membrane preparation follows the graft polymerization of styrene onto polymers and the subsequent sulfonation. The pioneering work of Chapiro on radiation-induced grafting led to interesting observations on the grafting process and opened up the route for several possibilities in radiochemical grafting of polymer films [47–50]. For most of the polymer–monomer systems, grafting proceeds by the grafting front mechanism, as proposed by Chapiro for grafting into polyethylene and FEP films [51–53]. The initial grafting takes place at the film surface and behaves as the grafting front. This grafted layer swells in the reaction medium and further grafting proceeds by the progressive diffusion of the monomer through this swollen layer and grafting front movement to the middle of the film. This mechanism of grafting has recently been the basis of several other investigations on membrane preparation based on polyethylene, FEP, and PFA films as the base matrix [54–57]. The following sections deal with the various parameters and factors that influence the DG.

Table 1 Common base polymer films used for the preparation of radiation grafted FC membranes [43]

Polymer	Abbreviation	Repeating unit
Perfluorinated polymers		
Polytetrafluoroethylene	PTFE	$-[-CF_2-CF_2-]_n-$
Poly(tetrafluoroethylene-co-hexafluoropropylene)	FEP	$-[-CF_2-CF_2-]_n[-CF_2-CF(CF_3)-]_m-$
Poly(tetrafluoroethylene-co-perfluoropropyl vinyl ether)	PFA	$-[-CF_2-CF_2-]_n[-CF_2-CF(OC_3F_7)-]_m-$
Partially fluorinated polymers		
Polyvinylidene fluoride[a]	PVDF	$-[-CF_2-CH_2-]_n-$
Poly(vinylidene fluoride-co-hexafluoropropylene)	PVDF-co-HFP	$-[-CF_2-CH_2-]_n[-CF_2-CF(CF_3)-]_m-$
Poly(ethylene-alt-tetrafluoroethylene)	ETFE	$-[(-CH_2-CH_2-)(-CF_2-CF_2-)]_n-$
Polyvinyl fluoride	PVF	$-[-CH_2-CHF-]_n-$
Hydrocarbon polymers		
Polyethylene	PE	$-[-CH_2-CH_2-]_n-$

Table 2 Monomers used for the preparation of radiation grafted FC membranes [43]

Styrene	(phenyl–CH=CH$_2$)	α,β,β-Trifluorostyrene (TFS)	(phenyl–CF=CF$_2$)
α-Methylstyrene (AMS)	(phenyl–C(CH$_3$)=CH$_2$)	Substituted trifluorostyrene (R = SO$_2$F, Me, MeO, PhO, ...)	(R-phenyl–CF=CF$_2$)

Table 3 Crosslinkers used as co-monomers in the radiation grafting process [43]

Divinyl benzene (DVB)	Bis(vinyl phenyl)ethane (BVPE)	Triallylcyanurate (TAC)

2.4.1 Nature of Base Polymer

The chemical nature of the base polymer is an important aspect in membrane development. There has been preference for the thermally stable fluorinated polymers over hydrocarbon polymers. Fluorine-containing polymers, characterized by the presence of carbon–fluorine bonds, are widely used as the base matrices owing to their outstanding chemical and thermal stability, low surface energy, and the ease of modification of various properties by the grafting method. Perfluorinated polymers and partially fluorinated polymers combining hydrocarbon and fluorocarbon structures are excellent candidates as base polymers. For instance, fluorinated FEP has drawn wide attention due to its reasonably good radiation stability [58].

The membranes, developed at the Paul Scherer Institut (PSI, Switzerland) for fuel cell applications, were initially based on FEP [59–61]. The use of ETFE as base material was revisited recently in this laboratory since ETFE is readily available in higher molecular weights and has desirable mechanical properties such as breaking strength and flexibility, which are enhanced with increasing molecular weight [62]. ETFE contains alternating structural units of ethylene and tetrafluoroethylene that confers a unique combination of properties imparted from both fluorocarbon and hydrocarbon polymers. Moreover, undesirable chain scission reactions occurring during preirradiation grafting can be minimized by using ETFE, especially in combination with electron beam irradiation under inert atmosphere [63].

The base polymer film type and its properties (such as film thickness, extent of orientation, and molar mass) have significant effect on both the degree of grafting and resultant membrane properties [64, 65]. Walsby et al. [65] have reported that under identical conditions, grafting of styrene onto different base polymers yielded different graft levels. The authors indicated that graft levels were 5% for PTFE, 56% for PVDF, 28% for FEP, and 62% for ETFE. It seems that the influence of the base polymer matrix on grafting is a complex scenario. The differences obtained in graft level may be due to the different radical concentrations, different structures of the radical centers, and different degrees of crystallinity. Since the grafting essentially takes place in the amorphous region, the high crystallinity of the polymer would provide lesser radicals in the amorphous region accompanied by low monomer diffusion for subsequent graft initiation and propagation. The glass transition temperature (T_g) may also contribute in terms of the mobility of the macromolecular chains in the amorphous region. If the grafting is carried out at a temperature higher than the T_g, the enhanced mobility of chains would favor mutual recombination of growing grafted chains, leading to the low graft levels [65]. The radical concentration in PTFE tends to be two orders of magnitude lower than in polyethylene and ETFE for an irradiation dose of 100 kGy and may be one of the reasons for low graft levels [66]. ETFE films are found to yield higher graft levels than that of FEP under identical grafting conditions. This behavior may be attributed to the greater number of reactive sites available for ETFE since more radicals are expected to be formed per kGy of radiation dose (lower bond strength of C–H than C–C and C–F) [67, 68].

Increasing the molecular weight of the base polymer film causes a decrease in the DG. Melt flow index (MFI) measurements are especially useful for obtaining both qualitative and quantitative information about the molecular weight of polymers, chain scission, and crosslinking. It was reported that MFI increases due to chain scission upon ETFE irradiation in air. Also, ETFE films tend to undergo crosslinking during irradiation at room temperature under inert atmosphere [64]. It is also observed that higher irradiation doses are required for thinner base films than for thicker ones to achieve comparable DG under identical grafting conditions. This may be attributed to the greater extent of orientation of polymeric chains in the machine direction in thinner films [63]. The extent of orientation has a significant effect on polymer permeability, which decreases as the orientation increases [64]. A negative dependence of grafting rate on film thickness for the grafting of acrylic acid onto PTFE has been observed [69]. However, other investigations have shown that the film thickness has no significant effect on grafting yield [70].

Another interesting development in membrane fabrication has been the use of porous base films [71]. The grafting of a monomer and subsequent sulfonation still leads to porosity in the membrane bulk. However, this membrane may be densified by impregnating it to substantially fill the porosity, or the porosity may be collapsed by the application of pressure and heat. The heating may be carried out to at least a melt flow temperature of the film but at a lower melting temperature (T_m) than grafted side chains.

The pregraft storage of irradiated films is an important aspect of membrane preparation. It has been observed that fluorinated polymers retain their grafting ability for a longer period, irrespective of their chemical structure [47, 72]. Horsfall et al. [73] have shown that irradiated ETFE and PVDF

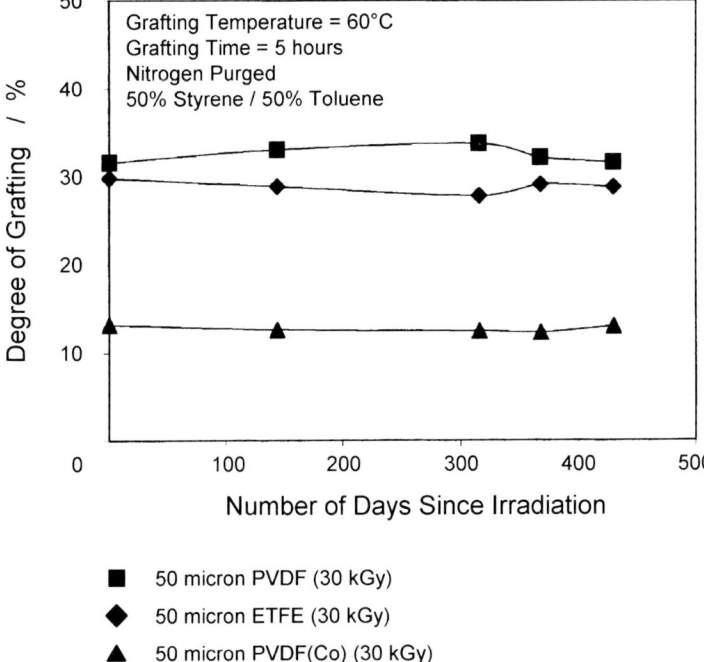

Fig. 3 Effect of low temperature storage on degree of grafting for the preirradiation grafting method [73]

films remain active even after more than a year of storage (Fig. 3). The storage of films may be accomplished at a low temperature of −18 °C or even less. The behavior of polyethylene films has shown to be quite different as they undergo considerable loss in the DG with storage [52]. This opens up an interesting aspect in the preirradiation grafting of monomers onto fluorinated polymers, where irradiation may be carried out once and the resultant films may be stored for subsequent membrane fabrication. It was reported that the storage of irradiated FEP films at −60 °C in the dark for 118 days had no significant effect on grafting [72].

2.4.2
Irradiation Dose and Dose Rate

The influence of the irradiation dose and dose rate on the grafting process has been the subject of detailed investigations. As the radiation dose increases, the number of radical sites generated in the grafting system also increases. This has been observed in the simultaneous radiation grafting of styrene into PTFE films, where the grafting increases almost linearly with the increase in the radiation dose and reasonably high graft levels up to 70% were achieved [74,75]. However, higher irradiation doses are not preferred due to the deterioration of mechanical properties [76].

Rager [77] has investigated the influence of irradiation dose on DG for grafting of styrene onto preirradiated FEP films (Fig. 4). Although DG increases as dose increases, it becomes more difficult to obtain higher degrees of grafting through a further increase in irradiation dose [77].

Chapiro [47, 48] demonstrated for the first time that the grafting yield increases with the total irradiation dose and is independent of the dose rate at low dose rates for simultaneous grafting of methyl methacrylate and styrene onto PTFE. It was emphasized that at low dose rates, the rate of polymerization was slow and grafting was diffusion controlled, whereas at high dose rates, the higher rate of polymerization exceeded the rate of diffusion and grafting was limited to the surface [47, 48]. As a matter of fact, the final DG increases with increasing dose and with decreasing dose rate for styrene grafting into PFA and PP [12]. It is important to note that a more efficient utilization of radicals is followed in simultaneous radiation grafting as compared to the preirradiation method. For the grafting of styrene onto Teflon–FEP films, a graft level of 40–50% is achieved using a radiation dose of 15 kGy in the simultaneous grafting method as compared to 100 kGy for similar graft levels in the preirradiation grafting method using gamma rays [72]. A significant fraction of radicals are deactivated during the course of preirradiation, and the polymer requires optimum activation by irradiation at additional doses to accomplish the high DG.

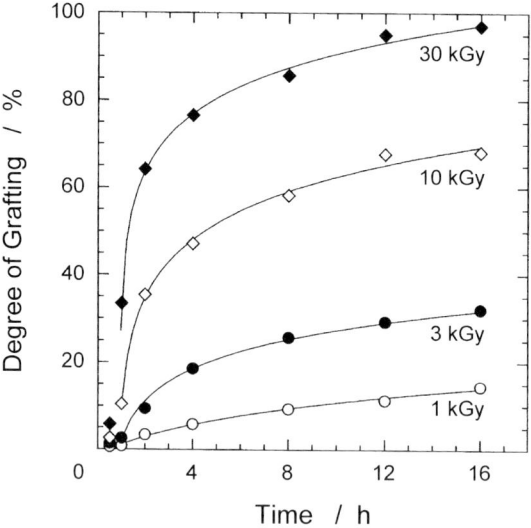

Fig. 4 Grafting kinetics as a function of preirradiation dose (grafting conditions: FEP 25 μm, 50% monomer concentration in isopropanol, 10% DVB, 60 °C) [76]

It is observed that gamma and electron beam irradiation lead to identical degrees of grafting in FEP-*g*-polyacrylic acid systems [53]. However, the grafting of acrylic acid into polyethylene films shows much higher grafting under gamma irradiation than under electron beam irradiation [52]. The difference in the behavior of FEP and polyethylene films lies in the ability of the polyethylene film to hydroperoxidize under the influence of irradiation. Moreover, gamma irradiation is carried out for a longer period than electron beam irradiation. Therefore, the hydroperoxide build-up is much higher in gamma irradiated films and offers much higher graft levels than are achieved in electron beam. Certainly, the influence of crystallinity and other factors needs to be considered, which will be over and above the influence of the chemistry of the polymers. This is what has been observed in the preirradiation grafting of styrene onto PVDF, where the graft levels are two to four times higher than for poly(vinylidene fluoride-*co*-hexafluoropropylene) [65]. Looking at the composition of this copolymer, there is only 7% hexafluoropropylene present in the copolymer matrix, but it diminishes the grafting drastically. Hexafluoropropylene not only enhances the plasticization of the matrix but also interferes with the crystallization process and results in low crystallinity. As a result, the mobility of chains is enhanced and radical–radical crosslinking dominates over the grafting process.

The radiation dose rate has a profound influence on the equilibrium grafting of styrene onto various polymers, both in the vapor phase and in solution, using the simultaneous grafting method [75, 78, 79]. The initial rate of grafting in such systems increases with the increase in the radiation dose. This is the outcome of the efficient utilization of radicals in graft initiation and subsequent chain propagation. It needs to be mentioned here that in the initial stages, homopolymer formation is very limited and the grafting proceeds smoothly with time. Owing to the faster homopolymerization, the grafting at higher dose rates reaches saturation much faster than at lower dose rates. However, for a constant radiation dose, the higher dose rate results in low graft levels and, maybe because the radical concentration is so high, the radical–radical recombination becomes the dominant reaction [75, 78]. Under such conditions, radiolysis reaches equilibrium with radical deactivation and the radical concentration does not increase further with a further increase in the dose rate [31]. Moreover, the higher rate of homopolymerization follows at higher dose rate and leads to an increase in viscosity and a depletion in monomer content. As a result, the monomer availability through the grafted layers is reduced [79–81].

The order of dependence, determined as 0.64 for styrene grafting into FEP [72], 0.58 for grafting of acrylic acid into FEP [82], and 0.53 for styrene–acrylic acid [83], is in agreement with the theoretical value of 0.5 for free radical polymerization. Momose et al. [70] reported that for the grafting of α,β,β-trifluorostyrene (TFS) into ETFE, the grafting rate and final percent grafting increase with increasing preirradiation dose, with the dose exponent of 0.3. The low dependence of grafting rate on the preirradiation dose may be attributed to the decay of trapped radicals due to the increased temperature during irradiation, radical decay during storage, or decay due to radical recombination. A similar trend has been reported for the radiation-induced grafting of acrylic acid onto PTFE [69, 84].

2.4.3
Monomer Concentration

Monomer concentration is the most dominant of the factors that significantly influence the grafting process. As long as the monomer accessibility to the propagating sites is facilitated, the grafting proceeds smoothly. This is the reason that an increase in the monomer concentration leads to an increase in the DG, which is observed for both the simultaneous and preirradiation grafting systems. The increase in grafting with increasing monomer concentration has been observed for the grafting of styrene and styrene–acrylic acid mixture into FEP films [55, 72]. Both the initial rate of grafting and equilibrium DG increase with the styrene concentration in the range of 20–100% [51]. This suggests that the grafting proceeds smoothly with the regular diffusion of monomer within the film. In contrast to the higher monomer dependence (1.9) observed for styrene grafting into FEP previously [72], a first-order dependence of the rate of grafting on the monomer concentration indicates that classical free radical polymerization kinetics operate in the system. However, the complexity arising from the extensive homopolymerization during the grafting may hinder monomer diffusion to the radical sites and may lead to diminishing grafting. This may lead to the maxima at specific monomer concentrations, beyond which the grafting would decrease rapidly. Liang et al. [85] have observed a maximum in the simultaneous radiation grafting of styrene into PTFE films, where the peak was observed at 70% monomer concentration in the grafting medium (Fig. 5).

Our group studied the influence of monomer concentration on styrene grafting into ETFE, using isoproponal/water as the solvent [80]. We found that the DG increases dramatically with an increase in the styrene concentration, until it reaches a maximum at 20% (v/v) styrene for reaction times above 2 h, and then decreases sharply as the concentration further increases. For grafting times below 2 h, this maximum is shifted to 50% (v/v) styrene. The increase in graft level was attributed to the increase in styrene diffusion and its concentration in the grafting layers. We determined the order dependence of the grafting rate on monomer concentration as 1.5. Nasef et al. [81] reported similar results for styrene grafting into ETFE in methanol as solvent. Moreover, these authors determined that the initial rate of grafting was significantly dependent on styrene concentration with an exponent as high as 2.0, which is not in agreement with a first-order dependence of free radical polymerization.

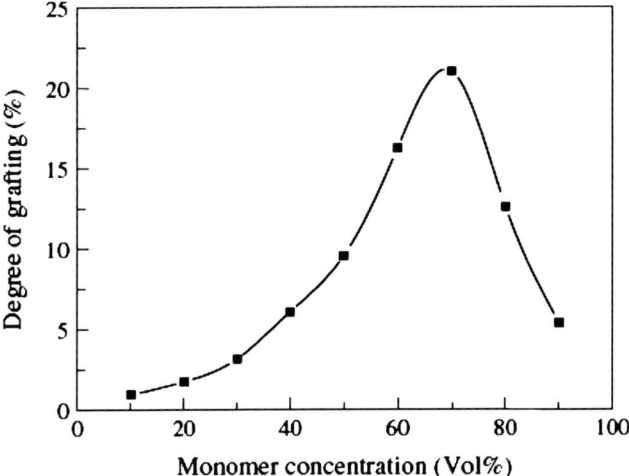

Fig. 5 Variation of DG with monomer concentration (grafting conditions: 20 kGy dose, 110 Gy min^{-1} dose rate, dichloromethane as the solvent, 50 μm film, ambient temperature, nitrogen atmosphere) [85]

It is important to see that a similar trend has been observed for the grafting of styrene into all three (PTFE, FEP, and PFA) films under identical conditions [75, 78, 86]. The DG increased dramatically with the increase in styrene concentration until it reached a maximum, and then decreased sharply as the concentration was increased further [74]. The authors emphasized that the DG of styrene in PTFE depends on both the number of radicals formed and the diffusion of styrene through the polymer matrix, and on its concentration in the grafting layers. Therefore, the increase in the DG in this system may be attributed to the increase in styrene diffusion and its concentration in the grafting layers. At very high concentrations of styrene, homopolymer formation was enhanced and the diffusion of styrene across the viscous medium was hindered. These studies are also supported by Cardona et al. [12] who observed that with increasing monomer concentration the DG reached a maximum and then decreased for styrene grafting into PFA and polypropylene.

The location of the maxima will be somewhat influenced by the nature of the solvent used in the reaction medium [56]. The initial rate of grafting should be largely dependent on the diffusibility of the monomer into the matrix and the grafting solvent must properly swell the grafted zone and make monomer diffusion possible. Such behavior has been proposed to be associated with styrene diffusion and its concentration within the grafted layers. It is stated that an increase in the monomer concentration up to 60% is accompanied by higher monomer availability within the bulk matrix, beyond which extensive homopolymerization leads to the depletion of monomer in the grafting medium and subsequent reduction of styrene diffusion into the film. The diffusion phenomenon has also been considered to be a decisive factor in the grafting of styrene into ETFE [87]. The grafting of styrene with acrylonitrile has been investigated recently [88]. It was observed that the graft yield is considerably enhanced by the addition of acrylonitrile as the co-monomer.

Our patent search of last 5 years shows that although most of the studies have been directed to the use of styrene-based monomers [89–92]. Some workers have tried to use substituted styrenes such as TFS to graft onto FEP [93–95]. The DG in fact remained lower than that observed for styrene grafting [93]. Momose et al. [96] has been granted a patent on the development of TFS-based graft copolymer membranes using both low density polyethylene and ETFE as the base polymers. Other patents describe grafting of TFS and trifluorovinyl naphthalenes onto ETFE film, which facilitates the introduction of more than one sulfonic acid group per monomer unit [97–102]. Considerably higher graft levels of $\sim 80\%$ and $\sim 44\%$ have been achieved for TFS and p-methyl trifluorostyrene, respectively [100]. A more recent patent describes the influence of the grafting mixture alcohol/water on the grafting of TFS derivatives [103]. Furthermore, a novel monomer combination, namely α-methylstyrene/methacrylonitrile, as grafting component is discussed in [104].

2.4.4
Grafting Temperature

The reaction temperature has a significant influence on the DG, irrespective of the nature of the polymer and the monomer. The general observation has been a decrease in the equilibrium DG as the reaction temperature increases. On the other hand, the initial rate of grafting increases with increasing temperature [72]. As a matter of fact, grafting is controlled by a cumulative effect of the monomer diffusion within the polymer bulk, termination of the growing polymer chains, and the deactivation of the primary radicals.

As the reaction temperature increases, the monomer diffusivity within the bulk also increases. This enhances the monomer accessibility to the grafting sites within the polymer bulk. As a result, the rate of initiation and propagation is enhanced. This is the reason that the initial rate of grafting increases with the increasing temperature. The other aspect of grafting is that the grafted zone remains swollen in the grafting medium, which leads to high mobility of the growing chains within the matrix. Therefore, termination of the two growing chains by mutual combination becomes dominant at higher temperatures. At the same time, the primary radical termination may also accelerate by the time the monomer reaches their vicinity. In spite of the higher rate of initial grafting, the final DG would decrease. A similar tendency has been reported for the grafting of styrene onto ETFE-based

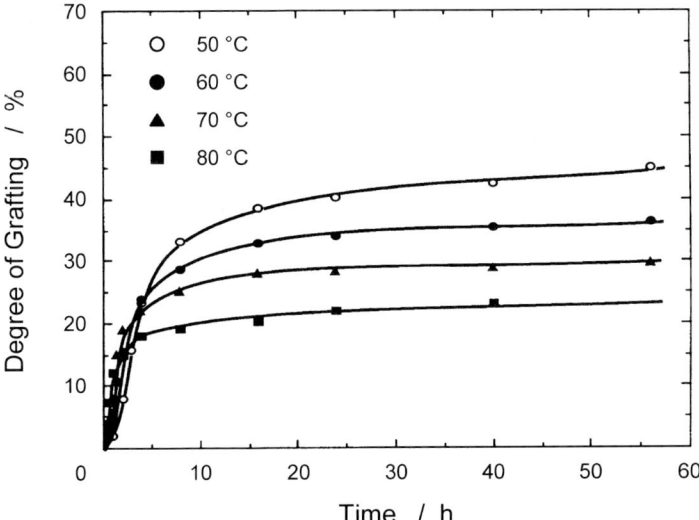

Fig. 6 Variation of DG with time at various temperatures (grafting conditions: 60 kGy dose, 60% monomer concentration, 50 μm film) [72]

films [64, 80, 81] and the grafting of TFS onto ETFE, FEP, PTFE, PFA, and LDPE films [105].

These observations are well in line with those of Rager [77]. The grafting studies were carried out at 50–85 °C and showed that the initial rate of grafting increases with the grafting temperature.

It may be mentioned that the T_g plays an important role in the grafting process. If T_g is lower than the grafting temperature, the mobility of chains is very high. Under such circumstances, the probability of primary radical termination becomes dominant. The final DG as a result may decrease. However, it may be overshadowed by the faster rate of chain initiation and higher monomer diffusivity at higher temperatures [72], as shown in Fig. 6. As a matter of fact, a sharp increase in the rate of grafting may be envisioned at the T_g of the specific polymer.

2.4.5
Grafting Medium

The graft copolymerization reaction is carried out by bringing the activated base polymer film into contact with the monomer in liquid or vapor form. The use of solvents in radiation grafting enhances the accessibility of monomer to the grafting sites due to the ability of the solvent to swell the base polymer. In poor swelling solvents, surface grafting occurs due to the slow down in monomer diffusion within the polymer. However, in good solvents, bulk grafting is highly favored and homogenous grafting is obtained across the film thickness.

The instantaneous swelling of the grafted matrix within the reaction medium is an important factor that governs the grafting process. With the progression of grafting, the polymer film is continuously being transformed into a grafted structure. It is, therefore, the swelling of the grafted matrix in the reaction medium of specific monomer composition that influences the monomer diffusion within the film. The swelling at 10 and 60% monomer concentration in the medium may be different than at higher concentrations and may, therefore, be reflected in the low graft levels, as observed in Fig. 5. This is further supported by the grafting of acrylic acid into polyethylene films, where a similar maximum was observed at 25% monomer concentration [51]. It was observed that the swelling of the grafted film is considerably reduced in a grafting medium containing monomer at higher than 25%, which diminishes monomer diffusion and hence the availability to the propagating chain within the bulk.

The nature of the solvent in the grafting medium is an interesting aspect of achieving efficient graft polymerization. The type of solvent and the composition of the monomer/solvent mixture may influence the grafting kinetics, the length of grafted chains, and polymer microstructure. Benzene, toluene, dichloromethane, and alcohols (methanol, ethanol, and propanol) have been employed as solvents for radiation grafting of styrene and styrene derivatives. It seems that a combination of the polarity (solubility parameter) and chain transfer constant of the solvent plays a major role in graft propagation. The use of dichloromethane has been observed to produce higher graft levels over benzene and methanol [56]. The radical yield in different solvent mediums has been established to be the reason behind such grafting behavior. The radical yields of irradiated styrene solutions in methanol, cyclohexane, and benzene have the order methanol < cyclohexane < benzene. The speculations of Nasef [78] and Dargaville et al. [13, 106] about the effect of viscosity changes in the grafting medium (due to the insolubility of polystyrene in methanol as medium) on decreasing the graft levels do not seem realistic. It may, in fact, be the lower swelling of the polystyrene-grafted matrix in methanol/styrene mixture as the medium that lowers the monomer diffusion within the film and results in a low DG. In such systems, the swelling of the original polymer matrix is not as important as that of the grafted matrix in the solvent medium [107]. This is achieved by using a solvent for the grafted component in combination with the monomer. The propagating graft chains become solvated in the surrounding medium. Since these chains are part of the matrix, the whole matrix exerts swelling. As the grafting proceeds, more polystyrene grafts are incorporated, leading to higher swelling of the matrix, which allows more and more monomer to diffuse into the polymer bulk for the propagation reaction. It is, therefore, the perfect matching of the solubility parameter of the solvent with the grafted polymer domain that would influence the swelling of the matrix during the grafting process. Benzene has a solubility parameter (18.6) much closer to that of styrene (19) as compared to dichloromethane (17.6) and methanol (29.7) [108]. The swelling of the polymer, therefore, would be higher in a solvent where the solubility parameters of the two are closer to each other. This would provide the least swelling in methanol but higher swelling of the grafted matrix in benzene medium for styrene-grafted films.

Cardona et al. [56] investigated the correlation of the efficiency of the grafting process with solubility parameters for polystyrene in various solvents. The authors reported that for grafting of styrene onto PFA in dichloromethane, the DG is higher than that of styrene in benzene and methanol. The chain transfer constants (0.15, 0.2, and 0.296 for dichloromethane, benzene, and methanol, respectively) were important parameters in this context. Low graft levels are obtained with solvents having a high chain transfer constant, since the growing chain will be quickly terminated, whereas solvents with low chain transfer constants enhance the propagation step and lead to higher grafting yields. The influence of solvent viscosity also plays an important role in surface graft–polymerization reactions [109].

An additional factor that originates from the use of a non-solvent medium, such as methanol, is the precipitation of the propagating chains and hindrance of diffusion of the monomer to the internal layers within the film, resulting in a decrease of the grafting [56]. However, recent investigations on the grafting of styrene onto PVDF and FEP films have exploited the use of alcohols as non-solvent for achieving higher graft levels [76, 107]. The pre-irradiation grafting of styrene/divinyl benzene (DVB) onto FEP films is accelerated in alcohols in the order methanol < ethanol < propanol. A fourfold increase in grafting kinetics was observed when toluene was replaced by isopropanol and has been attributed to the Trommsdorff effect, which can occur in chain polymerization when the increasing viscosity limits the rate of termination because of diffusion limitations operating in the system [110].

This certainly opens up an interesting route for achieving membranes with reasonable DG for relatively lower irradiation doses, which might be beneficial in retaining the mechanical properties of membranes to a large extent. Walsby et al. [111] reported the grafting of styrene into PVDF in both propanol and toluene, where not only the grafting kinetics but also the structural properties of the grafted films were dependent on the type of solvent. Higher grafting rates and saturation DGs were obtained in a propanol-based system, which was unable to swell the polystyrene grafts. On the other hand, the grafting in toluene yielded more homogenous films with better surface aspects and mechanical properties. Reduced elongation at break and much rougher surface with large cavities were observed for the films grafted in propanol. The authors reported that the film was swollen very little by the grafting solution, and that propanol served as a diluent without any contribution to the swelling of the polystyrene grafts. The authors attributed the higher grafting rate in propanol to the higher concentration of monomer in the reaction zone, whereas the higher saturation DG was due to the higher viscosity of the grafted zone, which prevents growing chain termination.

Some base polymers such as PTFE do not swell well in any common solvent. For this reason, the grafting reaction is performed in aqueous medium. Hegazy et al. [112] investigated the effect of various solvents on the radiation grafting of methacrylic acid onto PTFE film. The authors demonstrated that distilled water and methanol/water mixture (30/70 wt. %) are the most suitable solvents since the mixture swells the grafted regions. The increase in DG upon addition of water to isoproponal was emphasized for styrene grafting into FEP [76].

The radiation grafting of TFS onto various fluorine-containing base polymers, such as LDPE, ETFE, PFA, FEP, and PTFE has been accomplished by the pre-irradiation method [105]. A proper examination of the swelling properties and solubility parameters of these polymer films in pure TFS showed that LDPE yielded the highest, and PTFE led to the lowest graft levels. This is because of the fact that the sorption of liquid in polymer depends on the affinity between the liquid and the polymer film.

2.4.6
Additives

The influence of additives such as acids to the grafting systems has been explored for achieving higher graft levels [78]. The addition of sulfuric acid has been found to be effective in enhancing the DG of acrylic acid onto FEP and polyethylene films [18, 21]. Styrene grafting onto polyethylene films has also been observed to increase significantly in the presence of acids [113, 114]. However, there are contradicting reports where no influence of organic and inorganic acids was observed on the grafting of styrene into PTFE, PFA, and FEP films [78]. Different hypotheses have been postulated for the enhancement of the grafting but until today an exact mechanism of grafting in such systems has not been proposed.

2.5
Crosslinking

Crosslinkers are used in conjunction with the monomer to achieve certain desirable properties in the grafted membranes. The use of a crosslinker in the grafting medium has been investigated by different workers to obtain membranes that have improved stability in fuel cells [72, 115]. Lower graft levels are achieved as the crosslinker content in the grafting medium increases. This may be because the grafting starts at the film surface. In the presence of crosslinker, the very first polystyrene-grafted chains become crosslinked.

As a result, the mobility of chains is drastically lowered as compared to the crosslinker-free grafting reaction. Consequently, monomer diffusion to the grafting sites within the films is reduced. The higher the crosslinker content, the greater will be the crosslinking density of the grafted chains, which will hinder the monomer diffusion more and more, leading to comparatively low DG. However, it has been observed that crosslinkers may increase or decrease the grafting yield depending on their concentration [13]. At lower crosslinker concentration, the increased DG was attributed to enhanced branching reactions. At higher crosslinker concentration, on the other hand, a network structure was formed, which caused suppression in the swelling of the graft and an increase in viscosity of the grafting solution. This further resulted in a decrease of diffusion and in availability of the monomer and, consequently, the grafting yield was lower. These observations are well supported by the investigations of Rager [77] on styrene grafting onto FEP films. There was an initial rise in graft level for low a level of DVB content in the grafting medium and therefore the grafting decreased considerably. This has been attributed to the polyfunctional nature of the crosslinker.

The addition of crosslinking agents affects the kinetics of the grafting reaction. The addition of DVB decreased the initial rate of grafting and the limiting DG [116]. This is evident from the lower rates of grafting in crosslinked systems than in uncrosslinked ones. The rate of grafting for a crosslinker-free FEP–polystyrene system decreases from 3.6% per hour down to 2.2% and 1.4% per hour for 2 and 4% DVB content, respectively (Fig. 7). However, much higher values have been reported for the grafting of styrene/DVB onto PFA films using simultaneous radiation grafting, which may be attributed to the difference in the base matrix and the radiation dose rate. It was reported that the addition of DVB caused a significant decrease in the DG as a function of the DVB concentration for styrene grafting into PFA [115] and ETFE base films [117, 118].

The graft variation with the N,N,-methylene-bis-acrylamide as the crosslinker for grafting onto ETFE and FEP is quite different [119]. The grafting in fact did not show any specific trend with the increase in the crosslinker content.

The concept of double crosslinking has been examined previously by the use of DVB and triallylcyanurate (TAC) together for radiation grafting of styrene into FEP [72, 120, 121]. It was reported that TAC yielded improved mechanical properties and ionic conductivity [121]. Although it was found that TAC had a favorable promoting influence on the grafting kinetics, spectroscopic measurements failed to positively indicate that TAC was incorporated into grafted films and membranes [122]. Later, it was determined that TAC acted primarily as a graft-promoting additive rather than as a crosslinker [123].

The degree of crosslinking in the grafted film was found to be different from the composition of the grafting solution for FEP-based radiation grafted

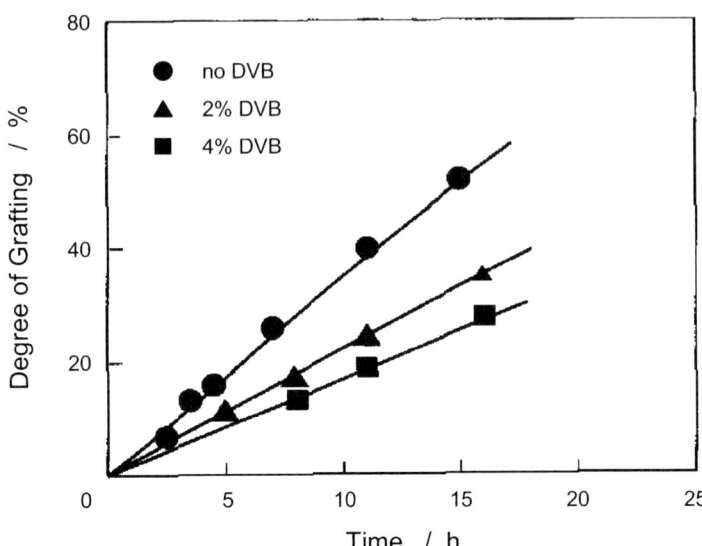

Fig. 7 Variation of DG with reaction time for different crosslinker contents [116]

films due to the different reactivity and diffusion coefficients of styrene and DVB in the film during the grafting process [124, 125]. It was observed that an increase in the degree of crosslinking decreases the membrane thickness, which means that crosslinking increases the structural density of the membranes. Moreover, the mobility of the protons in the membrane is reduced with increasing degree of crosslinking due to decreasing water uptake [125–127]. Moreover, Brack et al. [124] and Ben youcef et al. [118] reported that radiation grafted films are more highly crosslinked in their near-surface regions and thinner films are more extensively crosslinked.

Originating from the concept of crosslinking of fluoropolymers under irradiation at elevated temperature, grafting has been accomplished onto the crosslinked matrix so that the grafting-induced deterioration of mechanical properties may be compensated. As discussed in the preceding section, the crosslinking of PTFE is achieved in the molten state at a temperature of 340 °C. Surprisingly, the precrosslinked films (prepared under gamma irradiation doses of 60–320 kGy), lead to much higher polystyrene graft levels than the virgin one as given in Fig. 8 [128]. Such behavior is the result of two different factors operating in the system: (i) the availability of the amorphous area, and (ii) the radical site generation. It has been an established fact that grafting takes place predominantly within the amorphous region and on the crystal surfaces [127, 129]. The crystalline regions are impermeable structures and do not allow monomer diffusion and subsequent grafting with the radicals trapped within the crystallites [130]. Therefore, any process that leads to a decrease in the crystallinity would be expected to enhance the grafting

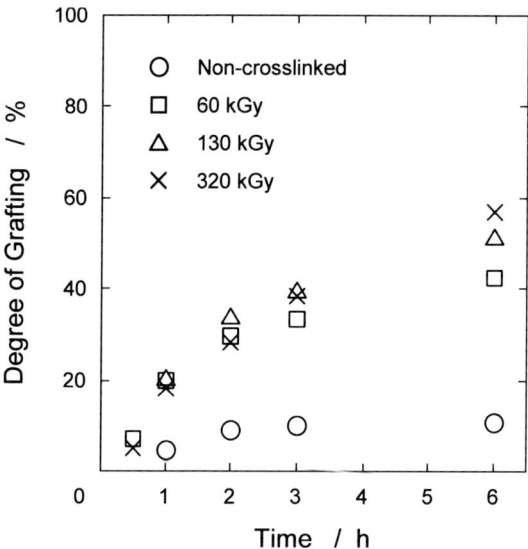

Fig. 8 Variation of DG with reaction time for styrene grafting into the PTFE films crosslinked with gamma rays at different doses (15 kGy preirradiation dose for the grafting reaction) [128]

reaction. The irradiation of PTFE is carried out in the molten state at a temperature of 340 °C where the crystallites are almost completely lost and the matrix behaves like the amorphous one. This state is achieved at irradiation at a high dose of 2 MGy, where the enthalpy of fusion in a differential scanning calorimetry reaches zero [131]. The irradiation at this stage would be favorable for the crosslinking reaction, providing a network structure due to the high mobility of chains. A crosslinked structure is more adapted to radical generation and has been found to have higher G values for the trapped free radicals than an uncrosslinked structure [32]. The radicals produced during the exposure of this crosslinked matrix would be more stable due to the reduced mobility of chains and would be available for graft initiation in contact with the monomer.

The precrosslinking of a polymer is an innovative approach to restoring mechanical strength. However, a proper monitoring of the precrosslinking dose has to be carried out to achieve reasonable graft levels. It is obvious that a precrosslinking dose that is too high may not bring about high graft levels [132]. It is observed that grafting enhances significantly with increasing dose but only up to a range of 50–500 kGy. Any further dose increase leads to loss in the grafting levels and very little grafting is obtained for film crosslinked at a dose of 2 MGy. This is because of the fact that the grafting ability of the polymer matrix is severely affected. The matrix is highly crosslinked to such an extent that the mobility of the molecular chains is suppressed. A crosslinked matrix may lead to lower diffusion of the monomer within the matrix and hence would have an adverse effect on graft propagation. However, it seems that the availability of the more amorphous region, along with the higher availability of radical sites, overpowers the impact of slow monomer diffusion. The temperature also has significant impact on the grafting reaction. An increase in the temperature brings about lower graft levels for films crosslinked at different doses. Here, the mobility of the growing chains at higher temperature increases to an extent that the bimolecular termination of chains is facilitated. The termination of the primary radicals would also be a dominant reaction and would contribute to the lower graft levels.

2.6
Sulfonation

Sulfonation is the final step for the preparation of polystyrene-based membranes for fuel cell applications. In this reaction a sulfonic acid group is added to the aromatic ring by electrophilic substitution. Sulfonation can be performed by several agents such as sulfuric acid, sulfur trioxide, sulfonyl chloride, acetyl sulfate, and chlorosulfonic acid.

Sulfonation conditions have a significant effect on membrane properties including ion exchange capacity, water uptake, and conductivity. Walsby et al. [111] demonstrated that the reaction time, concentration of the sulfonating agent, and reaction temperature have a considerable effect on sulfonation with chlorosulfonic acid. The authors reported that the sulfonation reaction proceeds by a front mechanism, that the grafts at the surface are sulfonated first, and that the rate of reaction depends on the diffusion of sulfonating agent within the membrane. An increase in the concentration of the sulfonating agent and in reaction temperature facilitates the reaction; however, side reactions, which cause a decrease in ion exchange capacity (IEC), water uptake, and proton conductivity, are favored at these conditions. This indicates that, although the use of harsher sulfonation conditions offers advantages in terms of speed of the sulfonation process and oxidative stability, the IEC, water uptake, and proton conductivity are decreased and the membrane becomes more brittle. Paronen et al. [6] emphasized that the rate of sulfonation increased with short sulfonation time, because with longer sulfonation time the hydrophilicity in the sulfonated regions governs the rate of sulfonation.

Sulfonation of FEP- and ETFE-based grafted films at PSI was performed by using 30% chlorosulfonic acid in dichloromethane (at 95 °C, 5 h) and membranes with reasonably good sulfonic acid content have been observed. Sulfonation conditions almost identical to those used at PSI have been used by others for the sulfonation of PFA-g-polystyrene films, i.e., a mixture of chlorosulfonic acid and 1,1,2,2-tetrachloroethane (30 : 70 v/v, 90 °C, 5 h) [133]. Phadnis et al. [83] performed the sulfonation of styrene–acrylic acid grafted FEP films in concentrated sulfuric acid (at room temperature). Concentrated sulfuric acid and refluxing under nitrogen (at 95 °C) has been used for PVDF-g-polystyrene films [134]. The attempts to sulfonate PVDF-g-polystyrene films in concentrated sulfuric acid at temperatures between 21 and 95 °C and in acetyl sulfate/dichloroethane solutions at 50 °C yielded low degrees of sulfonation, and the sulfonation was mainly restricted to the surface [111]. This may be due to the insufficient reactivity of these sulfonating agents. In addition, sulfuric acid may not be able to penetrate into the hydrophobic matrix.

The number of sulfonic acid groups in the membrane increases with the increase in the DG. At higher styrene concentrations more benzene rings are in contact with sulfonic acid groups, which results in more sulfonic acid groups in the membrane. However, the efficiency of the sulfonation reaction depends to large extent on whether or not the membrane is grafted through its thickness [111]. If the samples contained a core of ungrafted parts, sulfonation was incomplete at room temperature due to insufficient swelling of the samples and the difficulty of diffusion of the sulfonating agent. It was observed that full sulfonation of surface grafted samples can be achieved at higher temperatures.

3
Characterization and Structure of Grafted Films and Membranes

The characterization of membranes is essential for correlating their performance in fuel cells. It is the interface of the membrane that interacts with the electrode and hence a proper surface morphology may in fact improve the performance of the membrane electrode assembly. Membrane preparation involves the graft polymerization of a monomer, usually styrene, and subsequent sulfonation of the grafted matrix. This transforms a hydrophobic fluorinated structure into a hydrophilic ion exchange matrix. Therefore, the polymer film undergoes drastic modification in terms of the physicochemical properties and morphological nature, depending on the irradiation, grafting, and sulfonation conditions.

3.1
Graft Mapping

The most important requirement of the membrane is the homogeneous distribution of grafts across the membrane matrix. X-ray microprobe analysis (XMA) has been an effective way to monitor the graft distribution within the membrane matrix. The X-ray fluorescence for sulfur may be monitored across the membrane thickness and provides useful information about the distribution of the sulfonic acid groups and, hence, of the grafts across the matrix [127, 135, 136]. It was observed that the grafted phase was initially concentrated at the film surface. The low graft levels of ∼ 3% film shows a very high concentration of sulfur only on the surface, as presented in Fig. 9 [127]. The presence of sulfur in the middle of the membrane may be seen with a further increase in the DG. The two zones from both sides approach each other towards the middle and subsequently a homogeneous distribution of sulfur, or in other words polystyrene grafts, is achieved. This indicates that the grafting is a time-dependent process and that the homogeneous structure is possible only at a specific graft level and beyond a specified grafting time, irrespective of the grafting method used to produce the membranes. For instance, a homogenous distribution of grafts was achieved at DG higher than 20% for FEP-based films [87, 116].

This further substantiated the idea that grafting proceeds through a grafting front mechanism and that DG above 30–35% is required for two grafting fronts to meet and form a network for proton conduction [137]. It is also observed that an inhomogeneity, in the form of bubbles on the membrane surface, is created after sulfonation of grafted films with graft levels below 11%. The membrane inhomogeneity arises due to the presence of hydrophilic sulfonated polystyrene chains in the surface layer of the hydrophobic perfluorinated FEP matrix [138].

It was observed that the addition of crosslinker (2–4% DVB) to styrene considerably affected the homogeneity profile behavior [116]. The distribution became practically homogenous across the whole width of the film and the homogeneity increased at 4% DVB [116, 139]. That behavior was attributed to the decreased rate of diffusion in the grafted zone near the surface, an increase in the rate of termination of growing chains, and a decrease in the concentration of styrene in surface layers [116]. The observations for the TFS-

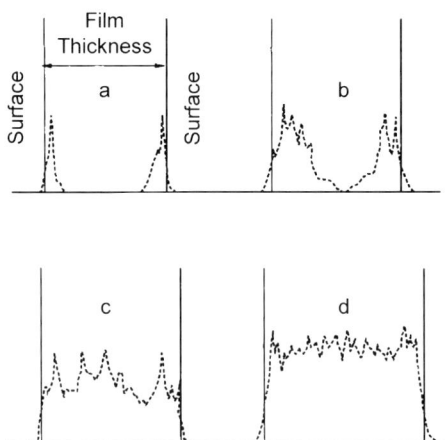

Fig. 9 Distribution of sulfur as determined by microprobe measurements in the transverse plane of FEP-based membranes with different DG: **a** 3.1%, **b** 5.9%, **c** 13.6%, **d** 27% [127]

grafted systems have been found to be completely different. It was observed that for TFS grafted onto PTFE and ETFE, although the graft chain distribution is almost constant over the range of film thickness for ETFE-based films, the grafted PTFE exhibited two peaks (XMA profile) located $\sim 10\,\mu\mathrm{m}$ inside the film surface. That was attributed to a better monomer diffusivity in an ETFE base film than in a PTFE base film [105].

Micro-Raman mapping is another interesting tool for analyzing the depth profile of the grafted component within the membrane matrix [12, 57, 79]. The ratio of the intensity of the Raman peaks associated with the aromatic band in polystyrene at 1601 cm^{-1} and in the fluorinated matrix, such as PFA at 996 cm^{-1} at the surface and along the cross-section, provides information about the distribution of the grafts. The graft penetration tends to be higher at higher radiation doses. Likewise, the vapor phase grafting has been observed to remain confined to the surface layers only [57, 79]. Hietala et al. [140] observed that for polystyrene-grafted PVDF films, although polystyrene distribution was homogenous on the surface at high graft levels, the surface became quite heterogeneous at low graft levels.

Hegazy et al. investigated the cross-sections of the poly(acrylic acid)-grafted FEP films [141] and PTFE films [142] by X-ray microscopy. It was observed that the monomer was limited to the surface at low graft levels. However, it penetrates the entire film and homogenous grafting throughout the entire film is observed for high graft levels.

It has been reported that the geometric dimensions of the styrene-grafted FEP films vary linearly, but not equally, with the increase in the DG. For instance, for a graft level of 52%, an increase of 25% in length as well as width and 45% increase in thickness have been obtained. Equal distribution of polystyrene within the FEP matrix prepared via simultaneous radiation grafting, at least for a graft level of 21%, has been monitored by Fourier transform infrared spectroscopy (FTIR) and attenuated total reflection spectroscopy (ATR) [126]. Similarly, FTIR-ATR was used to determine the surface grafting yields for styrene grafted onto ETFE by measuring the ratio of absorbance of the polystyrene peak at 699 cm^{-1} (C–C wagging band) to the ETFE matrix band at 1046 cm^{-1} (–CF$_2$ stretching vibration) [143].

Confocal Raman microscopy has been employed for the investigation of the changes in membrane composition after fuel cell experiments for PVDF-based radiation grafted membranes. In fact, severe degradation due to loss of polystyrene sulfonic acid (PSSA) was observed during the fuel cell run and only 5–10% of the initial content was found to be left behind. It has been reported that the degradation is an inhomogenous process that is different over the membrane surface and through the membrane depth [144]. It was proposed that the deterioration of fuel cell performance was because of the loss of entire PSSA chain segments rather than desulfonation [145] and is supported by the studies on ETFE-based membranes [146] and FEP-based membranes [125].

3.2
Surface Chemistry and Surface Morphology

The surface and wetting properties are known to influence the adhesive and bonding properties of materials [147]. The contact angle measurements of membranes provide useful information on the surface and interfacial behavior. Graft management within the membrane may take place in such a way that the surface is rendered hydrophobic in spite of the hydrophilic nature of the grafted component [148]. This could happen either during the grafting process or during the post-grafting treatments of the copolymer matrix. A fundamental investigation of the wetting and surface energy properties of commercial perfluorinated membranes and uncrosslinked radiation grafted membranes indicated that the surface properties of uncrosslinked radiation grafted membranes are similar to those of commercial perfluorinated membranes having similar ion-exchange capacities [148]. In addition, the contact angle of both the grafted and the sulfonated ETFE membranes shows distinct variations with different wetting agent [149]. The polystyrene-grafted films do not show any appreciable change with water as a function of graft level, but measurements with methylene iodide as a probing liquid indicate a decrease in the contact angle with an increase in graft level. At higher graft levels, the contact angle has been observed to behave identically to that for a pure polystyrene surface. This indicates that the surface of the membrane is rich in polystyrene. Sulfonation changes the wetting behavior drastically; the contact angle of water is significantly reduced to $32°$ for a graft level of 82%. This is an indication of the surface rendered hydrophilic due to the presence of sulfonic acid groups. However, absolute values of the contact angle have been observed to vary significantly in different investigations [85]. Maybe, the nature of the base matrix and the sulfonation process have some impact on the wetting behavior. The maximum degree of sulfonation in PTFE graft copolymer membranes has been reported to be 50% and may account for the higher contact angle in these membranes as compared to ETFE membranes [149].

Contact angle measurements on the fully swollen form of the radiation grafted membranes using several polar, non-polar, hydrogen-bonded, and non-hydrogen-bonded liquids have been performed by Brack et al. [149]. The high contact angle of water on the FEP-based membrane revealed the hydrophobic nature of the membrane due to the crosslinking and relatively low degrees of grafting. Moreover, crosslinking has a tendency to limit the mobility of chain segments. Due to restricted mobility it was difficult to undergo surface reconstruction to adjust the most favorable local structure at a surface or interface. The membrane cannot adapt a hydrophilic surface when it is exposed to water during an earlier swelling process [150].

X-ray photoelectron spectroscopy (XPS), provides quantitative information on surface chemical structure, chemical composition, and chemical bonding, and is one of the most extensively used methods for radiation grafted films and membranes. This method is useful for investigating the surface chemistry taking place during the grafting and sulfonation processes [151]. XPS has the ability to probe the surface within a few nanometers and, therefore, interesting information about the chemical composition at a few top layers is obtained. As a result, the polystyrene graft within and on the surface of the fluorinated matrix may be monitored [151]. The evolution of the C–F and the C–H:C–F ratio with respect to the DG or irradiation dose, as is evident from Fig. 10, indicates a high concentration of C–H, i.e., polystyrene chains on the surface [56]. Consequently, a significant loss of the fluorinated species in the PFA matrix is observed. A strong increase in the relative amount of C–H bonds at a dose of about 50 kGy is the indication of grafting taking place at the surface right from the beginning of the irradiation. As the radiation dose increases, more grafting takes place on the surface and in the bulk and, finally, the plateau beyond a dose of 250 kGy suggests that at least the top few nanometers of the surface can be considered to be the polystyrene grafts. Moreover, the matrix with lower crystallinity has a higher C–H:C–F ratio, suggesting more polystyrene grafts on the film. This, in principle, substantiates the earlier assumption that the lower crystallinity makes the matrix more amenable to monomer diffusion and subsequent grafting with the radical sites.

It was observed that the surface composition is strictly governed by the degree of crosslinking in FEP membranes [139]. The uncrosslinked FEP-g-polystyrene copolymer films show a well-defined C–H signal at ~ 286 eV, confirming the presence of polystyrene grafts on the surface. The absence of the –C–F signal in the uncrosslinked films is an indication of the abundance of the polystyrene on the surface. However, this signal is slowly lost in films prepared under increasing crosslinker content, while the C–F signal increases indicating that the polystyrene grafts are more and more confined to the bulk of the matrix. In addition, the sulfonated matrix shows a similar but weaker trend. The C–F signal was visible for the uncrosslinked membrane.

Nasef et al. [151, 152] investigated the structural changes enhanced by styrene grafting and subsequent sulfonation of PTFE film as well as a variation of the DG of PTFE-based membranes. It was reported that the membranes had side-chain grafts of polystyrene and structures composed of carbon, fluorine, sulfur, and oxygen. The authors determined that the base film undergoes structural changes in terms of chemical composition and shifting in binding energy. Although the binding energies of C1s, F1s, S2p, and O1s were found to be independent of DG, the amount of each component was shown to be dependent on DG.

It was observed that polystyrene grafted in a PVDF matrix under irradiation with γ-rays or heavy ion irradiation exhibited very large domains, when investigated using small angle X-ray and neutron scattering (respectively, SAXS and SANS) [153, 154]. The characteristic length of the ionic domains is observed at very low angles because of the large size of the domains. The

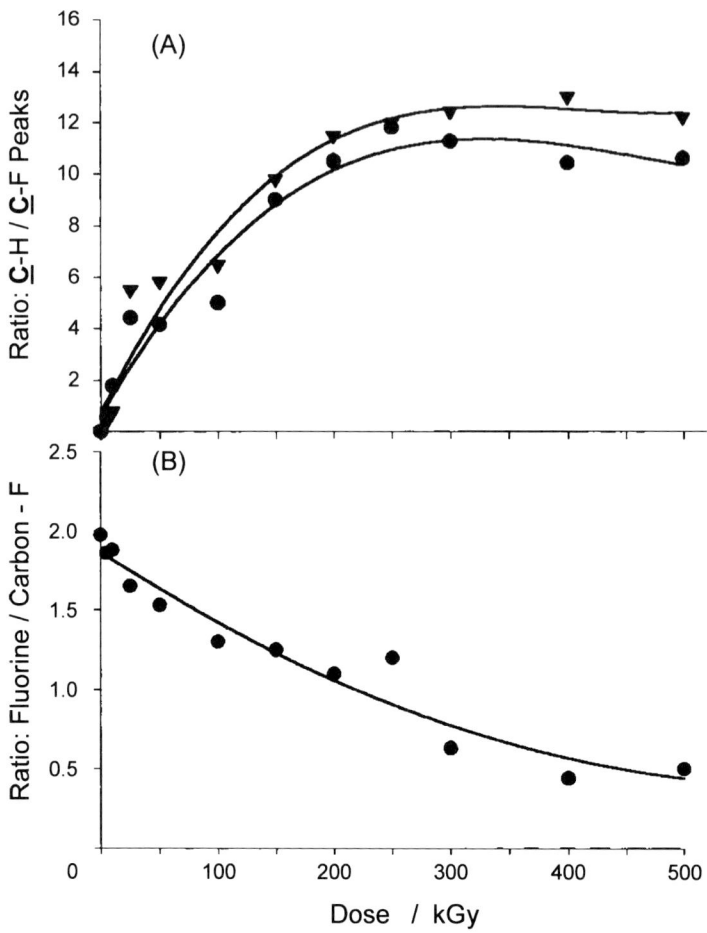

Fig. 10 Plots of **a** the atomic ratio of hydrogenated carbons (C–H) to fluorinated carbons (C–F) and **b** the ratio of fluorine (F1s) to fluorinated carbons (C–F) as a function of applied grafting dose: PFA-A (▼) and PFA-B (●). The irradiation was undertaken in nitrogen gas, with 50% styrene solutions in dichloromethane, and 6.5 kGy h^{-1} dose rate) [56]

broad maximum at large angles is only observable in membranes swollen in heavy water. The grafting in irradiated PVDF gives rise to a swelling on a microscopic scale, which is limited to low grafting levels (< 10%). The small-angle upturn observed for a water-swollen sulfonated sample was similar to that observed for the same sample before sulfonation, due to a dilution of the sulfonated groups by water swelling. Structural investigation of radiation grafted membranes by SAXS in the dry state of the membrane show a strong upturn in intensity, as observed over the investigated angular range. In the swollen state, a very broad maximum with low intensity was determined [155-157]. This difference was attributed to a characteristic distance between ionic domains.

Recently, the influence of crosslinking with DVB on the morphology of polystyrene-grafted FEP films was probed by SANS and a characteristic influence was observed. These results corroborate the interpretation of results obtained by DSC and TGA, namely the picture of a morphology for a two-phase semi-crystalline polymer, with the grafting component essentially being present in the amorphous phase (Mortensen et al. unpublished results).

Surface morphology is one the most important aspects of membrane design. The morphology is strongly influenced by the nature of the graft medium, which takes into account both the monomers and the diluents or additives. Scanning electron microscopy (SEM) has been an effective tool for visualizing the surface texture [158]. A distinct difference becomes visible in the styrene-grafted PVDF vis-à-vis the hexafluoropropylene copolymer of PVDF membranes. The PVDF membrane shows a much larger but wrinkled structure on the surface in comparison to the hexafluoropropylene-based PVDF membrane, which tends to be smoother. These results exhibit the importance of styrene diffusion within the films, as the monomer diffusion is faster in the latter film and the polystyrene-grafted layer formation becomes less pronounced, leading to the smoother surface. It should be mentioned here that the composition of the grafting medium has a strong influence over the surface morphology. The grafting of styrene onto PVDF introduces roughness, as is evident from SEM characterization [107]. The grafting in toluene as medium leads to some inhomogenous surface. However, isopropanol as the grafting medium introduces cavities of $\sim 10\,\mu m$ diameter. This is essentially due to precipitation of the polystyrene chains in isopropanol, which leads to phase separation within the grafted matrix and as a consequence, is reflected as cavity formation. It is important to mention here that a change in the opacity of the grafted films is observed in the presence of the crosslinker. These films turn light transparent at higher crosslinker concentration [77]. Cross-sections of the membranes may be visualized under SEM, where micrographs can be seen with distinct variation in the morphology of the membrane. A dark region in the middle and a clean region at one edge become evident for the ungrafted and grafted regions, respectively [152].

Atomic force microscopy is another interesting tool for investigating the surface morphology. A three-dimensional profile of the grafted structures may be achieved, which offers a more informative evaluation than SEM. The investigations on the surfaces of polystyrene-grafted PVDF films and membranes have revealed the heterogeneous character of membrane surfaces with alternation of PVDF and PSSA [140]. It was reported that after grafting the surfaces were found to be inhomogenous, and that blobs of polystyrene (domain size of $0.1-2\,\mu m$) were observed on the surface. Such a behavior arises due to the incompatibility of the grafted component and the base polymer films. As a matter of fact, the grafted components remain as distinct isolated phases within the fluorinated matrix and remain visible as inhomogeneity on the film surface. However, after sulfonation the blobs disappeared and the membrane surface became visually smoother [140]. Similarly, it was reported previously for polystyrene-grafted FEP membranes that the incompatibility between the hydrophobic perfluorinated backbone and the hydrophilic PSSA was overcome at high degrees of grafting and that the whole matrix behaved as a hydrophilic matrix. As a consequence, the film swells homogeneously in water leading to a smooth surface [138].

3.3
Thermal Characterization

Thermal behavior of radiation grafted films and membranes have been investigated mainly by using thermogravimetric analysis (TGA) and differential scanning calorimetry (DSC). It has been observed from TGA that a two-step degradation pattern is exhibited by styrene-grafted FEP-based films, indicating that the degradation of grafted polystyrene and that of the FEP base polymer occurred independently from each other [114, 159, 160]. In addition, the degradation pattern was found not to be much affected by the DG [101]. This shows that the polystyrene-grafted FEP copolymer films behave as a distinct two-phase system, where the polystyrene moiety forms a separate micro-domain within the FEP matrix. Similar observations have been made for the polystyrene-grafted FEP, ETFE, and PVDF films [161, 162], PFA films [115, 163], and PVDF films [164].

Sulfonation changes the stability pattern of membranes completely. The thermal degradation behavior of FEP-based membranes has been investigated previously by TGA in combination with FTIR and mass spectroscopy [160]. As presented in Fig. 11, unlike the two-step degradation pattern of the grafted films, a three-step weight loss pattern was observed for radiation grafted membranes and has been ascribed to dehydration of the membrane, desulfonation, and de-aromatization reactions, and finally degradation of the backbone [160]. A similar degradation pattern has been reported in the literature for the other radiation grafted membranes [135, 159, 161, 162, 164, 165].

It is important to understand that every step of membrane preparation, i.e., irradiation, grafting and sulfonation leads to certain changes in the crystalline structure. For instance, the incorporation of polystyrene grafts caused an increase in amorphous fraction and restricted the mobility of the chains, and T_g increased. Similarly, the incorporation of the sulfonic acid groups caused ionic interactions, and the mobility of the molecular chains and T_g increased. The slight decrease in T_m was attributed to the changes in original crystal size by styrene grafting and little disruption in the crystalline region was observed [165]. Moreover, the grafting process leads to a decrease in the heat of fusion with an increase in the DG in FEP-g-polystyrene copolymer films [114]. This arises because of the dilution effect on inherent crys-

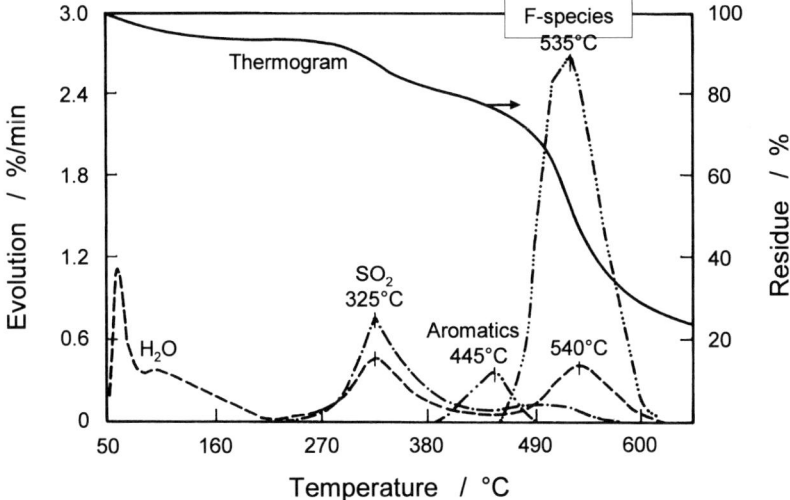

Fig. 11 Gaseous evolution pattern of FEP-based membranes [160]

tallinity of FEP by the incorporation of amorphous polystyrene grafts within the non-crystalline region of the film. According to the investigation of Cardona et al. [163] for PFA-based films, a relatively small decrease in inherent crystallinity after grafting has been observed since grafting occurred preferentially in the amorphous phase of the semi-crystalline polymer (diffusion was slow and radicals were less reactive in the crystalline phase). However, sulfonation of the grafted films leads to further decrease in the heat of fusion of the membranes (Fig. 12), and consequently decreases crystallinity [166]. It has been indicated that the loss of crystallinity in membranes is in addition to the changes induced by the dilution effect. These changes have been identified as crystal defects, as is evident from the loss of heat of fusion in Fig. 12. It is in fact the hydrophilic PSSA domains within the hydrophobic FEP matrix that absorb water and so strong hydrophilic–hydrophobic stresses develop in the water-swollen membrane and may be the reason for the distortion of the crystallite. This sounds reasonable considering the distortion of crystallites that has been observed in the sulfonation of polyethylene [167] and recently also in PVDF-g-PS [169].

The trend in the crystallinity of PTFE-based membranes seems to be different than for FEP membranes [151]. The grafting of styrene into PTFE film decreases the crystallinity from 43.2 to 32.1% for a graft level of 36%, which subsequently reduces to 21% on sulfonation. Although this trend accounts for the preservation of the inherent crystallites during the grafting and sulfonation processes, the authors attribute it solely to the dilution effect [169]. It seems that crystal distortion is also prevalent in this system because the crystallinity decreases more (21%) than if only the dilution effect persisted

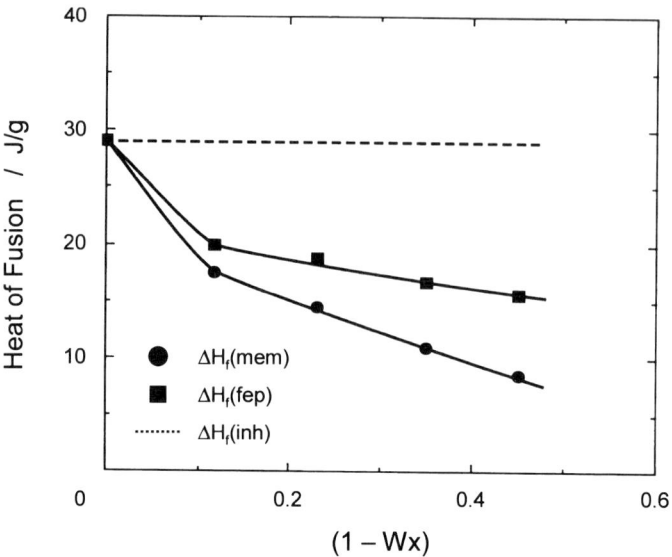

Fig. 12 Variation of the heat of fusion of the membrane $\Delta H_{f(mem)}$, FEP component in the membrane $\Delta H_{f(fep)}$, and inherent value $\Delta H_{f(inh)}$ with the cumulative weight fraction of acid group and water $(1 - W_x)$ [166]

(23.4%). However, there have been more investigations on the crystallinity variations of grafted films and sulfonated membranes based on poly(vinyl fluoride), PVDF, ETFE, and FEP using DSC [65]. It was observed that the dilution effect of grafted component is the only factor that influences the overall crystallinity, suggesting that the inherent crystallinity remains intact. This is supported by a decrease in crystallinity content with the increasing graft level of styrene-grafted PFA films, which is interpreted as indicating that this behavior is the dilution and partial destruction of the inherent crystallinity [115].

In a recent investigation, the influence of the irradiation and grafting processes on the crystallinity have been investigated for three base polymers by DSC [161]. The grafting process has been found to have the largest effect on base polymer crystallinity and resulted in a reduction of crystallinity in all cases. In addition, the authors reported as a result of TGA investigation that the extent of fluorination of the base polymer, the graft level, and the irradiation method all had important influences on the thermal degradation of the films and the activation energy for this process. These results were nicely confirmed for ETFE-g-polystyrene-based membranes [118, 162].

The X-ray diffraction studies have been interesting in supporting the observations on the crystallinity of membranes determined by DSC. The crystalline reflections in graft copolymer membranes with different degrees of grafting fall on identical angles. However, their intensity decreases, suggesting a decrease in their inherent crystallinity [74, 169]. A detailed characterization of a number of radiation grafted fluorinated films has been carried out to give a deeper glimpse of the crystal structure and orientation of the crystalline zone [170]. The grafting and subsequent sulfonation of the films led to a decrease in the crystallinity in each step again, because of the incorporation of amorphous polystyrene chains in the non-crystalline region of the film. The full width at half-maximum did not change, indicating the stability in the orientation. This shows that the grafted chains are bound to the amorphous region and do not disturb the crystalline region of PVDF films.

The effect of crosslinking on the degradation of the FEP-based grafted films and membranes have been investigated using TGA coupled to FTIR [171]. It was found that crosslinking causes a shift of the de-aromatization reaction to higher temperatures; however, the desulfonation reaction was shifted to lower temperatures. DVB increases the thermal stability of polystyrene grafts, facilitates the desulfonation process, and leads to a higher ash content.

3.4
Mechanical Properties

Mechanical integrity is one of the most important prerequisites for fuel cell membranes in terms of handling and fabrication of membrane electrode assemblies, and to offer a durable material. Robust fuel cell membranes are required because of the presence of mechanical and swelling stresses in the application [172]. Moreover, membranes should possess some degree of elasticity or elongation to prevent crack formation.

Typical mechanical properties of polystyrene-grafted FEP- and ETFE-based membranes have been investigated previously [62, 63, 146]. It has been reported that ETFE-based grafted films and membranes exhibit comparably better mechanical properties than FEP-based ones since ETFE films are available at higher molecular weight, which enhances breaking strength and flexibility. In addition, FEP undergoes a greater extent of chain scission reactions compared to ETFE. For both ETFE and FEP, the membranes from electron beam irradiation under inert atmosphere have better mechanical properties than the membranes from gamma irradiation under air. It is observed that thinner membranes possess poorer mechanical properties than the thicker membranes. Crosslinker also affects the mechanical properties and highly crosslinked membranes have poorer mechanical properties than the membranes with lower levels of crosslinker [62, 63]. The mechanical properties of FEP-based membranes are superior to those of the grafted films and may be due to the plastizing effect of water in the swollen membrane [62]. Similarly, the tensile properties of the grafted films and membranes are also reported [65].

The influence of irradiation dose and grafting solution on the mechanical properties of styrene-grafted FEP-based films has been investigated previ-

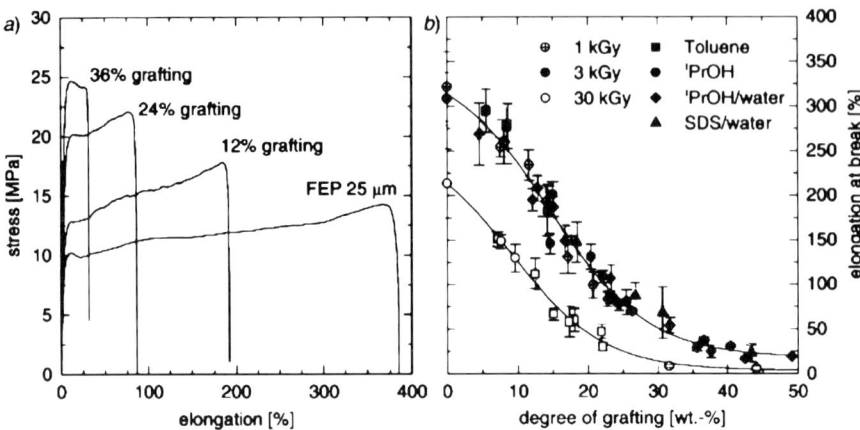

Fig. 13 a Stress–strain curves for pristine FEP and grafted films with different DG. **b** Elongation at break of grafted films as a function of DG, preirradiation dose, and type of solvent: FEP 25 μm, 10% DVB in solvents toluene, isopropanol (^{i}PrOH), isopropanol/water mixture (^{i}PrOH/water), and sodiumdodecyl sulfate/water (SDS/water) [76]

ously [76]. The elongation values of grafted films are lower than those of the unmodified base polymer (Fig. 13a). As presented in Fig. 13b, an increase of irradiation dose leads to considerable deterioration in the mechanical properties of pristine FEP and grafted FEP films. The loss in elongation at break with higher irradiation dose is attributed to an increased radiation damage to the trunk polymer. However, the type of solvent used during grafting has no significant effect on the elongation at break (Fig. 13b). Walsby et al. [107] has pointed out that the mechanical properties of PVDF-g-polystyrene films are seriously affected by the nature of the grafting medium. It was shown that better mechanical properties were obtained for the films in toluene compared to those in isopropanol. These and other authors have also reported that the mechanical properties of the base film in the machine direction and transverse direction differ significantly [118]. Although the film elongates several times compared to its initial length in the machine direction, elongation is negligible in the transverse direction.

4
Fuel Cell Application

4.1
Membrane Properties Relevant to Fuel Cell Application

In the polymer electrolyte fuel cell (PEFC), proton-conducting cation-exchange membranes are used as electrolyte, which consist of an organic polymer structure (crosslinked or uncrosslinked) containing pendant acid functional groups, e.g., sulfonic acid $-SO_3H$ [173]. Hydration of the membrane (i.e., incorporation of water molecules into the polymer structure) leads to dissociation of the acid groups into mobile $H^+(aq)$ and immobile anions fixed to the polymer backbone. The resulting nanophase-separated structure is an interpenetrating network of hydrophobic polymer backbone material providing structural integrity and aqueous domains allowing proton transport within water-containing channels. The proton conductivity of the material depends on the density of acidic groups, their dissociation constant (pK_a), and on the mobility of the proton, which is governed by the level of hydration (i.e., the water content of the membrane) and the geometry (dimensions, connectivity) of the hydrophilic channels.

4.1.1
Ion Exchange Capacity

The requirement of water within the polymer structure as a proton transport medium limits the operating temperature of such membranes to below 100 °C at moderate pressure. Alternative membrane concepts using anhydrous proton conduction are under development. Among the approaches, phosphoric acid-doped polybenzimidazole appears among the most promising. Here, protons are transported via a phosphoric acid network [174]. The technology of radiation grafting has not been adopted for the preparation of water-free membranes for high temperature operation, with the exception of the work mentioned in a patent by Toyota [175]. The method involves grafting of vinylpyridine onto an ETFE or PVDF backbone, followed by imbibition of the film with phosphoric acid. However, due to the limited references in this area, the following discussion will be concerned with radiation grafted membranes with a water-based proton transport mechanism.

In radiation grafted proton-exchange membranes, the structural integrity of the component originates from the base polymer film, and the proton conduction functionality is introduced with the graft component. Therefore, it can be expected that the proton conductivity will be a function of the number of exchange sites within a given membrane portion. The corresponding parameter is the ion exchange capacity (IEC), which is defined as:

$$\text{IEC} = \frac{n(SO_3H)}{m_{\text{polymer}}}, \quad (1)$$

where $n(SO_3H)$ is the number of exchange sites and m_{polymer} is the dry mass of the polymer. The IEC is determined by titration [149]. Obviously, the IEC increases as a function of DG (Fig. 14). For styrene-grafted membranes, the theoretical IEC, assuming one sulfonic acid group per aromatic ring, is:

$$\text{IEC}_{\text{th}} = \frac{\text{DG}}{M_S + \text{DG} \cdot M_{SSA}}, \qquad (2)$$

where $M_S = 104 \text{ g mol}^{-1}$ is the molar mass of styrene, and $M_{SSA} = 184 \text{ g mol}^{-1}$ is the molar mass of styrene sulfonic acid. It can be deduced from Eq. 2 that for high levels of grafting, the theoretical IEC approaches the value for pure sulfonated polystyrene, which is:

$$\lim_{\text{DG} \to \infty} \text{IEC}_{\text{th}} = \frac{1}{M_{SSA}} = 5.4 \text{ mmol g}^{-1}. \qquad (3)$$

Yet, the IEC value gives no indication about the distribution of the exchange sites across the membrane thickness, which is of course of paramount importance for the protons to be transported all the way from anode to cathode. It is possible that the conductivity of a membrane sample is low, even if the IEC is at acceptable levels. This happens when the grafting has not proceeded through the entire thickness of the base polymer film. Often, a threshold DG is observed, below which the conductivity is unmeasurably low, and above which acceptable conductivity is obtained [157, 176, 177]. The explanation is that at low degrees of grafting, the center of the membrane remains un-

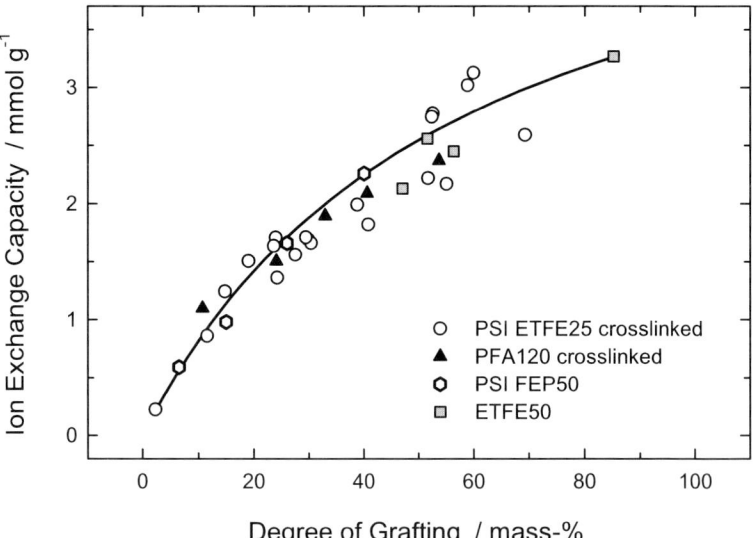

Fig. 14 The ion exchange capacity (IEC) of styrene-grafted and sulfonated membranes as a function of DG. The *solid line* represents the theoretical IEC for 100% degree of sulfonation, corresponding to one sulfonic acid group per aromatic ring (data for PFA120 crosslinked redrawn from [130]; data for ETFE50 redrawn from [213]; data for PSI FEP50 redrawn from [151])

grafted, and only above the threshold do continuous hydrophilic channels for proton transport exists through the membrane.

4.1.2
Water Uptake

As the acidic groups need to dissociate for the proton to become mobile, one can expect that the water content of the membrane will also have a strong influence on conductivity. Proton transport occurs either via hopping of protons from one water molecule to the next (Grotthus mechanism) or via the net transport of H_3O^+ or other aggregates of water and H^+ [178]. Evidently, as the DG increases and with it the number of ion exchange sites, so will the hydrophilicity of the material, resulting in an increase of the water uptake. The water uptake (ϕ) is expressed according to:

$$\phi = \frac{m_w - m_d}{m_d} 100\%, \qquad (4)$$

where m_w and m_d are the mass of the wet and dry membrane, respectively. A quantity that is often used to describe the water uptake of an ion exchange membrane is the so-called hydration number (λ), which is the number of

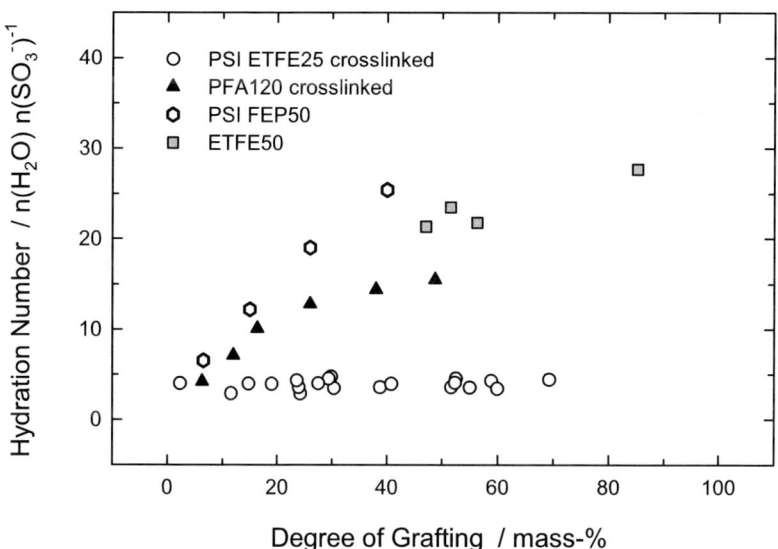

Fig. 15 Water uptake, expressed as the number of water molecules $n(H_2O)$ per sulfonic acid site $n(SO_3^-)$, as a function of DG (data for PFA120 crosslinked redrawn from [200]; data for ETFE50 redrawn from [213]; data for PSI FEP50 redrawn from [135])

water molecules $n(H_2O)$ per sulfonic acid site $n(SO_3H)$. λ is defined by:

$$\lambda = \frac{n(H_2O)}{n(SO_3H)} = \frac{m_{water}}{IEC \cdot M_{water}}, \tag{5}$$

where $M_{water} = 18\,\text{g mol}^{-1}$ is the molar mass of water. It is usually observed that the hydration number increases with the DG (Fig. 15), which points to the fact that as the membrane gets more hydrophilic upon incorporation of the graft component, the acidic sites become increasingly hydrated.

Monomers that act as crosslinking agents, such as DVB or bis(vinyl phenyl)ethane are introduced as co-monomers, in some cases to improve the dimensional and chemical stability of the membrane (as shown in Sect. 2.5). It is observed that the IEC of crosslinked membranes does not differ from that of uncrosslinked membranes with the same graft level (Fig. 14) [115, 118, 121, 125, 157]. This means that the introduced ionic sites are equally accessible through the hydrophilic domains in crosslinked membranes, regardless of the more constrained polymer framework, at least up to the level of crosslinking agent investigated, which is around 20%. The amount of swelling is substantially reduced upon crosslinking, which is the reason for the improved dimensional stability of crosslinked membranes (Fig. 15) [115, 125, 127]. Consequently, the hydration number decreases as the degree of crosslinking increases at a given graft level. We will see in the next section how this affects the conductivity of the material.

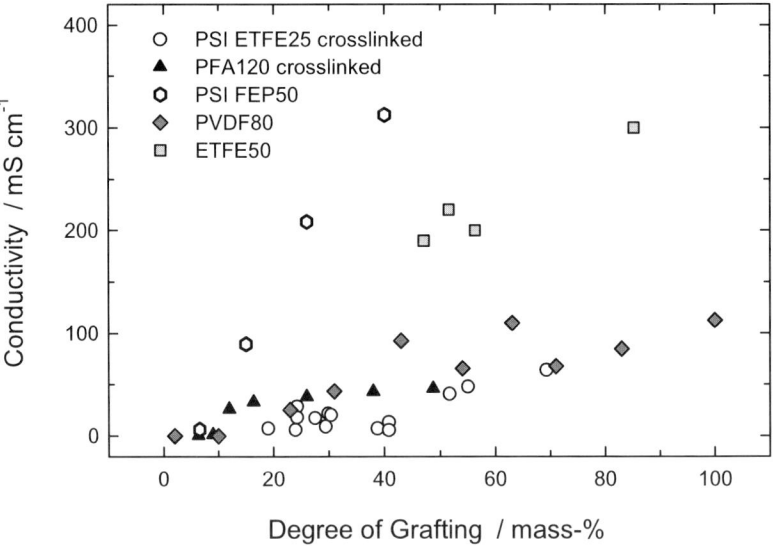

Fig. 16 Conductivity of various radiation grafted membranes as a function of DG at room temperature (data for PFA120 crosslinked redrawn from [200]; data for ETFE50 redrawn from [213]; data for PSI FEP50 redrawn from [135]; data for PVDF80 redrawn from [187])

4.1.3
Conductivity

As expected, the conductivity of radiation grafted ion exchange membranes increases with increasing DG, both for crosslinked and uncrosslinked membranes. There is, however, a tremendous range of conductivity values reported by different authors. The values range from < 1 up to 300 mS cm^{-1} at room temperature in fully hydrated (i.e., liquid–water equilibrated) state. The measured conductivity is governed or influenced by a number of parameters, above all by the distribution of the graft component across the membrane, as mentioned [157, 176, 177]. The base film thickness also appears to have an influence in some cases, thicker base films yielding a higher conductivity [63, 125, 146]. It is conceivable that this is a surface effect, i.e., that regions close to the surface of the irradiated film are less grafted, potentially due to loss of radical sites caused by exposure of the material to oxygen and water in the air. On the other hand, this thickness effect can also be observed for Nafion® [146], so it may also be a physical effect, presumably unfavorable aggregation or conformation of the ionophoric side chains close to the surface.

Proton conductivity (σ_{H^+}) can be related to the proton diffusion coefficient D_{H^+} using the Nernst–Einstein equation [179]:

$$\sigma_{H^+} = \frac{D_{H^+} c_{H^+} z^2 F^2}{RT}, \qquad (6)$$

where c_{H^+} is the volumetric density of protons and z, F, R, and T have the usual meaning. Proton diffusion in water and proton-exchange membranes is thermally activated, hence the quantity $\sigma \cdot T$ shows a temperature dependence of Arrhenius type. For perfluorosulfonic acid membranes such as Nafion®, activation energies between 12 and 15 kJ mol^{-1} are obtained [180]. As a comparison, 10.3 kJ mol^{-1} are found for pure water [181]. For radiation grafted membranes, only limited data is available. For the conductivity of uncrosslinked PVDF-based membranes in the temperature range between 20 and 70 °C, an activation energy similar to Nafion® 105 was found, yet quantitative values were not given [182]. Changes in membrane morphology and water uptake with temperature were put forward as further contributions to the increase in conductivity, in addition to the higher mobility of the protons. The resistance of membranes from Solvay, based on ETFE and crosslinked with DVB, was measured in situ during DMFC operation in a temperature range between 90 and 130 °C [183], and an activation energy of around 18 kJ mol^{-1} was calculated. A study carried out in the authors' laboratory, using water-swollen crosslinked and uncrosslinked FEP- and ETFE-based membranes with 20–25% graft level, showed higher activation energy for the crosslinked membranes (15.0–15.5 kJ mol^{-1}) compared to the uncrosslinked ones (14.0–14.5 kJ mol^{-1}), which may be a consequence of higher association of the protons with the counterions or polymer in crosslinked membranes, which have lower water uptake [184].

In addition to the conductivity in the water-swollen state, the conductivity of the fuel cell membrane under non-saturated water vapor conditions is of importance as, during cell operation, partial drying of the membrane and electrodes may occur. Also, fuel cell operation with partially humidified or even dry reactant gases is highly desirable to minimize system complexity. Walsby et al. [185] has investigated the influence of relative humidity on conductivity of radiation grafted membranes (Fig. 17). It was found that although the radiation grafted membranes displayed a superior conductivity at a relative humidity of 100%, the value dropped below that of Nafion® at relative humidities between 40 and 85%. Below 40%, all the membranes exhibited poor conductivity of around 1 mS cm^{-1} or lower. The different behavior could again be indicative of a dissimilar microstructure, polymer domain morphology, or extent of hydrophilic–hydrophobic phase separation [178]. There is no literature data on the polymer morphology of radiation grafted membranes; it is, however, likely that the microstructure will depend to a large extent on base film type, graft level, extent of crosslinking, and other design and process parameters. A sorption curve qualitatively similar to the data shown in Fig. 17 for radiation grafted membranes is observed for sulfonated poly(ether ketone) membranes. The strong drop in conductivity below 90% relative humidity is attributed to a less effective phase separation in polymer

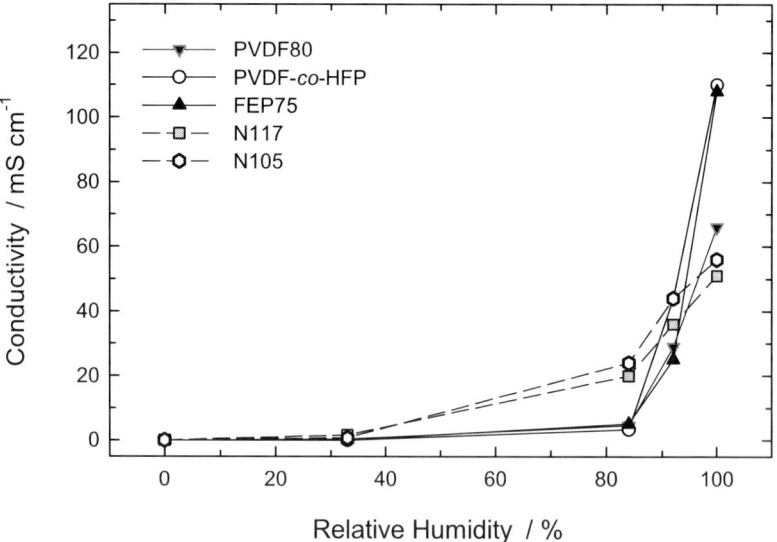

Fig. 17 Influence of relative humidity on conductivity at room temperature. Radiation grafted membranes are not crosslinked and have DG between 34 and 40% (redrawn from [206])

Table 4 Physical properties of radiation grafted membranes with different extent of crosslinking (redrawn from [121])

Base polymer	Degree of grafting (mass %)	Degree of crosslinking[a] (mol %)	Ion exchange capacity (mmol g^{-1})	Water content[b] (H$_2$O/SO$_3$H)	Conductivity[c] (mS cm^{-1})
FEP-50	19.1	0	1.39	27.2	98
FEP-50	18.8	3	1.07	25.9	93
FEP-50	19.6	6	n/a	11.9	63
FEP-50	19.0	12	1.27	7.0	28

[a] Determined in grafted films via FTIR
[b] Swollen in boiling water
[c] Determined in situ, fuel cell temperature of 40 °C, using equipment built in-house

backbone and proton conducting aqueous channels, a less favorable percolation of the hydrophilic domains, and higher localization of the protons due to the higher pK_a value compared to Nafion® [186].

For crosslinked membranes, the situation is somewhat different. Depending on the extent of crosslinking, excessive water uptake under fully humidified conditions is inhibited due to the network of covalent bonds in the polymer [184]. The effect of crosslinking on water uptake and conductivity has been investigated by Büchi et al. [125], as given in Table 4. For membranes of similar DG, it is observed that an increase in crosslinker content results in a decrease of water uptake and conductivity [118]. If the conductivity is plotted versus the water content, an approximately linear correlation is found, suggesting that the proton mobility is governed to a large extent by the hydration level of the material.

4.2
Performance in Fuel Cells

In PEFC, the membrane, together with the electrodes, forms the basic electrochemical unit, the membrane electrode assembly (MEA). Whereas the first and foremost function of the electrolyte membrane is the transport of protons from anode to cathode, the electrodes host the electrochemical reactions within the catalyst layer and provide electronic conductivity on the one hand, and pathways for reactant supply to and removal of products from the catalyst on the other hand. The components of the MEA need to be chemically stable for several thousands of hours in the fuel cell under the prevailing operating and transient conditions. PEFC electrodes are wet-proofed fibrous carbon sheet materials of a few 100 μm thickness. The functionality of the proton-exchange membrane extends to requirements of mechanical stability to ensure effective separation of anode and cathode, also under aggravated conditions such as operation on reactant gases below the water vapor saturation point, fuel cell start-up, and transient load. For a detailed review of fuel cell performance and in situ characteristics of radiation grafted membranes, the reader is referred to an article from the authors' laboratory published recently [43]. In this contribution, the insights are presented in a distilled manner with condensed facts and conclusions.

4.2.1
MEA Fabrication

The formation of an intimate contact between membrane and electrodes during MEA fabrication is of high importance to minimize interfacial voltage losses. When using radiation grafted membranes together with electrodes containing Nafion® as ionomer, it has been found that the membrane-electrode interface is of inferior quality compared to when Nafion® is used as membrane, resulting in a higher resistance and/or insufficient adhesion or delamination [61, 182, 187]. The likely reason for this is the mismatch in ionomer type between the membrane and electrode catalyst layer. Huslage et al. [60] and Gubler et al. [61] found that dip-coating FEP-based radiation grafted membranes in solubilized Nafion® prior to hotpressing leads to an improved fuel cell performance and lower impedance of the single cell. Furthermore, these authors showed that hotpressing with the membrane in wet state resulted in an improved membrane–electrode interface compared to when hotpressing with the membrane in dry state, which can be explained on the basis of the water acting as a plasticizer, allowing polymer flow during the hotpressing process.

4.2.2
Fuel Cell Testing

Generally, little fuel cell testing using radiation grafted membranes has been reported in the literature, compared to the total number of articles on the subject. Frequently, characterization is restricted to the membrane, and is not extended to include fabrication of MEAs and fuel cell testing. Important insights relating to electrochemical performance, membrane–electrode interface properties, membrane integrity, and lifetime are therefore missing. Of the studies published that include fuel cell test results, selected articles are reviewed in the following sections to highlight specific aspects.

4.2.3
Water States and Water Management

In the characterization of fuel cell membranes, there are a number of important materials and component properties that have to be assessed in order to determine the applicability and operability in the fuel cell environment Fig. 18. Since proton mobility within the polymer structure is a strong function of the water content, the water uptake and transport properties of the membrane are of paramount importance, determining the water profile through the thickness of the membrane as well as in-plane. Water transport mechanisms in the polymer are diffusion due to a gradient in water content, hydraulic permeation as a consequence of a pressure gradient between anode and cathode, and electroosmotic drag, i.e., water flux coupled to proton transport.

The states of water have an important role to play in determining the transport behavior of protons in membranes. The water directly associated with ionic sites in a membrane may behave in a way different from normal water due to its strong association in the form of hydrogen bonding or polar interactions with the functional sites within the membrane. Such water does not show any phase transition such as crystallization or melting in the temperature range 200–273 K. Using DSC, three different types of water molecules have been identified in sulfonated FEP-g-polystyrene membranes, which may be categorized as the freezing free, freezing bound, and non-freezing water [188]. The relative ratio of these three types of water molecules depends on the DG. The non-freezing water per ionic sites remains independent of the DG. However, the freezing free and freezing bound water per ionic site tends to increase with the DG (Fig. 19). The non-freezing water was evaluated to be six to eight water molecules per ionic site in membranes with DG in the range 15–40%. Recent investigations on membranes based on styrene grafting on different films showed that the non-freezing water remains almost the same, irrespective of the chemical nature of the membranes, and corresponds to ten water molecules per ionic site [185]. This is further supported by the studies on crosslinked PVDF membranes.

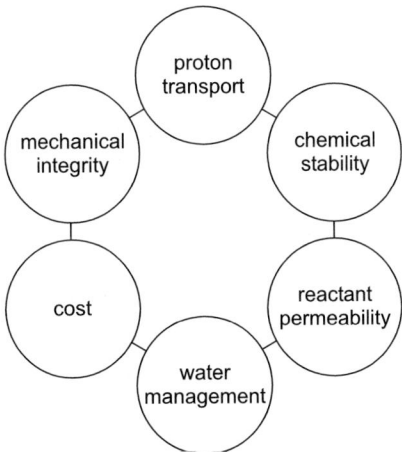

Fig. 18 Requirements for fuel cell membranes

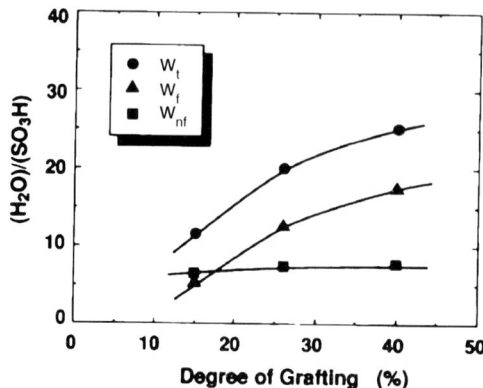

Fig. 19 Variation of water/ionic site ratio with DG for FEP-based membranes: W_t total water uptake, W_f freezing uptake, and W_{nf} non-freezing water uptake [188]

It may be stated that any increase in the water content with higher graft levels is associated with the incorporation of freezing water and should facilitate the ionic mobility [188]. With the increase in each grafting molecule, the hydrophilicity of the membrane matrix increases and crystallinity decreases. The structure as a result becomes more amenable to water penetration within the matrix. The crosslinking, however, influences the water uptake and its states. Highly crosslinked membranes developed from the DVB-styrene system do not show any freezing water and all of the water that accounts for the swelling of the membrane tends to be non-freezing in nature [129].

4.2.4
Reactant Permeability

Whereas uniform distribution of water within the membrane is desired, the permeability of the material to reactants (i.e., hydrogen or methanol and oxygen) has to be low to prevent direct chemical reaction between fuel and oxidant, which may lead to hotspots and, eventually, pinhole formation. Methanol permeability is a major challenge in the direct methanol fuel cell (DMFC), largely because methanol transport is strongly correlated with water transport, leading to significant penalties in fuel efficiency and poor cathode performance [189].

4.2.5
Chemical Stability

Chemical integrity of the polymer has to be maintained at the desired operating conditions for the designated operating time. The hostile fuel cell environment is a consequence of the simultaneous presence of H_2, O_2, H_2O_2

as intermediate, the noble metal catalyst, and possibly metallic contaminants such as Fe ions. It is widely accepted that radicals generated within this environment, such as hydroxyl (HO$^\bullet$) and hydroperoxyl (HO$_2^\bullet$) radicals, chemically attack the polymer, causing chain scission [190–195].

4.2.6
Mechanical Integrity

Furthermore, the material has to exhibit sufficient mechanical stability in order to fulfil its separator function. Not only tensile strength and elongation at break values have to be considered, but also dimensional stability upon swelling, and resistance to crack formation and propagation. Creep of the polymer is likely to occur because the water-swollen membrane is plasticized and the membrane is under a constant compaction force in the cell [196]. This may lead to membrane thinning and, eventually, puncturing a pinhole formation. An effect especially pertaining to swelling of the polymer upon water sorption is a fatigue-type phenomenon when the membrane electrode assembly is subjected to dry–wet cycles, which leads to periodic stress build up and relaxation in the membrane and, ultimately, to crack formation. This has been observed to be a membrane failure mode [197].

4.2.7
Fuel Cell Performance

Fuel cell characterization using radiation grafted membranes is mentioned in the work of Sundholm et al. [182, 198–200], Horsfall and Lovell [187, 201], Scott et al. [202], Nasef and Saidi [203], Hatanaka et al. [204], Aricò et al. [183], and Scherer et al. [61, 205–210] (in the last 5 years). In addition, in recent patent literature Ballard Power Systems [97, 211], Aisin Seiki [89, 91], and Pirelli [92] have filed inventions related to radiation grafted fuel cells membranes. The reported fuel cell performance characteristics span a substantial range, from unacceptably poor to values approaching or exceeding comparative samples based on Nafion® membranes [43]. It has to be emphasized at this point that direct comparison of fuel cell test data is not always straightforward and can be misleading. Occasionally, the grafted membranes used are thinner than the respective Nafion® comparison example. Consequently, similar fuel cell performance can be obtained although the conductivities of the two membrane materials are notably dissimilar. Crosslinked membranes have been used only in the minority of experiments by Nezu et al. [89], Aricò et al. [183], and in our own laboratory, e.g., [207]. The influence of DVB as crosslinker on ex situ membrane properties was discussed earlier. In the fuel cell, the level of crosslinking affects performance as well as durability [208] (also, Gubler et al. unpublished results). Optimum performance was found for 10% DVB content as a consequence of balanced membrane resistance and membrane–electrode interface characteristics. Stable performance was observed over a few hundred hours only for membranes with 10 and 20% DVB; the lower crosslinked membranes showed significant degradation. Membranes with high crosslinker content of 20% or more, however, suffer from poor mechanical properties. The increasing brittleness of the material can lead to membrane cracking during MEA fabrication or fuel cell operation.

One of the degradation modes observed using radiation grafted fuel cell membranes is correlated with reactant gas (i.e., H_2 and O_2) permeability through the membrane. Kallio et al. [182] found that oxygen diffusion and permeability increase with increasing water uptake and thus with DG. The open circuit voltage of the fuel cell was observed to be lower for membranes with higher water uptake, indicating a higher extent of mixed potential formation, especially on the cathode, due to gas permeation. Similar observations were made by Büchi et al. [125] as lower degrees of crosslinking at similar graft level yield membranes with higher water uptake, higher gas permeation, lower open circuit voltage, and shorter membrane lifetime in the fuel cell.

As an example for MEA performance, Fig. 20 shows the polarization behavior of an optimized radiation grafted membrane on the basis of FEP-25 film with 18% DG and 10% crosslinker content, compared against

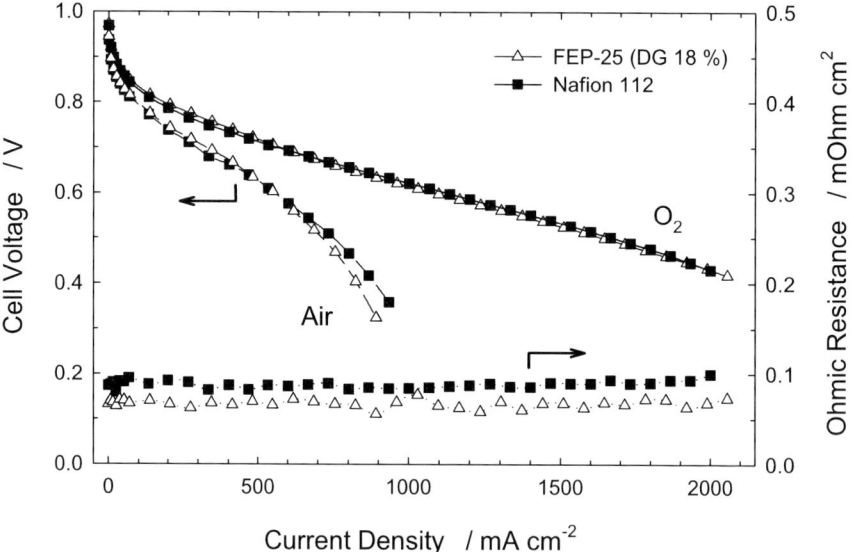

Fig. 20 Single cell performance comparison. Conditions: cell temperature 80 °C, H_2 stoichiometry 1.5, O_2/air stoichiometry 9.5/2.0, both fuel and oxidant reactant gases fully humidified, ambient pressure. Ohmic resistance was determined using auxiliary fast-current pulses according to [214]

a Nafion® 112-based MEA. The polarization curve of the two MEAs is similar, with a slightly lower ohmic resistance of the sample with the grafted membrane, which, however, is offset by a slightly higher interface resistance. This membrane is optimized for performance, durability, and mechanical stability. A membrane of this configuration was operated for over 4000 h at a cell temperature of 80 °C without loss in performance [61].

Very promising fuel cell performance results with respect to longevity were obtained with a novel monomer combination, namely a mixture of α-methylstyrene and methacrylonitrile as graft component [44]. Although these preliminary tests were carried out with non-crosslinked membranes, they nicely show the positive effect of substituting the α-H atom by a methyl group on stability under fuel cell test conditions. Testing of crosslinked membranes is ongoing (Gubler et al. unpublished results).

4.2.8
Performance in Direct Methanol Fuel Cells

The technology of radiation grafting of membranes is particularly interesting for the direct methanol fuel cell (DMFC), because the process parameters can be easily tuned to produce membranes with lower water and methanol

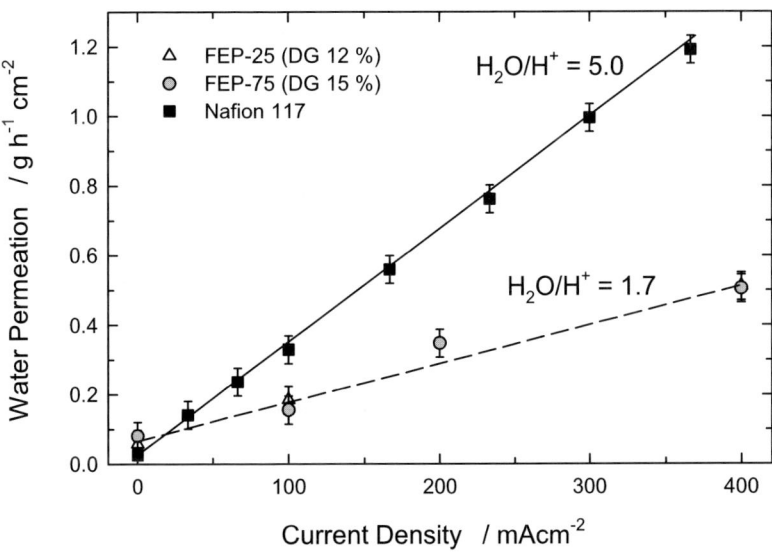

Fig. 21 Water permeation from anode to cathode in the direct methanol fuel cell for radiation grafted membranes based on FEP with different initial film thickness (25 and 75 μm) and Nafion® 117. The electroosmotic drag coefficient H_2O/H^+ is calculated from the slope of the regression line. Conditions: cell temperature 90 °C, pressure 2 bar, 20 mL min^{-1}, 0.5 M methanol, air stoichiometry is 2.0 for FEP and 3.0 for Nafion® 117

uptake, and with the desired transport properties [200, 212]. Compared to membranes used in the hydrogen fuel cell, optimized membranes for the DMFC are fabricated using thicker base film such as FEP 75 µm and having a lower DG [209]. With such membranes, identical performance is obtained compared to the Nafion® 117 standard, yet with methanol permeation reduced by 40%. In addition, lower water transport from anode to cathode leads to less cathode flooding. Water permeation data from anode to cathode for two radiation grafted membranes based on FEP-25 and FEP-75, and for Nafion® 117 are shown in Fig. 21. Water transport through these membranes appears to depend linearly on the cell current with comparably little permeation at zero current, indicating that the dominant mechanism for water transport is electroosmotic drag. From the slope of the curves, the electroosmotic drag coefficient can be calculated, yielding a value of 1.7 for the crosslinked radiation grafted membranes and 5.0 for Nafion® 117. Also, the reader may note that there is no marked difference in water permeation between the two grafted membranes of different thickness, reinforcing the conclusion that electroosmotic drag is dominant, and not diffusion.

5
Conclusions

This review demonstrates that radiation grafted membranes can be used successfully as solid polymer electrolytes for fuel cells. The membranes fabricated by radiation-induced grafting offer a cost-competitive option since inexpensive commercial materials are used and the preparation procedure is based on established industrial processes. Radiation-induced grafting is an attractive method to introduce desirable properties into a polymer owing to its simplicity in handling and its control over the grafting process. The method allows the use of a wide range of polymer–monomer combinations, such as various fluoropolymer films and vinyl and acrylic monomers. Partially fluorinated and perfluorinated polymers have been frequently used as base polymer to meet the requirements for chemically and thermally stable proton conducting membranes. Styrene and styrene derivatives have been extensively used as the monomer since grafted styrene can be readily modified to introduce a variety of functionalities.

Grafting parameters (irradiation dose, monomer concentration, grafting medium, temperature, etc.) have significant influence not only on grafting yield and grafting kinetics but also on resultant film and membrane properties. Crosslinkers are used in conjunction with the monomer to achieve certain desirable properties. For instance, the use of crosslinker is an effective means of enhancing the stability of styrene-grafted membranes in fuel cells.

Investigation of the structure, morphology, homogeneity, thermal and mechanical properties of both the grafted films and the membranes is important for understanding the grafting process and the operation mechanisms of the membranes. Several characterization methods are available to examine these properties.

The identification of membrane properties relevant to fuel cells (ion exchange capacity, water uptake, conductivity), aspects of membrane electrode assembly fabrication, and fuel cell performance are described in detail in this review.

References

1. Bungay PM, Lonsdale HK, Pinho MN (eds) (1986) Synthetic membranes: science, engineering and applications. Reidel, Dordrecht
2. Lemmons RJ (1990) J Power Source 29:251
3. Doyle M, Rajendran G (2003) Perfluorinated membranes. In: Vielstich W, Gasteiger HA, Lamm A (eds) Handbook of fuel cells: fundamentals, technology and applications, vol 3. Wiley, Chichester, p 351
4. Smitha B, Sridhar S, Khan AA (2003) J Membr Sci 225:63
5. Vie P, Paronen M, Stromgard M, Rauhala E, Sundholm F (2002) J Membr Sci 204:295
6. Paronen M, Sundholm F, Rauhala E, Lehtinen T, Hietala S (1997) J Mater Chem 7:2401
7. Kato K, Uchida E, Kang ET, Uyama Y, Ikada Y (2003) Prog Polym Sci 28:209
8. Schellekens MAJ, Klumperman B (2000) J Macromol Sci Rev Macromol Chem Phys C 40:167
9. Gupta B, Scherer GG (1994) Chimia 48:127
10. Hegazy EA, AbdEl-Rehim HA, Kamal H, Kandeel KA (2001) Nucl Instrum Methods Phys Res B 185:235
11. Rikukawa M, Sanui K (2000) Prog Polym Sci 25:1463
12. Cardona F, George GA, Hill DJT, Perera S (2003) Polym Int 52:827
13. Dargaville TR, George GA, Hill DJT, Whittaker AK (2003) Prog Polym Sci 28:1355
14. Nasef MM, Hegazy EA (2004) Prog Polym Sci 29:499
15. Kabanov VY (2004) High Energ Chem 38:57
16. Hickner MA, Ghassemi H, Kim YS, Einsla BR, McGrath JE (2004) Chem Rev 104:4587
17. Smitha B, Sridhar S, Khan AA (2005) J Membr Sci 259:10
18. Souzy R, Ameduri B (2005) Prog Polym Sci 30:644
19. Hickner MA, Pivovar BS (2005) Fuel Cells 5:213
20. Jagur-Grodzinski J (2007) Polym Adv Technol 18:785
21. Chapiro A (1962) Radiation chemistry of polymeric systems. Wiley-Interscience, New York
22. Chapiro A (1977) Radiat Phys Chem 9:55
23. Fischer NK, Corelli JC (1981) J Polym Sci A 19:2465
24. Lunkwitz K, Brink HJ, Handle D, Ferse A (1989) Radiat Phys Chem 33:523
25. Bürger W, Lunkwitz K, Pompe G, Petr A, Jehnichen D (1993) J Appl Polym Sci 48:1973
26. Iwasaki M (1971) Fluorine Chem Rev 5:1
27. Lappan U, Häussler L, Pompe G, Lunkwitz K (1997) J Appl Polym Sci 66:2287
28. Iwasaki M, Toriyama K (1967) J Chem Phys 47:559
29. Schlick S, Chamulitrat W, Kevan L (1985) J Phys Chem 89:4278
30. Hill DJT, Mohajerani S, Pomeri PJ, Whittaker AK (2000) Radiat Phys Chem 59:295

31. Dargaville TR, Hill DJT, Whittaker AK (2001) Radiat Phys Chem 62:25
32. Oshima A, Seguchi T, Tabata Y (1997) Radiat Phys Chem 50:601
33. Lunkwitz K, Lappan U, Lehmann D (2000) Radiat Phys Chem 57:373
34. Nasef MM, Dahlan KJM (2003) Nucl Instrum Methods Phys Res B 201:604
35. Oshima A, Ikeda S, Katoh E, Tabata Y (2001) Radiat Phys Chem 62:38
36. Tabata Y, Oshima A, Takashika K, Saguchi T (1996) Radiat Phys Chem 48:563
37. Oshima A, Tabata Y, Kudoh H, Seguchi T (1995) Radiat Phys Chem 45:269
38. Sun J, Zhang Y, Zhong X (1994) Polymer 35:2881
39. Zhong X, Sun J, Zhang Y (1992) Polymer 33:5341
40. Zhong X, Sun J, Wang F, Sun Y (1992) J Appl Polym Sci 44:639
41. Sun J, Zhang Y, Zhong X, Zhang W (1993) Radiat Phys Chem 42:139
42. Forsythe JS, Hill DJT, Logothetis AL, Seguchi T, Whittaker AK (1998) Radiat Phys Chem 53:611
43. Gubler L, Alkan Gürsel S, Scherer GG (2005) Fuel Cells 5:317
44. Gubler L, Slaski M, Wokaun A, Scherer GG (2006) Electrochem Commun 8:1215
45. Alkan Gürsel S, Yang Z, Choudhury B, Roelofs MG, Scherer GG (2006) J Electrochem Soc 53:A1964
46. Chen J, Asano M, Yamaki T, Yoshida M (2006) J Appl Polym Sci 100:4565
47. Chapiro A (1959) J Polym Sci 34:481
48. Chapiro A (1962) J Polym Sci 57:743
49. Chapiro A (1979) Radiat Phys Chem 14:101
50. Chapiro A (1987) Eur Polym J 23:255
51. Bozzi A, Chapiro A (1988) Radiat Phys Chem 32:193
52. Gupta B, Chapiro A (1989) Eur Polym J 25:1137
53. Gupta B, Chapiro A (1989) Eur Polym J 25:1145
54. Gupta B, Anjum N (2000) J Appl Polym Sci 77:1331
55. Phadnis S, Patri M, Hande VR, Deb P (2003) J Appl Polym Sci 90:2572
56. Cardona F, George GA, Hill DJT, Rasoul F, Maeji J (2002) Macromolecules 35:355
57. Cardona F, George GA, Hill DJT, Perera S (2002) J Polym Sci A 40:3191
58. Souzy R, Ameduri B, Boutevin B, Gebel G, Capron P (2005) Solid State Ionics 176:2839
59. Brack HP, Büchi FN, Huslage J, Scherer GG (1998) Proc Electrochem Soc 27:52
60. Huslage J, Rager T, Schnyder B, Tsukada A (2002) Electrochim Acta 48:247
61. Gubler L, Kuhn H, Schmidt TJ, Scherer GG, Brack HP, Simbeck F (2004) Fuel Cells 4:196
62. Brack HP, Büchi FN, Huslage J, Rota M, Scherer GG (2000) Development of radiation grafted membranes for fuel cell applications based on poly(ethylene-*alt*-tetrafluoroethylene). In: Pinnau I, Freeman BD (eds) Membrane formation and modification. ACS symposium series 744. Oxford University Press, New York, p 174
63. Brack HP, Scherer GG (1997) Macromol Symp 126:25
64. Brack HP, Bührer HG, Bonorand L, Scherer GG (2000) J Mater Chem 10:1795
65. Walsby N, Sundholm F, Kallio T, Sundholm G (2001) J Polym Sci A 39:3008
66. Lee W, Shibasaki A, Saito K, Sugita K, Okuyama K, Suyo T (1996) J Electrochem Soc 143:2795
67. Chen J, Asano M, Maekawa Y, Yoshida M (2005) J Membr Sci 277:249
68. Chen J, Septiani U, Asano M, Maekawa Y, Kubota H, Yoshida M (2007) J Appl Polym Sci 103:1966
69. Hegazy EA, Ishigaki I, Okamoto J (1981) J Appl Polym Sci 26:3117
70. Momose T, Yoshioka H, Ishigaki I, Okamoto J (1989) J Appl Polym Sci 37:2817
71. Stone C, Bonorand LM (2003) WO 03/018654 A1 (Ballard Power Systems Inc)

72. Gupta B, Büchi FN, Scherer GG (1994) J Polym Sci A 32:1931
73. Horsfall JA, Lovell KV (2002) Eur Polym J 38:1671
74. Nasef MM, Saidi H, Dessouki AM, El-Nesr EM (2000) Polym Int 49:399
75. Nasef MM, Saidi H, Nor HM (2000) J Appl Polym Sci 76:220
76. Rager T (2003) Helv Chim Acta 86:1
77. Rager T (2004) Helv Chim Acta 87:400
78. Nasef MM (2001) Polym Int 50:338
79. Dargaville TR, George GA, Hill DJT, Whittaker AK (2003) Macromolecules 36:8276
80. Alkan Gürsel S, Ben youcef H, Wokaun A, Scherer GG (2007) Nucl Instrum Methods Phys Res B 265:198
81. Rohani R, Nasef MM, Saidi H, Zaman K, Dahlan M (2007) Chem Eng J 132:27
82. Hegazy EA, Ishigaki I, Dessouki AM, Rabie A-GM, Okamoto J (1982) J Appl Polym Sci 27:535
83. Phadnis S, Patri M, Hande VR, Deb PC (2003) J Appl Polym Sci 90:2572
84. Septiani U, Chen J, Asano M, Maekawa Y, Yoshida M, Kubota H (2007) J Mater Sci 42:1330
85. Liang GZ, Lu TL, Ma XY, Yan HX, Gong ZH (2003) Polym Int 52:1300
86. Nasef MM, Saidi H, Nor HM, Dahlan KZM, Hashim K (1999) J Appl Polym Sci 73:2095
87. Elmidaoui A, Cherif AT, Brunea J, Duclert F, Cohen T, Gavach C (1992) J Membr Sci 67:263
88. Becker W, Bothe M, Schmidt-Naake G (1999) Angew Makromol Chem 273:57
89. Nezu S, Ito N, Yamada C, Kato M, Asukabe M (2001) US Patent 6 242 123 (Aisin Seiki Kabushiki Kaisha)
90. D'Agostino VF, Newton JM (2001) US Patent 6 225 368 (National Power PLC)
91. Asukabe M, Kato M, Taniguchi J, Morimoto Y, Kawasumi M (2004) US Patent US 6 827 986 B2 (Aisin Seiki Kabushiki Kaisha)
92. Dubitsky YA, Lopes Correia Tavares AB, Zaopo A, Albizzati E (2004) WO 2004/004053 A2 (Pirelli & C. SpA)
93. D'Agostino VF, Lee JY, Cook EH (1977) US Patent 4 012 303 (Hooker Chemicals & Plastics Corp)
94. D'Agostino VF, Lee JY, Cook EH (1978) US Patent 4 107 005 (Hooker Chemicals & Plastics Corp)
95. D'Agostino VF, Lee JY, Cook EH (1978) US Patent 4 113 922 (Hooker Chemicals & Plastics Corp)
96. Momose T, Tomiie K, Harada H, Miyachi H, Kato H (1986) US Patent 4 605 685 (Chlorine Engineers Corp)
97. Stone C, Steck AE (2002) US Patent US 6 359 019 (Ballard Power Systems Inc)
98. MacKinnon SM (2004) US Patent 6 828 386 (Ballard Power Systems Inc)
99. Sean MM (2004) US Patent 2004/0059015 A1 (Ballard Power Systems Inc)
100. Stone C, Steck AE, Choudhury B (2002) US Patent US 2002/0137806 A1 (Ballard Power Systems Inc)
101. Stone C (2003) WO 03/018655 A1 (Ballard Power Systems Inc)
102. Stone C, Steck AE (2001) WO 01/58576 A1 (Ballard Power Systems Inc)
103. Yang Z, Roelofs MG, Alkan Gürsel S, Scherer GG (2006) IP WO 2006/102672
104. Slaski M, Gubler L, Scherer GG (2006) EPA WO 2006/084591 A1
105. Momose T, Tomiie K, Ishigaki I, Okamoto J (1989) J Appl Polym Sci 37:2165
106. Dargaville TR, Hill DJT, Perera S (2002) Aust J Chem 55:439
107. Walsby N, Paronen M, Juhanoja J, Sundholm F (2000) J Polym Sci A 38:1512

108. Brandrup J, Immergut EH (1989) Polymer Handbook. Wiley-Interscience, New York, p 519
109. Farquet P, Kunze A, Padeste C, Solak HH, Alkan Gürsel S, Scherer GG, Wokaun A (2007) Polymer 48:4936
110. Odian G (1970) Principles of polymerization. McGraw-Hill, New York, p 255
111. Walsby N, Paronen M, Juhanoja J, Sundholm F (2001) J Appl Polym Sci 81:1572
112. Hegazy EA, Taher NH, Kamal H (1989) J Appl Polym Sci 38:1229
113. El-Assy N (1991) J Appl Polym Sci 42:885
114. Gupta B, Scherer GG (1993) Angew Makromol Chem 210:151
115. Nasef MM, Saidi H (2003) J Membr Sci 216:27
116. Gupta B, Büchi FN, Scherer GG, Chapiro A (1996) J Membr Sci 118:231
117. Chen J, Asano M, Yamaki T, Yoshida M (2006) J Appl Polym Sci 100:4565
118. Ben youcef H, Alkan Gürsel S, Wokaun A, Scherer GG (2008) J Membr Sci 311:208
119. Becker W, Schmidt-Naake G (2002) Chem Eng Technol 25:4
120. Rota M, Brack HP, Büchi FN, Gupta B, Haas O, Scherer GG (1995) Extended abstracts of 187th meeting of the Electrochemical Society, Reno, NV, 21–26 May 1995. 95-1:719
121. Büchi FN, Gupta B, Haas O, Scherer GG (1995) J Electrochem Soc 142:3044
122. Brack HP, Scherer GG (1998) Abstracts of papers of the American Chemical Society 216:281 – PMSE Part 2
123. Steuernagel L, Reich S, Kaufmann DE, Wokaun A, Scherer GG, Brack HP (2002) PSI scientific report. Villigen PSI, Switzerland, p 25; http://ecl.web.psi.ch/SciRep.html
124. Brack HP, Fischer D, Peter G, Slaski M, Scherer GG (2004) J Polym Sci A 42:59
125. Büchi FN, Gupta B, Haas O, Scherer GG (1995) Electrochim Acta 40:345
126. Rouilly MV, Kötz R, Haas O, Scherer GG, Chapiro A (1993) J Membr Sci 81:89
127. Gupta B, Büchi FN, Staub M, Grman D, Scherer GG (1996) J Polym Sci A 34:1873
128. Yamaki T, Asano M, Maekawa Y, Morita Y, Suwa T, Chen J, Tsubokawa N, Kobayashi K, Kubota H, Yoshida M (2003) Radiat Phys Chem 67:403
129. Gupta B, Anjum N (2001) J Appl Polym Sci 82:2629
130. Sakurai H, Shiotani M, Yahiro H (1999) Radiat Phys Chem 56:309
131. Oshima A, Tabata Y, Kudoh H, Seguchi T (1995) Radiat Phys Chem 45:269
132. Sato K, Ikeda S, Iida M, Oshima A, Tabata Y, Washio M (2003) Nucl Instrum Methods Phys Res B 208:424
133. Nasef MM, Saidi H, Nor HM (2000) J Appl Polym Sci 77:1877
134. Flint SD, Slade RCT (1997) Solid State Ionics 97:299
135. Gupta B, Büchi FN, Scherer GG, Chapiro A (1994) Polym Adv Technol 5:493
136. Alkan Gürsel S, Yang Z, Choudhury B, Roelofs MG, Scherer GG (2006) J Electrochem Soc 53:A1964
137. Mattsson B, Ericson H, Torell LM, Sundholm F (1999) J Polym Sci A 37:3317
138. Gupta B, Staub M, Scherer GG, Grman D (1995) J Polym Sci A 33:1545
139. Schnyder B, Rager T (2007) J Appl Polym Sci 104:1973
140. Hietala S, Paronen M, Holmberg S, Nasman J, Juhanoja J, Karjalainen M, Serimaa R, Toivola M, Lehtinen T, Parovuori K, Sundholm G, Ericson H, Mattson B, Torell L, Sundholm F (1999) J Polym Sci A 37:1741
141. Hegazy EA, Ishigaki I, Rabie A-GM, Dessouki AM, Okamoto J (1983) J Appl Polym Sci 28:1465
142. Hegazy EA, Ishigaki I, Rabie A-GM, Dessouki AM, Okamoto J (1981) J Appl Polym Sci 26:3871
143. Guilmeau I, Esnouf S, Betz N, Le Moel A (1997) Nucl Instrum Methods Phys Res B 131:270

144. Ericson H, Kallio T, Lehtinen T, Mattsson B, Sundholm G, Sundholm F, Jacobsson P (2002) J Electrochem Soc 149:A206
145. Mattsson B, Ericson H, Torell LM, Sundholm F (2000) Electrochim Acta 45:1405
146. Brack HP, Büchi FN, Huslage J, Scherer GG (1998) In: Gottesfeld S, Fuller TF (eds) Proton conducting membrane fuel cells II. Electrochemical Society PV 98-27:52
147. Good RJ (1992) J Adhes Sci Technol 6:1269
148. Gupta B, Anjum N (2002) J Appl Polym Sci 86:1118
149. Brack HP, Wyler M, Peter G, Scherer GG (2003) J Membr Sci 214:1
150. Brack HP, Slaski M, Gubler L, Scherer GG, Alkan Gürsel S, Wokaun A (2004) Fuel Cells 4:1
151. Nasef MM, Saidi H, Nor HM, Yarmo MA (2000) J Appl Polym Sci 76:336
152. Nasef MM, Saidi H (2006) Appl Surf Sci 252:3073
153. Gebel G, Ottomani E, Allegraud JJ, Betz N, Le Moel A (1995) Nucl Instrum Methods Phys Res B 105:145
154. Gebel G, Diat O (2005) Fuel Cells 5:261
155. Hietala S, Holmberg S, Näsman J, Ostrovskii Paronen M, Serimaa R, Sundholm F, Torell L, Torkkeli M (1997) Angew Makromol Chem 253:151
156. Jokela K, Galambosi S, Karjalainen M, Torkkkeli M, Serimaa R, Eteläniemi V, Vahvaselkä S, Hietala S, Paronen M, Sundholm F (2000) Mater Sci Forum 321–324:481
157. Elomaa M, Hietala S, Paronen M, Walsby N, Jokela K, Serimaa R, Torkkeli M, Lehtinen T, Sundholm G, Sundholm F (2000) J Mater Chem 10:2678
158. Aymes-Chodur C, Betz N, Porte-Durrieu MC, Baquey C, Le Moel A (1999) Nucl Instrum Methods Phys Res B 151:377
159. Gupta B, Scherer GG (1993) J Appl Polym Sci 50:2129
160. Gupta B, Highfield JG, Scherer GG (1994) J Appl Polym Sci 51:1659
161. Brack HP, Rüegg D, Bührer H, Slaski M, Alkan Gürsel S, Scherer GG (2004) J Polym Sci B Polym Phys 42:2612
162. Alkan Gürsel S, Schneider J, Ben Youcef H, Wokaun A, Scherer GG (2008) J Appl Polym Sci 108:3577
163. Cardona F, Hill DJT, George GA, Maeji J, Firas R, Perera S (2001) Polym Degrad Stabil 74:219
164. Hietala S, Koel M, Skou E, Elomaa M, Sundholm F (1998) J Mater Chem 8:1127
165. Nasef MM, Saidi H, Nor HM, Foo OM (2000) J Appl Polym Sci 78:2443
166. Gupta B, Haas O, Scherer GG (1994) J Appl Polym Sci 54:469
167. Zevin L, Messalem R (1982) Polymer 23:601
168. Nasef MM, Saidi H (2006) Macromol Mater Eng 291:972
169. Nasef MM (2002) Eur Polym J 38:87
170. Jokela K, Serimaa R, Torkkeli M, Sundolm F, Kallio T, Sundolm G (2002) J Polym Sci A 40:1539
171. Gupta B, Scherer GG, Highfield J (1998) Angew Makromol Chem 256:81
172. Scherer GG, Brack HP, Büchi FN, Gupta B, Haas O, Rota M (1996) Hydrogen energy progress XI. Proceedings of the 11th world hydrogen energy conference, Stuttgart, Germany. 2:1727
173. Scherer GG (1990) Ber Bunsenges Phys Chem 94:1008
174. He R, Li Q, Xiao G, Bjerrum NJ (2003) J Membr Sci 226:169
175. Taniguchi T, Morimoto T, Kawakado M (2001) JP 2001/213987A2 (Toyota Central Research & Development Laboratory Inc)
176. Nasef MM, Saidi H, Nor HM, Foo OM (2000) J Appl Polym Sci 76:1
177. Lehtinen T, Sundholm G, Holmberg S, Sundholm F, Björnbom B, Bursell M (1998) Electrochim Acta 43:1881

178. Kreuer KD (2001) J Membr Sci 185:29
179. Choi P, Jalani NH, Datta R (2005) J Electrochem Soc 152:E123
180. Halim J, Büchi FN, Haas O, Stamm M, Scherer GG (1994) Electrochim Acta 39:1303
181. Kreuer KD (1992) In: Colomban P (ed) Proton conductors. Cambridge University Press, Cambridge, p 474
182. Kallio T, Lundström M, Sundholm G, Walsby N, Sundholm F (2002) J Electrochem 32:11
183. Aricò AS, Baglio V, Cretì P, Blasi AD, Antonucci V, Brunea J, Chapotot A, Bozzi A, Schoemans J (2003) J Power Source 123:107
184. Gubler L, Finsterwald T, Keller M, Scherer GG (2005) International conference on solid state ionics, SSI 15, Baden-Baden, Germany, 17–22 July 2005. Oral contribution no 78
185. Walsby N, Hietala S, Maunu SL, Sundholm F, Kallio T, Sundholm G (2002) J Appl Polym Sci 86:33
186. Kreuer KD (1997) Solid State Ionics 97:1
187. Horsfall JA, Lovell KV (2001) Fuel Cells 13:186
188. Gupta B, Haas O, Scherer GG (1995) J Appl Polym Sci 57:855
189. Jiang R, Chu D (2004) J Electrochem Soc 151:A69
190. LaConti AB, Hamdan H, McDonald RC (2003) Mechanisms of membrane degradation. In: Vielstich W, Lamm A, Gasteiger H (eds) Handbook of fuel cells: fundamentals, technology, applications, vol 3. Wiley, Chichester, p 647
191. Chen J, Asano M, Yamaki T, Yoshida M (2006) J Mater Sci 41:1289
192. Yamaki T, Tsukada J, Asano M, Katakai R, Yoshida M (2007) J Fuel Cell Sci Technol 4:56
193. Chen J, Septiani U, Asano M, Maekawa Y, Kubota H, Yoshida M (2007) J Appl Polym Sci 103:1966
194. Mitov S, Vogel B, Roduner E, Zhang H, Zhu X, Gogel V, Jörissen L, Hein M, Xing D, Schönberger F, Kerres J (2006) Fuel Cells 06:413
195. Gubler L, Scherer GG (2008) Durability of radiation grafted fuel cell membranes. In: Inaba M, Schmidt TJ, Büchi FN (eds) Proton exchange fuel cells durability. Springer, New York (in press)
196. Makharia R, Kocha SS, Yu PT, Gittleman C, Miller D, Lewis C, Wagner RT, Gasteiger HA (2006) Abstracts of the 208th meeting of the Electrochemical Society, Los Angeles, 16–21 Oct 2005. 502:1165
197. Gasteiger HA (2005) International conference on solid state ionics, SSI-15, Baden-Baden, Germany, 17–22 July 2005. Oral contribution no 72
198. Gode P, Ihonen J, Strandroth A, Ericson H, Lindbergh G, Paronen M, Sundholm F, Sundholm G, Walsby N (2003) Fuel Cells 3:21
199. Kallio T, Kisko K, Kontturi K, Serimaa R, Sundholm F, Sundholm G (2004) Fuel Cells 4:328
200. Saarinen V, Kallio T, Paronen M, Tikkanen P, Rauhala E, Kontturi K (2005) Electrochim Acta 50:3453
201. Horsfall JA, Lovell KV (2002) Polym Adv Technol 13:381
202. Scott K, Taama WM, Argyropoulos P (2000) J Membr Sci 171:119
203. Nasef MM, Saidi H (2002) J New Mater Electrochem Syst 5:183
204. Hatanaka T, Hasegawa N, Kamiya A, Kawasumi M, Morimoto Y, Kawahara K (2002) Fuel 81:2173
205. Huslage J, Rager T, Kiefer J, Steuernagel L, Scherer GG (2000) Proceedings of 197th Electrochemical Society Meeting, Toronto, Canada. 14–18 May 2000.

206. Geiger AB, Rager T, Matejcek L, Scherer GG, Wokaun A (2001) In: Büchi FN, Scherer GG, Wokaun A (eds) Proceedings of 1st European PEFC Forum. Lucerne, Switzerland, 2–6 July 2001. p 124
207. Gubler L, Beck N, Gürsel SA, Hajbolouri F, Kramer D, Reiner A, Steiger B, Scherer GG, Wokaun A, Rajesh B, Thampi KR (2004) Chimia 58:826
208. Schmidt TJ, Simbeck K, Scherer GG (2005) J Electrochem Soc 152:A93
209. Gubler L, Gürsel SA, Slaski M, Geiger F, Scherer GG, Wokaun A (2005) Proceedings of 3rd European PEFC Forum, Lucerne, Switzerland, 4–8 July 2005. Oral presentation B113
210. Gubler L, Prost N, Alkan Gürsel S, Scherer GG (2005) Solid State Ionics 176:2849
211. Stone C, Steck AE, Choudhury B (2004) US Patent 6 723 758 (Ballard Power Systems Inc)
212. Geiger AB, Rager T, Huslage J, Scherer GG, Wokaun A (2001) PSI scientific report, vol 5. Villigen PSI, Switzerland, p 99; http://ecl.web.psi.ch/SciRep.html
213. Chuy C, Basura VI, Simon E, Holdcroft S, Horsfall J, Lovell KV (2000) J Electrochem Soc 147:4453
214. Büchi FN, Marek A, Scherer GG (1995) J Electrochem Soc 142:1895

Advances in the Development of Inorganic–Organic Membranes for Fuel Cell Applications

Deborah J. Jones (✉) · Jacques Rozière

Institut Charles Gerhardt UMR 5253, Agrégats, Interfaces et Matériaux pour l'Energie, Université Montpellier II, Place E. Bataillon, 34095 Montpellier cedex 5, France
Deborah.Jones@univ-montp2.fr

1	Introduction	220
2	Use of Ionomer Membranes as Templates for Inorganic Particle Growth	222
2.1	Morphology and Microstructure of Ionomer Membranes	222
2.2	Preparation and Characterisation of Hybrid Inorganic–Organic Membranes Using Pre-Formed Membranes	225
3	Preparation of Composite Membranes by Addition of an Inorganic Component to a Polymer Solution or Dispersion	242
4	Preparation of Composite Membranes Using Polymer Solutions or Dispersions as Reaction Medium for In Situ Formation of the Inorganic Component	252
5	Conclusions and Outlook	259
	References	260

Abstract Inorganic–organic membranes are characterised by the presence of a certain amount of inorganic solid within an organic polymer that serves as the matrix component. From its origins some 15 years ago as a means of conferring proton conduction properties to an insulating polymer matrix by addition of a powdered inorganic proton conductor, the methods of preparation have developed and currently include a range of approaches that allow control over the localisation of the inorganic component preferentially in the hydrophilic or hydrophobic regions of an ionomer, and incite development of morphologies ranging from nanoparticulate to extended network forms. The presence of an inorganic phase is effective in enhancing interaction between components, in limiting dimensional change and in improving fuel cell performance under high temperature, low relative humidity conditions. These approaches have enabled the field to develop from the stage of different concepts of inorganic–organic fuel cell membranes to their implementation in fuel cell stacks. Novel approaches make use of further degrees of organisation of the organic and inorganic components, for example by use of nanoporogens or bimodal/spinodal transformations.

Keywords Hybrid ionomeric membrane · Inorganic particles and networks · Nanocomposite membrane · PEM fuel cell membrane

1
Introduction

Proton exchange (polymer electrolyte) membrane fuel cells (PEMFCs) hold great promise in the future diversification of energy supply for applications ranging from the portable electronics market, through automotive use and stationary power generation to other "niche" areas related in particular to military purposes. Real impact of the potential in terms of reduced emissions and alternative fuels allowed by fuel cell technology awaits introduction of a mass-market application, and arguably the greatest bearing will be brought by fuel cell powered cars, an essential driving force for the hydrogen economy. However, despite progress, the fuel cell is not yet an established technology under "real-life" conditions of use, where thermal and load cycling, high and low temperature and relative humidity (RH) environments, in particular, all inflict their own ageing mechanisms on the component materials.

One of the bottlenecks to fuel cell implementation is the proton conducting electrolyte membrane, the lifetime of which is not yet compatible with the target applications [1]. For example, efficiency requirements for fuel cell systems for automotive applications [2] have driven the need for a high-performance membrane that operates under conditions of low RH and temperatures exceeding the boiling point of water. These conditions put a severe constraint on the amount of water that can be maintained in the fuel cell membrane electrode assembly (MEA), limiting conductivity in conventional membrane materials, amongst other effects. Membranes must in addition have adequate strength and stability, be compatible with electrodes and have sufficient fuel cell performance: for automotive applications across the full operating range of temperature from start-up (sub-zero temperatures) to full power (up to 120–130 °C); for stationary application at higher temperatures > 150 °C, and this over lifetimes of 5000 h (automotive) to 50 000 h (stationary power generation).

Many significant advances have been made in the design and development of a broad range of polymers and membranes as "alternatives" to the conventional perfluorosulfonic acid (PFSA) type, of which Nafion®, generated by copolymerisation of a perfluorinated vinyl ether comonomer with tetrafluoroethylene, is by far the best known, most fully studied [3] and applied [4]. New materials have primarily explored radiation-induced graft polymerisation on per- or partially fluorinated aliphatic polymers [5], and sulfonated aromatic or heterocyclic polymers having more facile preparative routes [6], and that are less costly, in particular polysulfones [7, 8], polyetherketones [9, 10], polybenzimidazoles [11, 12] (PBIs) and polyimides [13, 14], in which the protogenic groups are covalently bound either to the polymer backbone or via a spacer. Acid-doped PBI [15–19] represents a second class of proton conducting membrane which, at the present time, is the only real contender for use in MEAs operating above 150 °C. Strategies have been developed for non-fluorinated polymers that limit membrane swelling, for example by blending with a polymer of opposite polarity [20], by introduction of cross-links, or by the introduction of a hierarchical organisation into gas-tight microporous arrangements during the membrane casting process [21], or that enhance hydrophobic–hydrophilic phase separation [22], with the aim of improving proton conduction properties.

However, a different approach pre-dates the above, which makes use of the proton conduction properties of inorganic solid acids and the development of inorganic–organic membranes [23]. From its origins as a means of conferring the proton conduction properties of a powdered inorganic material to a composite system incorporating non-functionalised (insulating) polymers prepared by screen-printing [24], the preparation methods for inorganic–organic membranes have progressed and now embrace a range of approaches including the in situ precipitation of inorganic particles in pre-cast ionomeric membranes or in functionalised polymer solutions [25], and the formation of interpenetrating inorganic and organic networks [26]. We have clearly shown that this is an effective route to enhance interaction between components and to limit swelling of sulfonated polyaromatic membranes that generally suffer even more severely than the PFSA type from excessive dimensional change and plastification under conditions of high water uptake, and yet require a higher hydration number (number of water molecules per sulfonic acid group) for equivalent conductivity [27]. Other properties of prime interest of inorganic–organic membranes include reduction of methanol crossover [28], of relevance to direct methanol fuel cells (DMFCs), improved conductivity [29], and opportunities in the direction of higher temperature operation that are attributed to improved mechanical properties and to the water-retentive characteristics of the inorganic phase, and which are of clear advantage when operating a fuel cell at low or no humidification of reactant gases. These observations have provided major impetus to the field, and recent years have witnessed a largely increased effort in the development of inorganic–organic membranes for fuel cell applications. Earlier reviews by Jones and Rozière [26, 30] and by Alberti and Casciola [31] are available, as well as a broader review by Savadogo [28] that also includes organic–organic composite membranes for fuel cells.

Inorganic–organic membranes are characterised by the presence of a certain amount of inorganic solid within an organic polymer that serves as the matrix component. The term "composite" is used here as a general term to describe all such membranes, while the designation "hybrid" is set aside for nanocomposite membranes with nanoscale integration of the organic and inorganic components. Hybrid membranes can be most readily generated by in situ precipitation of the inorganic material within an ionomer membrane or in a polymer solution [26]. Based on our previous reports, the present work aims to develop a general approach to the preparation of nanocomposite membranes in which specific interactions (ionic interaction, or weaker, physical or hydrogen-bonding interactions) are favoured, and to review recent literature on this overall scheme.

2
Use of Ionomer Membranes as Templates for Inorganic Particle Growth

2.1
Morphology and Microstructure of Ionomer Membranes

Charge-driven self-assembly between material components has been known and used for the preparation of nanocomposites for many years. It has been utilised to assemble organic and inorganic species into layered hosts for example, with applications in the fields of preparation of polymer-layered host hybrids, materials with non-linear optical properties and pillared layered solids [32, 33]. Polyelectrolytes of sufficiently high acidic (or basic) characteristics, such as acid-functionalised ionomeric membranes, have a distinct morphology arising from self-organisation into hydrophobic and hydrophilic regions, and which is related to that shown by amphiphilic systems. Various models describing this microstructure/nanostructure have been proposed, the enduring conceptual basis of which is the cluster-network model of hydrated ionic aggregates of diameter ≈ 4 nm (Fig. 1a), proposed by Gierke [34] some 25 years ago on the basis of the presence of a single small-angle X-ray scattering peak, a more recent example of which is given in Fig. 1b [35], and the behaviour of this peak with membrane swelling. In this model, sulfonic acid end groups of the perfluoroalkyl ether side chains are organised as inverse micelles that are connected via channels of length ≈ 1 nm, through which ions and polar solvents permeate, and are embedded in a semi-crystalline, hydrophobic matrix that ensures mechanical integrity. As more extensive structural studies have been conducted, alternative microstructures have been suggested [36-39] (Fig. 1c and d), all of which, however, retain the central recognition that ionic groups aggregate to form a network of clusters, although debating the form and spatial distribution of these clusters, as well as the structure and distribution of crystallites in the fluoropolymer backbone. As recently pointed out by Mauritz and Moore [3], by recognising the particular perspective by which a particular model was developed, certain disparities between the different models may be reconciled. There now seems to be agreement that the arrangement of the ionic domains is less well-ordered than originally proposed, with anisotropy of individual clusters superimposed on heterogeneity in spatial organisation, and transitional interphases between hydrophobic and hydrophilic regions. In sufficiently hydrated samples, and as described by Kreuer, water is an extended phase [27].

All of the above is relevant to our understanding of the properties of inorganic–organic membranes obtained from films. Other studies are signifi-

(a)

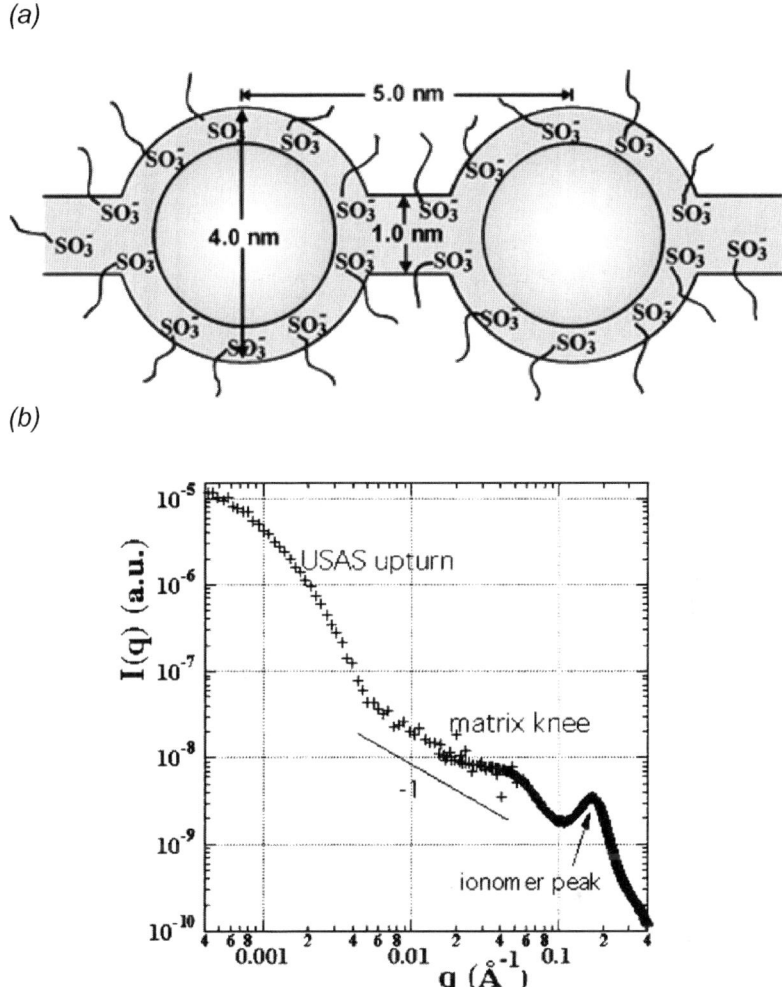

(b)

Fig. 1 Morphological models for the microstructure of Nafion®. a Cluster network model for the morphology of hydrated Nafion®. Reproduced with permission from [3]. © (2004) American Chemical Society. b Typical X-ray scattering spectrum of a water swollen Nafion®-1100. Reproduced with permission from [35]. © (2004) American Chemical Society

cant in the context of composite membrane preparation using other routes, in particular using solvent casting. For example, small-angle X-ray scattering (SAXS) profiles of films of Nafion® cast from low-boiling solvents (ethanol/water), such as typically used at the membrane/electrode interface, differ from those prepared from high-boiling solvents with casting temperature > 160 °C (and which are similar to those of Nafion® films prepared by extrusion). While ionic domains gave their characteristic feature in all

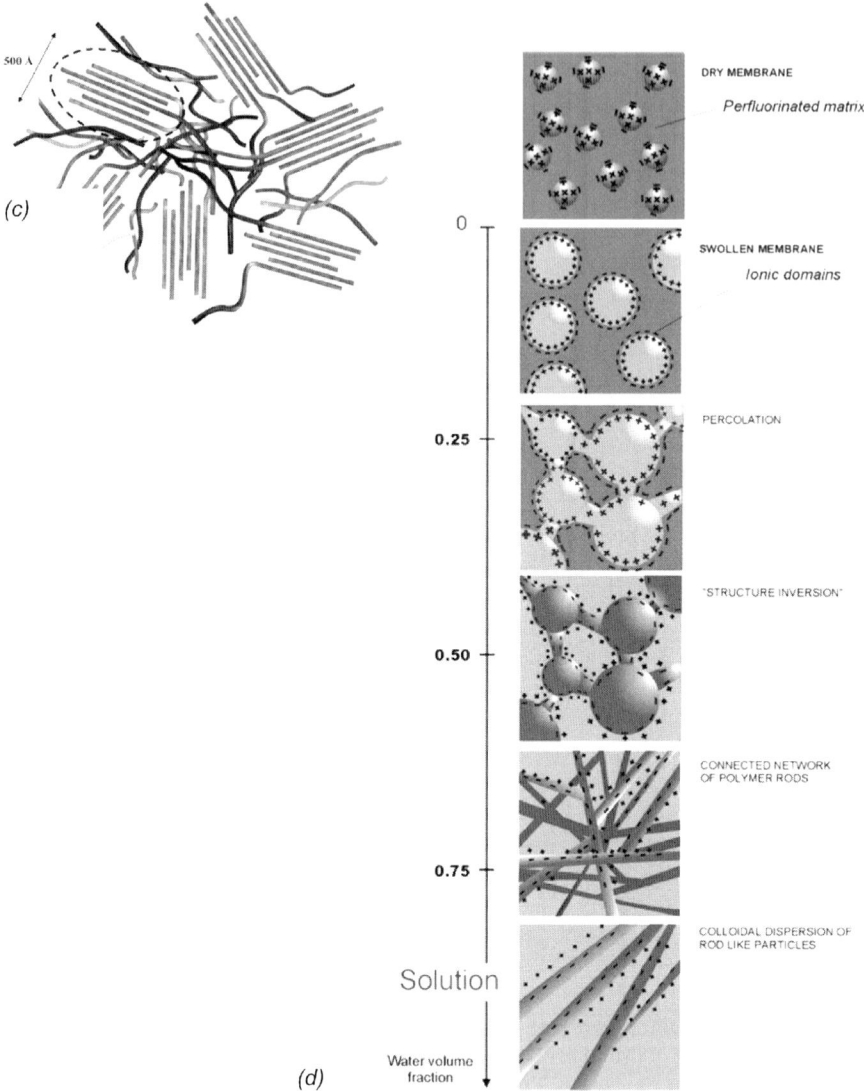

Fig. 1 c Entangled network of rodlike aggregates in Nafion®. Reproduced with permission from [35]. © (2004) American Chemical Society. The position and orientation within an aggregate is characterised by the ionomer peak, and the correlation length by the USAS upturn in (**b**). **d** Conceptual model for reorganisation and continuity of the ionic domains as the dry membrane is swollen with water to the state of complete dissolution. Reprinted from [39]. © (2000), with permission from Elsevier

SAXS profiles, the low angle maximum associated with scattering from well-organised crystal domains was absent in the former, the structure being essentially amorphous. Nafion® recast from low-boiling solvents retains a colloidal morphology with little or no entanglement between polymer aggregates [40, 41] (Fig. 1c). Only a history of high-temperature processing enabled the polymer chains to reorganise and entangle, and ultimately to give a sufficiently robust membrane. For detailed background on this topic, the reader is referred to the excellent review on the state of understanding of Nafion® by Mauritz and Moore [3]. In addition, some 10 years ago a description of Nafion®-117 membranes was formalised according to whether they had been submitted to heat treatment at 80, 105 or 120 °C (normal, shrunken and fully shrunken forms, respectively), or to no pre-treatment (expanded form) [42], and their water content and associated proton conduction properties elucidated for these various forms. No equivalent description has been agreed upon, however, for recast Nafion®.

Studies of the morphology of non-fluorinated ionomers are far less extensive, and although description of their microstructure is made in the same general terms, there are central differences of detail [1]. In hydrocarbon-based ionomers, the sulfonic acid groups are less acidic than those attached to a perfluorocarbon structure, the polymer chain is less hydrophobic and the backbone is less flexible, due to the presence of aromatic groups. These factors all contribute to reducing the extent of nano-phase separation between ionic cluster regions and hydrophobic domains, which has a direct effect on proton transport properties [43, 44], in particular by a stronger dependence of proton conductivity on water content than is observed for PFSA membranes [9]. With regard to inorganic–organic membranes prepared with perfluorinated versus non-fluorinated polymers and polymer membranes, the narrower ionic/water-containing channels and the higher degree of branching of these channels, as well as the presence of cul-de-sac tunnels in non-fluorinated systems [27], will all impact the distribution of inorganic material throughout the nanocomposite structure and, ultimately, the extent to which co-continuous inorganic and organic networks may be produced.

2.2
Preparation and Characterisation of Hybrid Inorganic–Organic Membranes Using Pre-Formed Membranes

The preparation of nanocomposite membranes by intra-membrane growth within a proton exchange membrane was first described by Mauritz et al. [45–47]. The then novelty of this approach and the breadth and depth of these studies warrant the following discussion of the results, which in many ways laid the foundation for future work in this area. This group made use of the hydrophilic ionic cluster regions of Nafion® for confined, sulfonic acid group catalysed, hydrolysis/condensation reactions of impregnated alkoxides. Nafion® membranes were first swollen in ethanol/water, then tetraethoxysilane (or aluminium, titanium and zirconium alkoxides) permeated from one side of the membrane. In addition to the concentration profile of inorganic oxide across the thickness of the membrane created by this approach, tailoring of compositional gradients was achieved at the nano-scale by shell-like construction of ceramic particles by successive permeation of metal alkoxides [48]. Characterisation of the nanocomposite membranes by FTIR [49], thermal analysis, dielectric relaxation and, in particular, SAXS is compatible with the hypothesis that Nafion® acts as a morphological template for growth of the inorganic phase. Following from the discussion of Sect. 2.1, investigation using SAXS is of special interest. The earliest studies showed that the ionomer peak is still observed for Nafion®–$SiO_{2(1-x/4)}(OH)_x$ membranes, and the original Bragg spacing, most often associated with the average inter-cluster spacing, is unchanged, despite the invasion of the polar regions by the sol–gel derived silicon oxide phase [50]. It was further concluded that although the inorganic phase was to be found predominantly as isolated clusters, some of these were linked by silicon oxide bridges, presumed to be formed in the later stages of reaction. Pursuing the idea of using the "filling" of the ionic cluster and channel regions to probe the Nafion microstructure, it was further reported that when incorporated silicon oxide (with many surface silanol groups) is post-reacted with monofunctional ethoxytrimethylsilane or with difunctional diethoxydimethylsilane, the SAXS spectrum obtained for the latter no longer shows the presence of an ionomer peak, owing to the lowered electron density contrast resulting presumably from the formation of an inorganic phase co-continuous with the perfluoropolymer network [51]. Unfortunately, no transmission electron microscopic (TEM) images are available of these first nanocomposite membranes.

Of relevance to fuel cell use is the observation that water uptake from liquid water is higher for the Nafion®–silicon oxide membranes than for unmodified Nafion, which was attributed to the presence of a large number of accessible $\equiv$SiOH groups. These nanocomposites exhibit progressive material strengthening with decreasing elongation to break, followed by a ductile to brittle transformation that occurs with increasing silica content. Finally, in membranes permeated simultaneously with titanium and silicon alkoxides and from both membrane surfaces, a glassy zone of titania located in the immediate sub-surface regions was considered responsible for the inferior mechanical properties of these composites compared with those containing only silica.

Despite the careful examination of the structure and properties of Nafion®–metal oxide and ORMOSIL nanocomposites, only partial electrical characterisation had been reported until more recently, some of which was in a patent dating from the same period [52]. The proton conductivity of Nafion®–silicon oxide membranes prepared according to the above protocols has been described as being slightly lower at room temperature than that of Nafion®-115 (undefined RH), and decreasing with increasing silica content [53], while the methanol permeability also decreases. Another measurement on a sample in which the silica content was not specified was higher than that of Nafion® at 80 °C [54]. Characterisation in DMFCs [53] using a 2 M methanol feed suggested that membranes containing 12 wt.% silicon oxide gave better performance than those with lower or higher amounts, giving increasing power density at temperatures up to 125 °C, but the study was unfortunately flawed by the lack of any comparison with unmodified Nafion®-115. More complete fuel cell characterisation has been performed by Bocarsly et al. [55, 56], who compared the performance of various PFSA-type–silicon oxide membranes prepared using the above methodology: Nafion®-115, -112, -105 and Aciplex-1004, allowing assessment also of the role of membrane thickness and PFSA equivalent weight (EW) (Aciplex-1004, Nafion®-105: EW 1000 g mol^{-1}; Nafion®-112, -115: EW 1100 g mol^{-1}; Nafion®-115, -105: thickness 125 µm; Aciplex-1004: 100 µm; Nafion®-112: 50 µm). At operating conditions of 3 atm, cell temperature 140 °C and humidifier temperatures at the anode and cathode of 130 °C, the order of performance for the unmodified membranes was Aciplex-1004 > Nafion®-105 > Nafion®-112 > Nafion®-115. This tendency can be rationalised on the basis of the above physical/chemical characteristics, the resistivity of the cell decreasing as the EW and the thickness of the membranes decreases, and the former playing a larger role in maintaining proton conductivity at high temperature than membrane thickness. Using nanocomposite PFSA–silicon oxide membranes, the trend becomes Aciplex-1004 > Nafion®-112 > Nafion®-105 > Nafion®-115 (at both 130 and 140 °C, with humidifiers at 130 °C), suggesting that, since silicon oxide assists water management in all systems under high-temperature PEMFC operation, more emphasis is placed on membrane thickness than the density of sulfonic acid groups. In addition, the current density of all the silica-containing membranes was stable over 50 h when operated at constant 0.65 V potential, while that of the reference unmodified Nafion®-115 dropped rapidly and gave no current after 1 h, indicating that the positive influence of the inorganic component was not only transitory.

Impregnation relies on the ability of molecular species to diffuse into the polar regions of ionomeric membranes, and is driven by the affinity of polar/charged aggregates for the ethanol/water environment of the pre-swollen films. The acidic sites act as catalysts for hydrolysis/condensation, but do not otherwise participate in the reaction. In a different scheme, further advantage may be derived from the specific character of ionomer membranes by using the sulfonic acid sites in ion-exchange reaction with a medium containing metal cations. On conditioning the ion-exchanged membrane in a solution of an appropriate counter-ion, the inserted metal ions act as centres for local particle growth. In this scheme, the existing clustered morphology of the membrane is still used as a template, confining development of the inorganic "sub-lattice" to the hydrophilic regions of the film. This approach was first extensively described for the preparation of sulfonated poly(ether ether ketone) (sPEEK)–metal(IV) phosphate nanocomposite membranes [57], following the first report by Grot and Rajendran on its use to prepare composites with Nafion® [52]. Process variables include the concentrations of metal salt (and the nature of the salt) and phosphoric acid solutions used for ion-exchange reaction and precipitation of the metal phosphate, respectively, membrane pre-swelling in alcohol/water mixtures, as well as the temperature and reaction times. A particularly crucial point is to choose experimental conditions that favour metal(IV) phosphate formation in the membrane, over reverse ion-exchange and return of the membrane to its acid form.

Metal hydrogen phosphates and phosphonates have been extensively studied for their proton conduction properties [58–61], and their surface acidity, chemical stability in an acidic environment, insolubility in water, the versatility of preparation routes and understanding of the influence of preparation conditions on crystallite size and morphology are all salient characteristics that have led us to suggest that these are probably the most appropriate inorganic proton conductors for PEMFC and DMFC applications [25]. Zirconium phosphate exists in different crystalline arrangements, but only the so-called α structure is formed under the conditions prevailing in nanocomposite membrane preparation. α-Zr(HPO$_4$)$_2$·H$_2$O is formed by reaction of an aqueous solution of M(IV) ions with phosphoric acid. In general, the size of the crystallites increases with the concentration of phosphoric acid used, the temperature of the reaction medium and the duration of the reaction. α-ZrP can be prepared by heating the amorphous form in concentrated phosphoric acid [62], by direct precipitation from a solution of Zr(IV) in a mixture of phosphoric and hydrofluoric acids [63], or by complexation of zirconium propoxide (or other metal organic compound) by phosphoric acid, followed by heat treatment in concentrated phosphoric acid [64]. Similar routes may be followed to prepare α-titanium and α-tin phosphates. Crystalline α-ZrP, α-TiP and α-SnP have a layer structure, in which hydrogen phosphate groups are organised on the upper and lower surfaces of a plane formed by the metal atoms. Of particular consequence in the context of ZrP as a component of hybrid membranes is the fact that its conductivity depends upon the degree of crystallinity and particle size; for the crystalline compound, the conductivity is far too low for interest (10^{-7} S cm^{-1}), but in the amorphous material it reaches 10^{-3} S cm^{-1} [58]. Proton conductivity strongly depends upon RH.

In our work, sPEEK membranes of polymer equivalent weight ca. 770 g mol^{-1} (ion-exchange capacity 1.3 meq g^{-1}) and Nafion®-117 membranes were immersed in an aqueous solution of either 5 M zirconyl or tin(IV) chloride for 6 h at 80 °C, rinsed and transferred to phosphoric acid baths of 1–14 M at 80 °C where they were kept for a duration of 10 min to 7 days, in order to determine the influence of these parameters on the metal(IV) phosphate formed. After washing with boiling water to remove surface-bound phosphoric acid, the ion-exchanged membranes sPEEK-Zr and sPEEK-Sn contained 11 wt.% Zr and 17 wt.% Sn, respectively. Just as for the bulk metal(IV) phosphates, the concentration of H$_3$PO$_4$ used at the precipitation stage affects both the amount of metal(IV) phosphate formed and the mole ratio P/M(IV)—expected to be 2 from the above chemical formula (Table 1). It is observed that this ratio is

Table 1 Composition (elemental analysis) of sPEEK–metal phosphate membranes in terms of mass percent of metal phosphate formed in situ, and the metal to phosphorus ratio

Concentration of phosphoric acid (mol L^{-1})	Duration of acid treatment at 80 °C (h)	Amount of M(HPO$_4$)$_2$·H$_2$O in hybrid membrane (wt. %)	Mole ratio Zr : P
sPEEK–zirconium phosphate membranes			
1	15	25	1 : 2.1
5	15	10	1 : 2.1
7	15	12	1 : 2.6
14	15	0.5	1 : 10.8
Expected			1 : 2.0
sPEEK–tin phosphate membranes			
1	60	25	1 : 0.86
7	60	ppm	–
1	100	21	1 : 1.7
7	100	ppm	–
Expected			1 : 2.0
Nafion®-117 membranes			
1	15	18	1 : 2.1

attained in the hybrid membranes only when phosphoric acid of 1–5 M is used, and with use of acid of higher concentration, the final membranes are greatly enriched in phosphorus, symptomatic of a reverse ion-exchange process. Zirconium diffuses from the membrane and into the surrounding acid medium, while the membrane becomes progressively swollen with phosphoric acid. The reverse ion-exchange process no longer dominates when lower concentrations of phosphoric acid are used, and it is under these conditions that hybrid membranes most highly loaded in metal(IV) phosphate are formed: up to 25 wt. % zirconium phosphate when immersed in 1 M H$_3$PO$_4$. Membranes ion exchanged with Sn(IV) show similar trends, but the rate of formation of tin phosphate is much slower (as for the equivalent bulk materials) [65]; longer reaction times are needed for formation of a phosphate with P/Sn ratio close to 2, and after 100 h of immersion the membrane contains 21 wt. % SnP. It should be noted that P/M(IV) ratios < 2 correspond to hydroxyphosphate derivatives of metal(IV) phosphates, and since these are less stable hydrolytically and, in consequence, less suitable for fuel cell application, it is indispensable to understand the relation between the preparation conditions and the chemical nature of the inorganic material formed.

The presence of an inorganic phase broadens the range of applicable characterisation techniques to those sensitive to heavier atoms and/or that are specific probes for the environment of a given atom. ^{31}P MAS NMR is one such method that probes the degree of molecular connectivity within the

 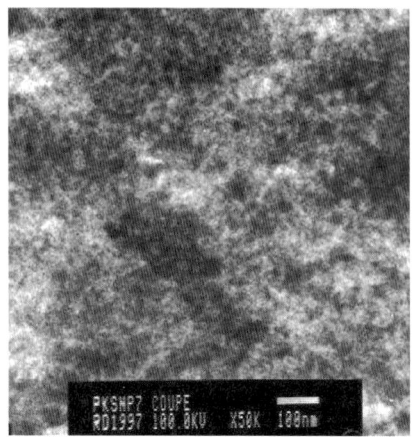

Fig. 2 Preparation of hybrid membranes by ion exchange/precipitation: transmission electron micrographs of **a** sPEEK incorporating zirconium phosphate particles (20 wt.%), **b** hybrid membrane of sPEEK incorporating tin phosphate particles (20 wt.%). Magnification 50 000 ×

metal(IV) phosphates. Applied to sPEEK–ZrP and sPEEK–SnP, it reveals differences in the phosphorus environment in the two nanocomposites that can be related to particle size and morphology. A single resonance at – 18 ppm in the spectrum of sPEEK–ZrP is characteristic of an environment HO–P–O_3, while the appearance of three signals (at – 13.4, – 8.4 and – 6.0 ppm) in the spectrum of sPEEK–SnP indicates the presence also of $(HO)_2$–P–O_2 groups [66]. This conclusion is compatible with the chemical analyses described above, and the availability of pendant hydroxyl groups also implies that SnP particles have greater surface/interface than ZrP particles. From direct observation of the nanocomposite membranes using TEM (Fig. 2), the particle sizes of ZrP and SnP are estimated as 15–30 and 5–10 nm, respectively, with particles of SnP being more isotropic in shape than the platelet morphology of ZrP. Diffraction lines from zirconium phosphate are broader in Nafion®–ZrP membranes than in sPEEK–ZrP, and the first diffraction line corresponding to the interlayer spacing is very weak or absent. Reflections (100) and (101) from crystalline regions of the PFSA backbone are also observed. The particle size observed in Nafion®–ZrP by TEM is smaller, ca. 10 nm (Fig. 3).

X-ray absorption spectroscopy has been very little applied as a tool for the characterisation of proton exchange membranes, and yet the element-specific information on the nature and number of nearest neighbours and interatomic distances it is capable of providing on disordered materials is unique. Rozière and Jones et al. [67, 68] made use of extended X-ray absorption fine structure (EXAFS) and X-ray absorption near edge structure (XANES) spectroscopies to investigate the local environment of Zr in the

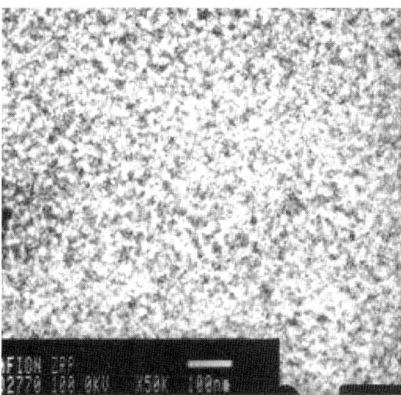

Fig. 3 Preparation of hybrid membranes by ion exchange/precipitation: transmission electron micrograph of Nafion®-117–ZrP. Magnification 50 000 ×

ion-exchanged membranes, and follow the evolution of the local structure in sPEEK–ZrP and Nafion®-117–ZrP membranes as a function of duration of reaction of the ion-exchanged membranes with phosphoric acid (1 M, 80 °C). The amount of metal(IV) phosphate that can be formed in a one-step process depends on the uptake of metal ions in the ion-exchange step, itself dependent on the ion-exchange capacity (IEC) of the ionomer membrane, and the charge on the metal ions. In this context, the first result of interest concerns the nature of the species exchanged from aqueous $ZrOCl_2$ to the sulfonated polymer membrane. In its crystalline form, zirconyl chloride adopts a structure in which zirconium is bound into complex tetranuclear cations $[Zr_4(OH)_8(H_2O)_{16}]^{8+}$ [69] (Fig. 4). Studies using

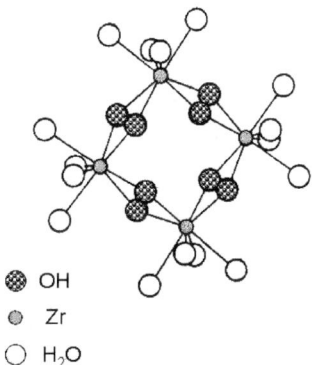

Fig. 4 Tetrameric zirconium species, formally $[Zr_4(OH)_8(H_2O)_{16}]^{8+}$, solid $ZrOCl_2$. A Zr species of this geometry is identified by EXAFS spectroscopy in sPEEK and Nafion®-117 membranes ion-exchanged from $ZrOCl_2$ solution

Table 2 Coordination numbers and interatomic distances determined from EXAFS spectroscopy of ion-exchanged membranes Nafion®-117-Zr, sPEEK-Zr and crystalline ZrOCl$_2$

		N_i^a		R_i^b (Å)
Nafion®-117-Zr	N_1	7.9	R_1	2.21
	N_2	2.3	R_2	3.61
sPEEK-Zr	N_1	8.6	R_1	2.20
	N_2	1.6	R_2	3.59
ZrOCl$_2$	N_1^c	4	R_1	2.29
	N_2	4	R_1' ^{d}	2.15
	N_3	2	R_2	3.59

^{a,b} Coordination number and interatomic distance of ith shell of oxygen atoms. R_1, R_2 are the interatomic distances from zirconium to the first and second shells of oxygen atoms, respectively
^{c} Four atoms at 2.29 Å and four at 2.15 Å give an average of eight atoms at 2.22 Å
^{d} In ZrOCl$_2$, there are two Zr–O distances in the first coordination shell, denoted R_1 and R_1'. Estimated errors are 0.02 Å and 20% on R_i and N_i, respectively.

SAXS conclude that this tetrameric species also exists in aqueous solutions of Zr(IV) [70]. EXAFS spectroscopy of sPEEK-Zr and Nafion®-117-Zr membranes shows the immediate environment of Zr to comprise two distinct shells of oxygen at 2.20 and 3.59 Å. To within the accuracy of the EXAFS method, this local structure matches that in crystals of ZrOCl$_2$ (Table 2) and it is concluded that tetrameric zirconium species are exchanged into the ionomer membranes. In solution the charge on this tetramer can deviate from the formal 8+, and an estimate can be made using the polymer IEC and the weight of inorganic material formed in situ. In the case of sPEEK with IEC of 1.3 meq g^{-1}, an average charge per zirconium of 2+ (as in [Zr$_4$(OH)$_8$(H$_2$O)$_{16}$]$^{8+}$) would allow formation of a maximum amount of 16 wt. % of ZrP in a one-step process. Higher amounts can be formed using multiple exchange–precipitation steps with regenerated exchange sites. Experimental results indicate that, in a single exchange–precipitation step, up to 25 wt. % of ZrP is formed, which implies that the average charge is lower, closer to 1.3+. In Nafion®-117 membranes the amount of ZrP formed in a single step is ca. 18 wt. %, which corresponds to the expected amount for the membrane IEC if the average charged species is the same in the two ionomeric membranes. In conclusion, zirconium is not exchanged as simple Zr^{4+} ions (as is written in some recent reports), neither in polyaromatic sPEEK nor in polyfluorosulfonic acid Nafion®-117 membranes, but as a part of a tetranuclear species in which the average charge per Zr is ca. 1.3.

Figure 5a displays spectra in the X-ray absorption near edge region given by sPEEK-ZrP samples after immersion in H$_3$PO$_4$ for periods of 10 min–

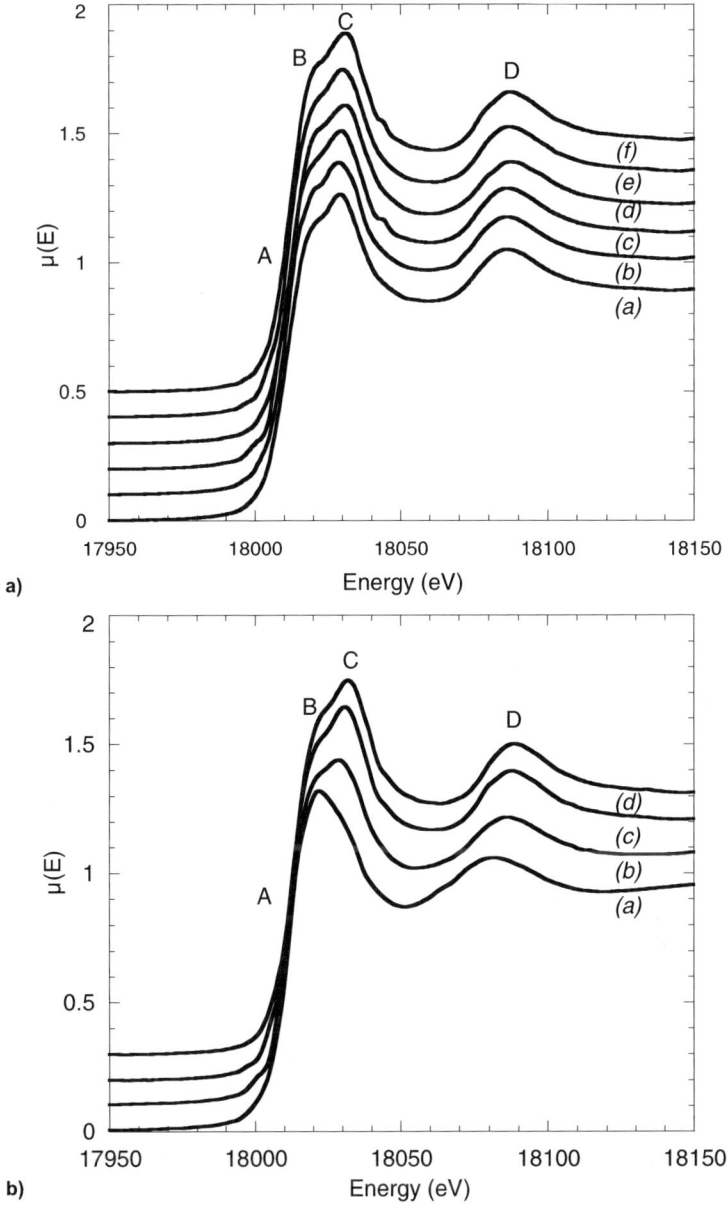

Fig. 5 a X-ray absorption spectra in the near Zr-edge region of sPEEK–Zr membranes after immersion in H_3PO_4 (1 M, 80 °C) for various times: (a) 5 min, (b) 10 min, (c) 40 min, (d) 60 min, (e) 4 h, (f) 15 h. *A* Rising edge; *B* near-edge feature; *C* absorption maximum; *D* first post-edge oscillation. **b** X-ray absorption spectra in the near Zr-edge region of Nafion®–Zr membranes after immersion in H_3PO_4 (1 M, 80 °C) for various times: (a) 10 min, (b) 40 min, (c) 4 h, (d) 15 h. *A* Rising edge; *B* near-edge feature; *C* absorption maximum; *D* first post-edge oscillation

15 h. While the absorption edge (A) does not change in position, the absorption maximum (C) and the first oscillation (D) both increase in intensity at longer reaction times, and a shoulder (B) on the rising absorption edge becomes more marked. These features are the expression in XAS of the transition from an eight-coordinate arrangement (eight nearest neighbours in $[Zr_4(OH)_8(H_2O)_{16}]^{8+}$) to the six-coordinate environment of α-$Zr(HPO_4)_2 \cdot nH_2O$, the onset of which begins very rapidly, even after 5 min immersion in H_3PO_4. Before discussing the corresponding EXAFS region of the spectra for sPEEK–ZrP membranes, a short detour will be made to examine the near-edge spectra of membranes based on Nafion®-117 (Fig. 5b). Here the absorption maximum (C) and the first oscillation (D) both move to higher energies (and become more intense) at longer reaction times, and the shoulder (B) grows in on the rising absorption edge; it is clear that with Nafion® the evolution from eight to six coordination is noticeably slower, beginning after ca. 40 min, while no further evolution in the spectrum is seen after 4 h. These differences could simply arise from the difference in membrane thickness (sPEEK, 50 μm; Nafion®-117, 250 μm), but it is more likely that re-protonated sulfonic acid sites in sPEEK exert less influence by Donnan exclusion than the superacid sites in Nafion®, and allow faster diffusion kinetics.

From numerical analysis of the EXAFS region of the spectra given by sPEEK–ZrP and Nafion®-117–ZrP (15 h of reaction in H_3PO_4), it is deduced that zirconium is surrounded by six oxygen atoms at 2.05 Å ($r(Zr-O)$ = 2.07 Å in α-ZrP), a second shell of oxygen at 3.59 Å ($r(Zr-O)$ = 3.43 Å in α-ZrP) and a shell of zirconium atoms at 5.35 Å ($r(Zr-Zr)$ = 5.29 Å in α-ZrP). Although the coordination environment corresponds to that of α-ZrP, the slightly longer interatomic distances indicate the structure to be more relaxed, with less long-range interaction, which is consistent with the presence of nanoparticulate zirconium phosphate. An earlier EXAFS study of zirconium phosphates in X-ray amorphous and (partially) crystalline forms [71] is of relevance in this context. Further strong evidence for relaxation from the positions expected in an extended organisation, as in ZrP crystals, is deduced from the emergence of an intense signal from the shell of phosphorus atoms ($r(Zr-P)$ ca. 3.5 Å). For crystalline α-ZrP, the intensity of this maximum in the radial distribution-like function is vanishingly small, due to destructive interference of the EXAFS oscillations singly backscattered from phosphorus, with oscillations scattered in a multiple scattering path involving the first oxygen atom shell (Zr–O–P–O–Zr). Since the extent of this destructive interference strongly depends upon the Zr–O–P angle, the intensity of the maximum due to back-scattering from phosphorus in the radial distribution function allows us to calculate approximately that the Zr–O–P angle in ZrP confined in sPEEK is ca. 170° [68], which can be compared with 155° in crystalline α-ZrP. The radial distribution-like functions obtained by Fourier transformation of the EXAFS spectra of nanocomposite membranes sPEEK–

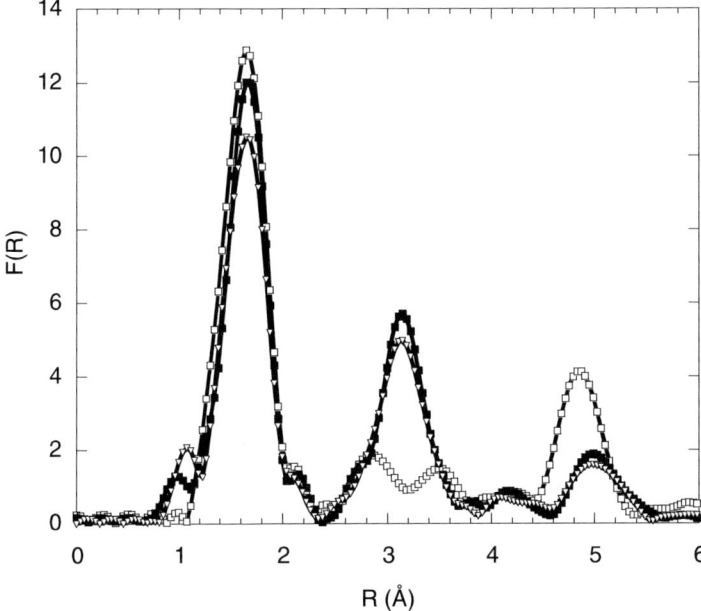

Fig. 6 Fourier transformed EXAFS spectra of sPEEK–ZrP membranes prepared by contact of ion-exchange membranes with H_3PO_4 (1 M, 80 °C) for 5 min (▽) and 15 h (■), compared with that of well-crystallised α-ZrP prepared using the "HF" method (□)

ZrP and Nafion-117–ZrP obtained after initial immersion (5 min) and 15 h of immersion in phosphoric acid are compared in Fig. 6 with those given by bulk, well-crystallised α-ZrP prepared in an H_3PO_4/HF medium. These spectra have not been corrected for phase shifts, and thus the interatomic distances are skewed to lower values by about 0.3 Å.

The proton conductivity of sPEEK–ZrP membranes (measured across the film thickness) has low temperature dependence at 100% RH, increasing from 2 to 5×10^{-2} S cm^{-1} in the range 20 to 100 °C. Using membranes containing up to 25 wt. % ZrP in MEAs operated under hydrogen/oxygen and hydrogen/air fuel cells [25, 29, 57], slightly higher cell voltages were achieved at 100 °C with hybrid systems than with unmodified sPEEK membranes of the same thickness, the current density (H_2/O_2) at 0.6 mV being > 1 A cm^{-1} (Fig. 7).

Following the above preparation protocol, Nafion®–ZrP membranes have also been extensively characterised by Yang, Bocarsly et al. [72–74] and Willert-Porada [75, 76], in particular with regard to fuel cell performance and microstructural properties. In these studies either commercial Nafion®-117 [68, 75] or -115 [72], or recast Nafion® [73] films have been used, Nafion being re-formed from alcoholic solution and thermally treated at 160 °C before use. The starting membranes in all studies are thus expected to be simi-

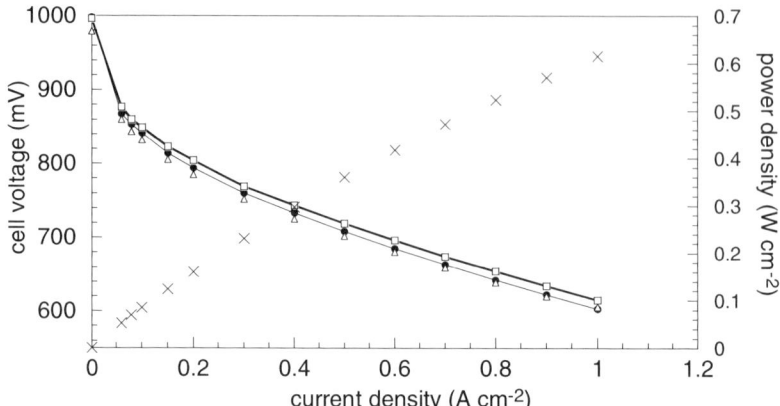

Fig. 7 Polarisation characteristics of sPEEK and sPEEK-ZrP (50 μm) prepared by ion-exchange/precipitation with sPEEK membranes. The electrodes employed in these MEAs were standard 1 mg Pt cm^{-2} E-TEK, Nafion® dispersion was used at the membrane/electrode interface, and the cell was pressurised to 2.6 bar abs. sPEEK at 85 °C (△); sPEEK-ZrP at 85 °C (●) and sPEEK-ZrP at 100 °C (□)

lar, since high-temperature processing re-establishes a microstructure that is comparable to that of commercial extruded films [41].

The morphology of hybrid Nafion®-ZrP membranes has been investigated recently using SAXS [74]. From these studies, it was deduced that ionic clustering persists in the hybrid system, and that the characteristic spacing between clusters in fully hydrated Nafion®-115 and fully hydrated Nafion®-115-ZrP is similar. On the other hand, whereas with decreasing hydration number the inter-cluster spacing in Nafion®-115 diminishes (scattering maximum shifts to smaller spacing with decreasing water content), it remains unmodified in the hybrid membrane. This would imply that the ZrP particles act as an internal prop to bolster and stabilise the ionic cluster volume, which is then less susceptible to changes in degree of hydration. Observations at this length scale can be related to the well-documented macroscopic dimensional change in Nafion® with water uptake. Swelling/contraction effects rapidly induce mechanical fatigue, and finding reliable means of limiting hydration/dehydration is a key challenge in the development of new membrane materials. Finally, Nafion®-ZrP membranes in the dry state give no SAXS signal; the electron densities of the perfluorocarbon backbone and the inorganic inclusion are sufficiently similar that no scattering peak is observed, from which it is concluded that the formation of zirconium phosphate particles extends throughout the hydrophilic regions of the ionomer membrane. However, such conclusions relate only to membranes in which the above preparation protocols have been rigorously followed. In another recent study, Nafion®-117 ion exchanged with ZrOCl$_2$ was immersed in phosphoric acid for only 1 h [75], which leads to membranes displaying a distinct concentration profile that decreases from the membrane surfaces to the centre, even when successive exchange/precipitation cycles were made, as determined from SEM-EDX. In addition, no analysis of the phosphorus to zirconium ratio was provided; the chemical analyses described above indicate that a zirconium oxophosphate is the dominant inorganic species in Nafion®-117 based membranes for immersion times < ca. 4 h [68]. In any case, the properties of these membranes are expected to be different from those where zirconium phosphate is precipitated throughout the membrane thickness. From other recent SAXS studies, it was concluded that the presence of ZrP does not modify the structure of the pre-formed Nafion® membrane [77].

The hydration number (λ) under saturation conditions is identical for Nafion® and Nafion®-ZrP membranes for water uptake from the liquid phase ($\lambda = 25$ at 25 °C), while a significant difference was reported for uptake from the vapour phase, with λ of 11 for Nafion®-115 and 19 for Nafion®-115 at 80 °C, with no change at higher temperatures up to 140 °C. As for sPEEK-ZrP membranes, RH affects the proton conductivity of Nafion®-ZrP membranes to a much larger extent than temperature. In all cases, the hybrid membranes had a somewhat lower conductivity than unmodified Nafion®, and this despite the increased total IEC, although this distinction is blurred above a water activity of ca. 0.8 (Fig. 8). On this basis alone, the promise of Nafion®-ZrP for fuel cell use would appear to be mediocre, and it provides a good illustration of the importance of in situ characterisation to fully assess a membrane's potential. Yang, Bocarsly, Srinivasan et al. [73, 74] have compared the fuel cell performance (H_2/O_2) of Nafion®-115-ZrP membranes with

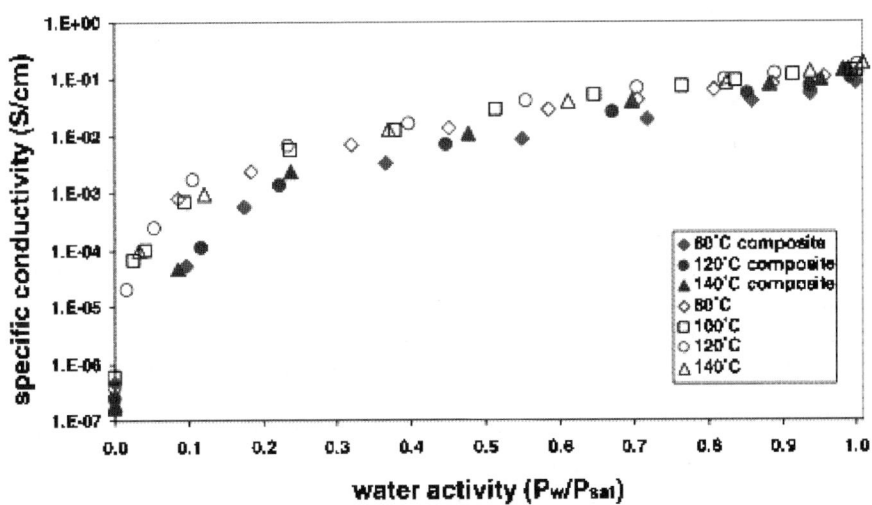

Fig. 8 Comparison of conductivity at 80–140 °C of Nafion® (*open symbols*) and Nafion®-ZrP membranes (*filled symbols*). Reprinted from [74]. © (2004), with permission from Elsevier

that of Nafion®-115 under various conditions of cell and gas humidification temperatures, and pressure. The results clearly show superior performance for the hybrid membrane at high temperature, under conditions where RH is < 100%, whereas at 80 °C and full humidification, the performance was similar or slightly lower (Fig. 9). Determined from the fuel cell data, the membrane resistance for Nafion®-115 increased by a factor 50 when the RH was reduced from 100 to 70%, very much more than the increase from 0.20 to 0.22 Ω cm^2 that was expected from ex situ resistance experiments. In contrast, the membrane resistance of Nafion®-115–ZrP derived from fuel cell tests only increased by a factor of 3. It was suggested that for Nafion®-115, swelling of the membrane in the fuel cell generates internal pressure that can cause a change of state of the membrane [78]. The presence of zirconium phosphate provides mechanical support which resists compression in the fuel cell MEA and the nanocomposite membrane can take up water and swell [74]. Nafion®–ZrP or, more generally, inorganic–organic membranes, are able therefore to gain more advantage from available water in an operating fuel cell than corresponding polymer-only membranes, the underlying key factor being the reinforcement effect of the inorganic phase. The effect of mechanical constraint is amplified at higher temperature in Nafion®-based membranes because the elastic modulus decreases with increasing temperature, and the pressure needed is greatly reduced above the glass transition temperature. Nafion®–ZrP membranes prepared by ion-exchange/precipitation have also been operated in DMFCs up to 150 °C [72, 76], and the results are compared in Sect. 4 with those obtained using nanocomposites containing ZrP and elaborated by in situ preparation in a polymer solution.

A further degree of organisation and structuring of the inorganic and organic components has been introduced in recent work when porous ionomer membranes have been used as the host for nascent zirconium phosphate [21, 79–81]. Amongst other methods, nanosized pores may be generated by solvent extraction of a dispersed porogen from a cast membrane, or by phase inversion of a cast polymer solution, and since the pores formed increase the free volume available for formation of ZrP, this is one means of increasing the proportion of the inorganic phase in a composite system. However, it should be considered also that the nanostructure of the porous film formed will depend on the method used for pore generation; a hydrophobic porogen, such as dibutyl phthalate, will locate preferentially in the hydrophobic regions of a sulfonated ionomer and lead to pores in these regions [80], while the use of spinodal or binodal phase transformation of a cast sulfonated ionomer by water vapour induced phase separation (VIPS) will lead to preferential pore formation in hydrophilic regions [21, 82], and ultimately the properties of the membranes are not expected to be the same. However, in both cases, immersion of porous sulfonated ionomers in aqueous $ZrOCl_2$ leads to ion exchange with the protons of sulfonic acid groups and, in addition, impregnation into the micron-sized pores induced by processing, and on subsequent phosphoric

Development of Inorganic–Organic Membranes for Fuel Cell Applications

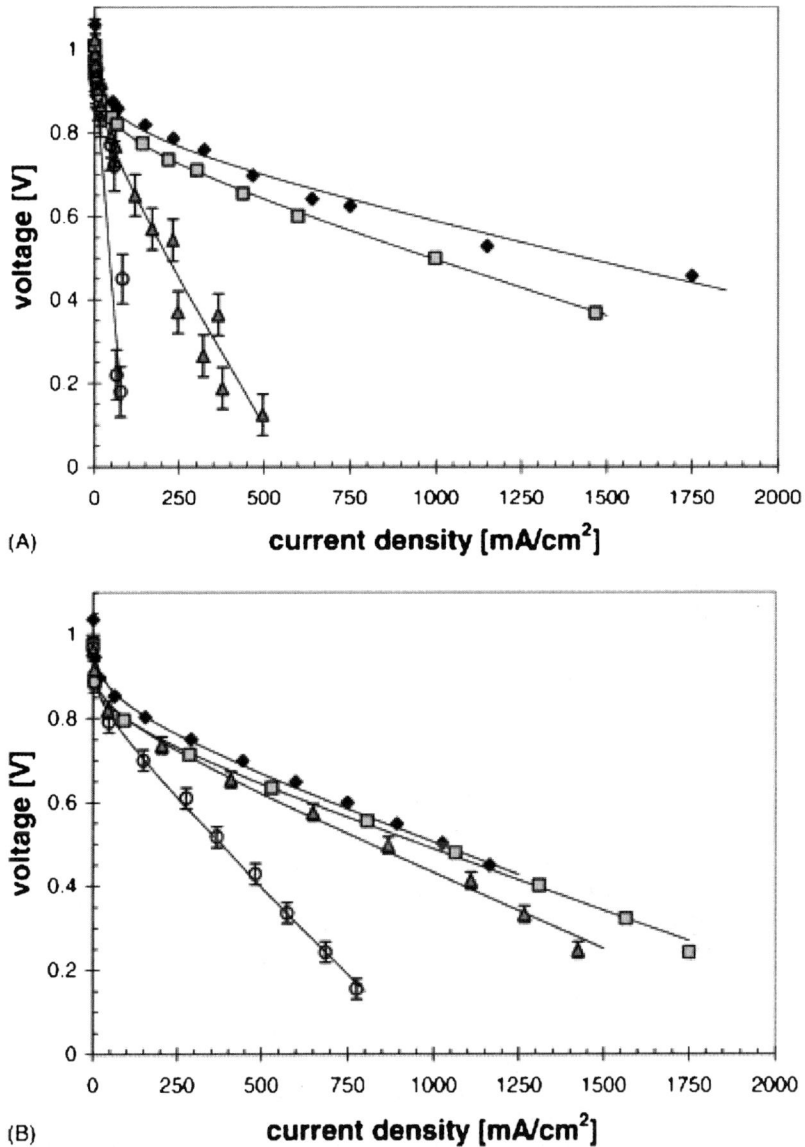

Fig. 9 Comparison of fuel cell performance given by Nafion-115 (**a**) and Nafion-115–ZrP (**b**) membranes. The cell voltage is plotted as a function of the average current density in the fuel cell. Operating conditions are: (♦) $P = 1$ bar, $T_{anode}/T_{cell}/T_{cathode} = 90/80/88\,^\circ$C; (□) $P = 3$ bar, $T_{anode}/T_{cell}/T_{cathode} = 130/120/130\,^\circ$C; (▲) $P = 3$ bar, $T_{anode}/T_{cell}/T_{cathode} = 130/130/130\,^\circ$C; (○) $P = 3$ bar, $T_{anode}/T_{cell}/T_{cathode} = 130/140/130\,^\circ$C. Reprinted from [74]. © (2004), with permission from Elsevier

acid treatment, ZrP is formed both in the hydrophilic regions of the polymer and precipitated into to polymer micropores. In the case of porous sPEEK prepared via VIPS (*por*-sPEEK), ZrP is located in the hydrophilic pore walls and, depending (amongst other factors) upon the membrane pore size, either completely fills the pores or coats their internal surface to give an eggshell-like arrangement (Fig. 10).

On a macroscopic level, the most significant observation is that composite membranes of *por*-sPEEK–ZrP show no dimensional change with water uptake, and the polymer swelling that occurs on water uptake is absorbed by changes in pore volume and not by an increase in membrane dimension. The spongy character of porous ionomer membranes allows very high water uptake that is no longer related only to hydration of sulfonic acid groups. These microstructural properties of *por*-sPEEK–ZrP and porous Nafion®–ZrP, different from those of dense nanocomposite membranes incorporating ZrP, lead to different conductivity trends. At 130 °C, the conductivity of *por*-sPEEK–ZrP determined on fully humidified membranes remained unchanged as the RH was lowered to 25% over a 24-h period [81, 83]. This

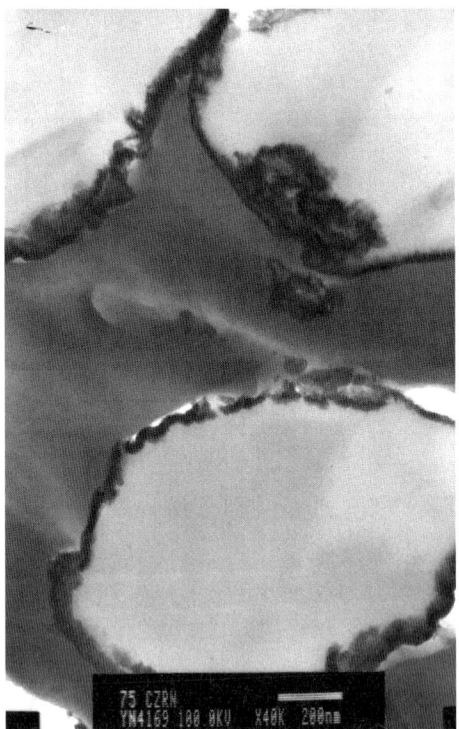

Fig. 10 Transmission electron micrograph of *por*-sPEEK–ZrP, pore diameter ca. 1–2 μm. ZrP of thickness 30 nm coats the internal surfaces of the pores in an eggshell arrangement

delayed response to low ambient RH is of high relevance in the context of automotive fuel cell application, where driving conditions (idling, acceleration, deceleration) lead to load and temperature cycling that generate intermittent low- and high-RH conditions. Membranes that respond rapidly to high RH but have delayed response to low RH are an alternative option to the as yet elusive candidates of target conductivity under equilibrium over the whole range of RH values. The above observations are corroborated by those on porous Nafion® containing 20 and 40 wt. % ZrP where the conductivity at 130 °C after initial saturation of the membrane was $> 10^{-2}$ S cm^{-1} [80]. They are further confirmed by fuel cell characterisation, where the maximum power density of the above porous Nafion®–ZrP was twice that of Nafion®-115 at a cell temperature of 110 °C and humidifier temperature 80 °C, due to reduced areal resistance of the composite membrane. A similar trend is seen for *por*-sPEEK–ZrP membranes, which have better performance than reference Nafion®-115 at 95 °C, even though at lower temperatures their performance was inferior. More extensive in situ characterisation over a fuller range of temperature and RH is required, as well as durability testing, so as to fully explore the potential of inorganic–organic composites based on porous membranes.

In a modification of the Gore-type methodology for composite membrane preparation [84], Si and co-workers have impregnated porous (0.5 μm) Teflon with $ZrOCl_2$ dissolved in a commercial Nafion® dispersion, and treated the resulting imbibed membrane with phosphoric acid [85]. No chemical characterisation of the final membrane is provided, so it not possible to assess the extent of formation of zirconium phosphate or the possible presence of other reaction products under the conditions used. However, the resulting composite has a membrane resistance a factor of 2 lower than that of the corresponding *por*-Teflon-Nafion® prepared in a similar way (0.24 Ω cm^2 at 120 °C/31% RH), and its H_2/O_2 fuel cell performance, although the same as that of reference Nafion®-112 at 80 °C, was higher at 120 °C, even after correction for the difference in membrane thickness. In a related approach, Alberti and Casciola described successive impregnation/drying cycles of porous (0.5 μm, 85% porosity) Teflon. The removal of solvents leads to a composite membrane in which ca. 70–80% of the pore volume is filled with zirconium phosphate sulfophenylphosphonate (see Sect. 4) [86], complete occlusion occurring when the anhydrous inorganic phase is hydrated. Finally, successive impregnation cycles of porous PTFE by Nafion® and then by tetraethoxysilane (TEOS) gives membranes containing up to 3 wt. % of silica that have been characterised in DMFCs [87].

In a different approach again, the open porosity of an ordered inverse silica opal material has been filled by multiple impregnation steps with sulfonated polysulfone (sPSU). In a highly polymer-charged composite, the polymer and silica phases are co-continuous, and the conductivity of a material with mass ratio of polymer to silica of 70:30 was ca. 0.1 S cm^{-1} at 80 °C and 90% RH, using sPSU of IEC > 2 meq g^{-1} [88, 89]. Water swelling/solubility and eventual polymer elution should be addressed, as well as the mechanical properties, but the degree of organisation achieved using this methodology makes the approach an interesting reference for in situ methods to co-continuous network materials (described in Sect. 4).

3
Preparation of Composite Membranes by Addition of an Inorganic Component to a Polymer Solution or Dispersion

In this approach, an inorganic solid, generally in the form of a powder, is added to a polymer dispersion or solution, followed by film casting and solvent removal. Different associations of functionalised/non-functionalised components may be envisaged. Most studies have been carried out on the addition of metal oxide particles and inorganic proton conductors, mainly phosphates and phosphonates, to ionomers, although it is recalled that non-functionalised polymers were used in the first work in this area, necessitating rather high proportions of inorganic proton conductors to achieve percolation, with an ensuing detrimental effect on the membrane mechanical properties. More recently, a class of composites developed by Peled et al. is based upon non-functionalised polymers (polyvinylidene difluoride (PVDF), polytetrafluoroethylene) and ceramic particles (silica, alumina, titania), with impregnated acid as the source of protons [90]. Although dependent on a system capable of providing acid at moderate concentration (3 M) at the anode, these so-called nanoporous proton conducting membranes have given high performance in DMFCs, with a report of 0.5 W cm^{-2} at 130 °C in direct methanol/air [91, 92]. Other composites of PVDF with solid acids show high proton conductivity, without recourse to imbibed acid. Membranes cast from a dispersion of DMF-swollen gels of amorphous zirconium phosphate sulfophenylphosphonate (ZrSPP) and PVDF containing 5–25 wt. % ZrSPP have conductivity values of 0.3–20 × 10^{-4} S cm^{-1} at 120 °C and 90% RH [93], but loss of mechanical stability above 20 wt. % loading prevents higher conductivity from being attained. Water uptake by these membranes is the same, within experimental error, from the liquid and the vapour phase at 90% RH, being ca. 7.3 H$_2$O/sulfonic acid group. The microstructure of PVDF-based composite membranes certainly differs from those with an ionomer matrix, and such behaviour, different from that of ionomer membranes in this respect, is not unexpected. Other composite membranes have been reported incorporating antimonic acid, sulfonated polystyrene and PVDF [94].

The main problem encountered using methodology of a straightforward dispersion of a pre-formed inorganic material in a polymer solution followed by casting is that particles tend to aggregate and form agglomerates of varied size that are distributed non-homogeneously throughout the membrane. This problem can be alleviated in the presence of a specific interaction (hydrogen bonding, ionic interaction) between inorganic and organic components that is favourable for high dispersion and enhances the extent of the inorganic–organic interface. In this context, the nature of the ionomer itself is also a key factor, since the degree of formation of a solution, as opposed to a dispersion, is different for sulfonated polyaromatic polymers and for PFSA-type polymers. Initial "as-made" Nafion® dispersions have a high molar mass shoulder on the mass distribution determined from size exclusion chromatography—low angle light scattering as a result of a process-dependent aggregation phenomenon. This aggregate structure is irreversibly broken down on heating the dispersion to high temperature ($> 230\,^\circ$C) to give a molar mass of ca. 10^5 g mol^{-1}. The Nafion® dispersion particle shape is considered anisotropic, possibly having a rod and/or ribbon form, generally several hundreds of nanometres in size. In fact, current models for Nafion® dispersions make use of elongated, charge-stabilised particles distributed on a three-dimensional lattice with the particle centre of mass at the lattice points [95]. Thus, the addition of nanometre-sized silica to a PFSA dispersion can be considered as giving a mixed suspension of two colloids, one roughly spherical, one of elongated morphology, and with an order of magnitude (at least) difference in elementary "particle" size. The extent of interpenetration of the organic network with the inorganic phase depends upon the strength and the nature of interactions that can be developed: interactions between the hydrophilic sulfonic acid rich regions considered as being present as fringes on the outside of Nafion® rods or ribbons in solvent dispersion and hydroxyl groups at the surface of metal oxide particles, interaction between unsaturated coordination sites on the metal oxide surface and sulfonated groups, and non-specific interactions of the metal oxide with the hydrophobic perfluorocarbon core. In the absence of any specific, few or weak interactions, high-temperature processing is essential, as for recast Nafion® membranes [40], to establish entangling of the PFSA chains with each other and with the inorganic particles, and thus the nature of the polymer in solution or dispersion can strongly influence the microstructure of the ultimate composite membrane.

Some of the first work on Nafion®–metal oxide composites was performed by Watanabe et al. [96, 97] who demonstrated that the addition of small amounts of colloidal titania or silica to a dispersion of Nafion® in alcohols, followed by recasting, led to membranes giving higher current density in an H_2/O_2 fuel cell than Nafion®-112 or recast Nafion® of the same thickness. This observation was attributed to an improvement in water retention properties as a result of interaction of water with the metal oxide surface, both reducing electro-osmotic drag and enhancing back-diffusion of water. In related work, subsequent thermal treatment of dried composite Nafion®–SiO_2 (aerosil 200) and Nafion®–TiO_2 (anatase, rutile) membranes was made at 160 °C [98, 99]. Other authors have developed this approach [100], in particular for application in DMFCs where, although the presence of oxide particles can lead to higher cell resistance, particularly at normal temperatures of operation, this effect can be compensated by a lower rate of methanol permeation due to a barrier effect and increased tortuosity path for methanol crossover.

Water physisorption properties depend on the surface characteristics of metal oxides, surface functional (hydroxyl) groups acting as hydrophilic centres available for hydrogen bonding and dipole–dipole interactions. An approach to quantification of these characteristics can be made via surface acidity properties and the particle size and surface area. Various studies have compared the efficacy of Al_2O_3, TiO_2, ZrO_2 and SiO_2 as additives in composite membranes prepared with Nafion® [99, 101, 102] or other sulfonated polymers such as sPEEK. It has been reported that for membranes based on Nafion®–alumina, the DMFC voltage at low current densities was significantly lower than for those containing zirconia or silica, which suggests inhibition of the electrochemical reactions at the interface by alumina species. This is possibly due to local release of aluminium ions from the oxide at low pH. Water self-diffusion coefficients determined from pulsed field gradient spin-echo ^{1}H NMR experiments from ambient temperature to 150 °C are slightly lower for Nafion®–SiO_2 than for Nafion® below 100 °C, and slightly higher than for Nafion® above 100 °C, while those of Nafion®–Al_2O_3 are more markedly lower throughout the temperature range [103]. At 145 °C, the maximum power densities were also appreciably different being, in decreasing order, $SiO_2 > ZrO_2 >$ neutral $Al_2O_3 >$ basic Al_2O_3, which is the same order as the point of zero charge on the oxide surface, i.e. directly related to the density of charged sites available at the low pH environment of Nafion® [101, 104]. Incorporation of zeolites chabazite and clinoptilolite in Nafion® substantiates these results [105], by showing that whereas the DMFC resistance of unfilled recast Nafion® is lower than that of composite membranes below 100 °C, between 120 and 150 °C the cell resistance of the Nafion®–zeolite composites is significantly lower (anode back-pressure up to 3.5 atm abs. at 140 °C). These observations are reflected in the values of maximum power density obtained with recast Nafion® and Nafion®–zeolite-based MEAs.

It has recently been argued that effects induced by the presence of metal oxide particles other than water retentive properties are also, if not more, relevant to high-temperature PEMFC operation. In particular, the glass transition temperature T_g of dried composite membranes was observed to increased in Nafion®–SiO_2, Nafion®–ZrO_2 and Nafion®–TiO_2 (100–120 °C) compared to dried recast Nafion® (92 °C). Although exact values might differ from those in situ in a fuel cell environment depending on hydration conditions, the trend is to a higher T_g in the composite membranes. This increase in T_g reflects increased polymer stiffness, as expected if a specific interaction exists between components. Interface interactions are potentially more abundant as the metal oxide particle size is reduced, and it is noteworthy (although is not mentioned by the authors) that the composite membranes in this study giving higher cell potential at 130 °C and RH below 90% than recast Nafion® have a particle size in the nanometre range (membranes of inferior fuel cell properties having particles in the micron range). Coordination bonding between sulfonate groups and unsaturated coordination sites on the metal oxide surface was considered as being the most likely interface interaction, in particular with SiO_2 and TiO_2 particles [102]. Hydrogen-bonding interactions between the sulfonate groups and surface hydroxyl groups are, however, probably more prevalent at the interface, although less specific in nature.

More specific interaction between the polymer and inorganic components can be introduced by surface functionalisation of metal oxide particles in an approach first described by Rozière, Jones and co-workers [67, 68]. Colloidal silica was reacted with aminophenyltrimethoxysilane to give particles coated by aminophenyl groups. In this system, protons are transferred from sulfonic acid to amine groups. The specific acid–base (sulfonic acid–amine) or ionic ($SO_3^--NH_3^+$) interactions favour charge-driven dispersion of basic silica throughout the polymer, self-assembling and cross-linking polymer chains and inorganic particles. Furthermore, once protonated, the charged silica particles tend to repel each other, and their agglomeration is avoided. The influence of the degree of functionalisation on silica dispersion can be seen from the scanning electron micrographs shown in Fig. 11 for membranes based on sPEEK and containing 20 wt. % silica. Without functionalisation, dispersion of an aqueous colloidal silica suspension in sPEEK dissolved in NMP gives membranes in which silica is aggregated and heterogeneously distributed. On addition of 10–20 wt. %. of aminophenyltrimethoxysilane (with respect to the amount of colloidal silica), the distribution of silica becomes progressively more and more homogeneous, and individual silica particles are seen at high magnification, being 10 nm in diameter. It was observed that

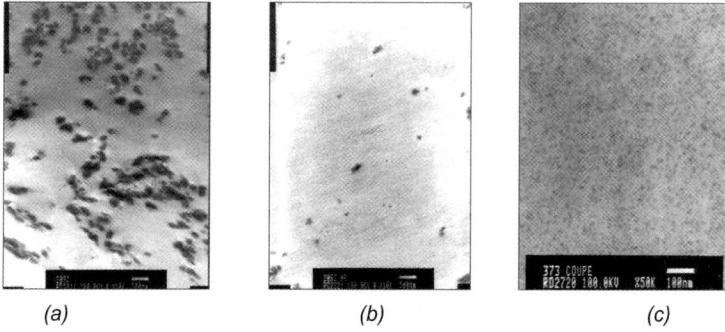

(a)   (b)   (c)

Fig. 11 Scanning electron micrographs of sPEEK–silica membranes containing 20 wt. % colloidal silica surface functionalised with aminophenyl cross-linking groups: **a** no surface functionalisation, **b** colloidal silica/aminophenyltrimethoxysilane 90/10 wt. %, **c** colloidal silica/aminophenyltrimethoxysilane 80/20 wt. %

the mechanical properties of non-functionalised sPEEK–silica and functionalised sPEEK–silica membranes are quite different. In the absence of surface functionalisation, the maximum strength drops from 40 MPa for sPEEK to 10 and 5 MPa for sPEEK–SiO$_2$ with 10 and 20 wt. % silica, respectively. However, for ionically cross-linked membranes, the maximum strength reaches values of ca. 35 and 30 MPa, respectively, for nanocomposite membranes with 10 and 20 wt. % silica.

Proton transfer from sPEEK to aminophenyl groups reduces the effective IEC of the nanocomposite membrane and ultimately the proton conduction properties, and a compromise must be found between the charge-driven particle dispersion and the need to maintain an adequately high IEC. In the above membranes, an increasing difference was observed between the effective IEC determined by titration and the value calculated assuming proton transfer to each amino group, as the number of aminopropyl groups increased. This suggests that particle/polymer spatial organisation is an important factor, and it is possible that protons are transferred preferentially where sulfonic acid sites of sPEEK are in close proximity to the silica particle surface. In agreement with this, the observed proton conductivity of the nanocomposite membranes is only slightly lower than that of sPEEK in the studied temperature range to 100 °C.

It is recalled that ionic cluster sizes in Nafion®–silicon oxide membranes prepared by permeation of TEOS into Nafion®-117 were reported to be of similar size to those of an unmodified membrane [50]. It is thus of particular interest that SAXS investigations of Nafion®–SiO$_2$, Nafion®–TiO$_2$ and Nafion®–Al$_2$O$_3$ membranes prepared by recasting with metal oxide particles show the Bragg spacing attributed to cluster-to-cluster spacing to be smaller than that of comparison recast Nafion®, for the same hydration number (Fig. 12) [102]. It has been proposed that a decrease in the cluster-to-cluster distance correlates with cluster size, smaller separations being associated with larger cluster sizes [106]. On this basis, hydrated silica- and titania-containing membranes prepared by recasting from colloidal suspensions are characterised by larger ionic clusters, and it is postulated that these larger ionic clusters enable lower cell resistance and improved fuel cell current–voltage response under elevated temperature, low humidity conditions in which the polymer is placed under significant external stress by the cell test frame [102].

Heteropolyacids (tungstophosphoric, silicotungstic, tungstomolybdic acids) containing Keggin ions [$(M^1M^2_{12}O_{40})^{3-}$] or lacunary Keggin-type structures have attracted considerable interest as the inorganic component of composite membranes because of their low cost, commercial availability (in some cases), and high proton conductivity when fully hydrated. In the solid state, the above heteropolyacids (HPAs) exist as distinct 6, 14, 21 and 28(29) hydrates that are increasingly hygroscopic and increasingly proton conducting. Above 80 °C under ambient RH, the conductivity drops by up to a factor

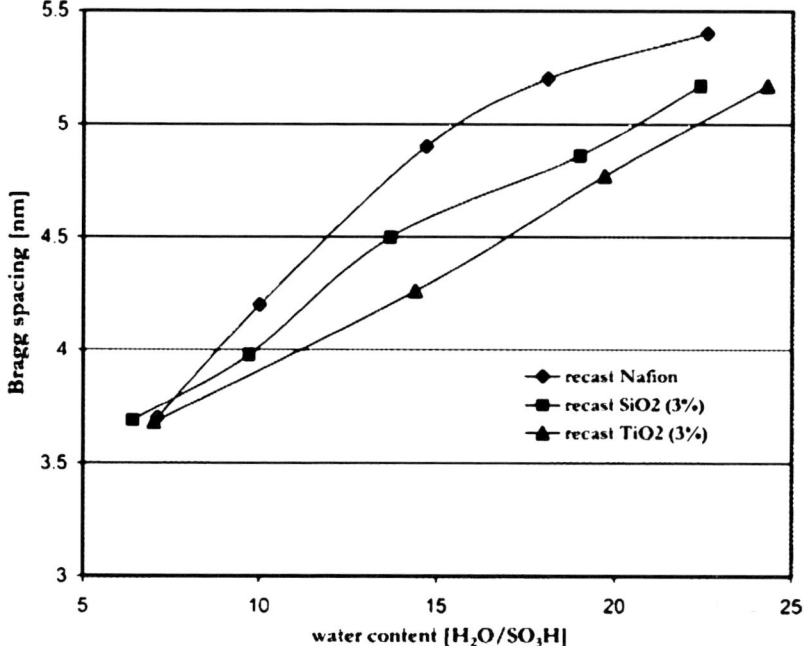

Fig. 12 Bragg spacing from SAXS as a function of water content in recast Nafion®, Nafion®-SiO$_2$ and Nafion®-TiO$_2$ membranes. Reproduced with permission from [102]. © (2006) American Chemical Society

of 10^3 [107–109] and thus such systems need to operate under locally high RH to be effective in increasing the proton conductivity. It is possible that the high temperature/low RH conductivities of composite systems incorporating HPAs described in some reports [110] could result from the presence of phosphoric acid formed by decomposition of phosphorus-containing HPAs. The main drawback, however, to the use of HPAs as a component of PEMFC membranes is their solubility in water, which means that, in principle, membranes incorporating HPAs should not be allowed to come into contact with liquid water, such as is generated on load or temperature cycling in situ. It is considered that even if HPA moieties interact with the sulfonic acid/sulfonate groups of the polymer host, the strength of these interactions will not be high enough to avoid elution in the presence of a water flux.

Savadogo [111, 112] and others [113, 114] have developed composite membranes based on Nafion® and HPAs introduced either by soaking Nafion® membranes in a solution of HPA, or by casting a suspension containing Nafion® dispersion and dissolved HPA. These reports concur in the observation that silicotungstic acid (STA) gives more highly conducting composite membranes and superior fuel performance than those incorporating either phosphomolybdic or phosphotungstic acids (PTA), both at full humidification, and at higher temperature/lower humidity (sample temperature 120 °C, gas humidification 90 °C) operation conditions. It is reported that water uptake from the liquid phase by all Nafion®–HPA composites is greater than that of Nafion®-112, -115 or -117 [111], while uptake from the vapour phase by composite and non-composite membranes is similar. For Nafion®-based membranes containing PTA, the inorganic particles were of micron size, even at low (10 wt. %) loading [114] and, after immersion in H_2SO_4 at 85 °C for 3 h, followed by water washing (85 °C, 1 h), PTA was almost completely leached out [115]. Partial exchange of the protons of PTA for NH_4, Cs, Rb or Tl significantly alleviated this loss to water, and Nafion®–Cs/H-PTA membranes were reported to have a conductivity of 0.016 S cm^{-1} at 120 °C/35% RH, the same as unmodified Nafion® determined under the same conditions. Although this result seems encouraging, the longer-term stability of the conductivity needs to be considered, since the metal ions can potentially exchange with protons of sulfonic acid groups of Nafion®, and could also lead either to polarisation and/or to an increased rate of membrane or MEA degradation, and the advantages of this stabilisation should be more clearly shown. Composite membranes using molybdenum-based HPAs are electrochemically unstable due to redox transitions in the fuel cell operating range, with migration of the HPA to the cathode catalyst layer [114].

In sulfonated poly(arylene ether sulfone) (BPSH) composite membranes with PTA, small shifts in IR signals arising from the stretching vibrations $v(W=O)$ and $v(W-O-W)$ in the Keggin structure of $(PW_3O_{40})^{3-}$ ions when incorporated in a composite membrane were interpreted as resulting from interaction with polymer sulfonate groups via H_3O^+ ions [116]. As developed above, these interactions are certainly responsible for the homogeneity of the distribution of the inorganic phase, which is present as particles 30–50 nm in diameter; in a control polymer without sulfonic acid functions, particles were in the 3 μm range. Membranes containing 30–50-nm particles had higher tensile strength than a reference BPSH membrane without inorganic component, as is expected for a material reinforced mechanically by hydrogen-bonding interaction. The loss of PTA from these membranes after contact with water vapour (100 °C, 15 h) depended upon the degree of sulfonation of the polymer, being 4, 0, 2, 3 and 18 wt. % from membranes with 0, 20, 30, 40 and 60% sulfonation, respectively. These data illustrate both the importance of polymer functionalisation in reducing extraction of PTA, and the need for judicious choice of the degree of sulfonation so as to avoid excessive water uptake and dissolution of the HPA. HPAs have also been used as a support for Pt particles, and incorporated into sulfonated PEEK [117] to provide sites for potential catalytic recombination of permeated hydrogen and oxygen [96].

Formation of HPA composites with a basic polymer such as polybenzimidazole (PBI) could in principle be favourable, since stronger hydrogen bonding between the two components could delay elution of the HPA by water. On the other hand, the HPA will tend to be more susceptible to hydrolysis in this environment, possibly leading to local formation of phosphoric acid that can enhance conductivity at high temperature/low RH, but be eluted in the presence of water. Composites of HPAs with AB-PBI doped with phosphoric acid have also been reported [118]. Other efforts have been made to stabilise HPA, in particular by grafting to the surface of (modified) metal oxide particles, generally either by reacting the HPA with an alcoholic solution of TEOS, or with an organosilyl derivative, and varying the ratio of PTA to SiO_2. PBI [119] and sPEEK-based composite membranes [120] with silica-supported HPA have been described. The approach seems to only partially resolve the problem; in the above study, sPEEK membranes containing STA supported on an organo-silica and aerosil led to a reduction in the amount of HPA bled out (water, 55 °C, 24 h) to around 15 wt. % from around 80% in the reference sPEEK-HPA [121]. However, the sPEEK was highly sulfonated, with high water uptake, indeed favouring not only loss of the inorganic material but also dissolution of the polymer itself. This approach may represent a way forward, but more convincing results are required to show that supported HPAs are effectively stabilised against water solubility in an ionomer environment. Other studies, in particular the work of Honma and co-workers, have developed the approach of composite membranes based on the incorporation of HPAs into cross-linked hybrid silicate–polyether matrices [122–127], with covalent bonding between HPA and the organic skeleton [128]. These are considered beyond the scope of this article, and readers are referred to a recent overview [129].

Preparation of composite membranes with natural, synthetic and modified clays is attracting renewed interest. So-called swelling clays delaminate in water and give objects of high surface/volume ratio which, if oriented parallel to the membrane surface, will provide physical barriers to fuel crossover. Other solvents also authorise this exfoliation phenomenon, in particular when the organophilicity of the interlayer regions of the clay is increased. This is generally achieved by ion exchange with alkylammonium species, the presence of which will also enhance ionic interaction with a sulfonated polymer matrix, although it will tend to lower proton conductivity. Such properties are observed with Nafion®–organo-montmorillonite composite membranes containing only 1 wt. % clay [130], for which methanol permeability was reduced by an order of magnitude compared with Nafion®-117, and which allowed operation in a DMFC using 10 M methanol, when a power density of ca. 100 mW cm^{-2} was obtained at 70 °C. The loss of sulfonic acid groups was delayed to higher temperature, while maximum strength and elongation at break were reported to be higher for membranes containing up to 15 wt. % montmorillonite than for reference recast Nafion®, observations that are compatible with enhanced cohesion at the polymer/clay interface. Similarly, mechanical reinforcement in hexadecyltrimethylammonium-exchanged montmorillonite–sPEEK is revealed by significant reduction in water uptake [131]. Ionic interactions via positively charged alkylammonium ions link together the macroanionic montmorillonite layers and macroanionic polymer in these systems. In other work, synthetic and natural clays modified to covalently graft imidazole to their surfaces via alkyl or alkyl ether spacer groups have been used as inorganic components in composite membranes with sPEEK [132]. Imidazole and other nitrogeneous heterocycles, in particular when immobilised on a polymer backbone, have attracted interest over recent years as amphoters able to act as centres for proton transfer in the same way as water [133]. Although interesting as a reference system, the above grafting to clays [132] occurs via a nitrogen atom of imidazole, which is less favourable for proton transfer, and in addition the long-term stability of the spacer groups in a fuel cell environment is uncertain. It has also been reported that surface functionalisation of montmorillonite to give pendant sulfonic acid groups followed by dispersion in Nafion® leads to membranes and corresponding MEAs [134, 135] that give a maximum power density of ca. 70 mW cm^{-2} at 40 °C with 2 M methanol, and it is interesting to note that this is lower than the performance obtained with Nafion®–montmorillonite composites having surface functionalisation with basic groups [130].

Despite the low-cost advantage of natural clays, the dependence of exact composition on source and the possible presence of metal ion contaminants effectively exclude them from use as fuel cell membranes. As described in Sect. 2, metal(IV) phosphates have a layer structure which, under particular conditions (solvent, temperature) spontaneously delaminates to give a colloidal suspension of nominally single layers. In the particular case of α-ZrP, a procedure of intercalation of propylamine to 50% saturation of the ion-exchange sites, followed by de-intercalation gives a colloidal suspension when the recovered proton form ZrP is dispersed in water [136]. A recent report describes the formation of composite membranes by addition of α-ZrP dispersed in a solution of propylamine/PBI/DMF to sPEEK/DMSO [137]. These conditions are not expected to favour delamination/dispersion, although the presence of both PBI and micronic ZrP effectively reduces methanol crossover. It was earlier observed that colloidal suspensions of α-ZrP with particles of ca. 100 nm thickness and a few μm^2 surface area, when dispersed in DMF and used to prepare composite membranes with sulfonated polyetherketone (sPEK) [138], led to a decrease in conductivity with increasing zirconium phosphate loading up to 20 wt.%. This result, and those of other authors on composites of exfoliated α-ZrP in Nafion® [139], is in contrast with that obtained when α-ZrP is precipitated in situ (Sects. 2 and 4), and is certainly a consequence of the orientation of rather large particles having a layer surface perpendicular to the direction of proton transport. Much smaller particle size, X-ray amorphous ZrP, when exfoliated, forms gels with organic solvents in which the average thickness of the particles is 25–80 nm [140] and that can readily be dispersed in ionomer solutions and cast to form nanocomposite membranes. Using Nafion® as polymer component, a "stability map" identifying the temperature and RH conditions under which the conductivity value is stable for 150 h indicates that there is a high temperature gain for Nafion®–ZrP nanocomposite membranes of 20 °C over the polymer-only membranes, attributed to the greater stiffness of membranes incorporating ZrP [78].

Zirconium (titanium) phosphate sulfophenylphosphonates are members of the same family. Here, some of the phosphate groups are replaced by sulfophenylphosphonate, giving a general formula:

$$Zr(O_3POH)_{2-x} \cdot (O_3PC_6H_4SO_3H)_x \cdot nH_2O, Zr(SPP)_x.$$

Hybrid membranes based on sPEEK containing 40 wt. % of exfoliated zirconium phosphate sulfophenylphosphonate ($x = 1$ and 1.5 in the above formula) and added to sPEEK solutions show conductivities lying between those of the individual polymer and inorganic components, being $> 10^{-2}$ S cm^{-1} at 90% RH and 100 °C [29]. In comparison, composite Nafion®–titanium phosphate sulfophenylphosphonate membranes prepared by bulk mixing were reported to give conductivities lower than either of the components separately [141]. Amorphous zirconium phosphate sulfophenylphosphonate, $Zr(SPP)_x$ (with $x = 1$ to 1.3 in the above formula) can also be obtained as DMF-swollen gels that can be dispersed in a DMF-based polymer solution to give membranes in which the inorganic domains are 20–200 nm in size [93, 142]. The conductivity of composite membranes sPEK–$Zr(SPP)_{1.3}$ containing 10 and 20 wt. % of inorganic material was higher than that of sPEK (IEC 0.95 meq g^{-1}) in the range 80–110 °C, and at 90% RH, by a factor of 5 and 10, respectively. Interfacial effects between the polymer and $Zr(SPP)_{1.3}$ in these systems were assumed to be negligible, and it was proposed that inorganic-rich or continuous inorganic component conduction pathways run through the membrane [142]. Zirconium phosphate sulfophenylphosphonate contributes very effectively to proton conductivity in composite and nanocomposite membranes, but is less stable to hydrolysis than zirconium phosphate, and the benefits it brings in terms of conductivity enhancement may be lost or reduced in a fuel cell with time of operation. Another report has been made of Nafion–Zr(SPP) composite membranes, but comparison with other reports [31] and with the foregoing results is difficult since the composition and characterisation conditions are somewhat incomplete [143].

Other layered phosphates have also been used in composite membrane formation. For example, layered phosphoantimonic acid ($H_3Sb_3P_2O_{14} \cdot xH_2O$) dispersed in sulfonated polysulfone has been reported to increase conductivity, lower permeability to oxygen and reduce water uptake, properties that were attributed to interaction between the inorganic and polymer components [144]. Finally, boron phosphate has interesting characteristics that merit special mention. Although best known for its properties as an acid catalyst, BPO_4 is a proton conductor under certain conditions that allow local formation of (di)hydrogen phosphate groups and/or phosphoric acid. Best conduction properties are observed in samples of low degree of crystallinity, but this parameter requires careful control since amorphous BPO_4 is highly hygroscopic and tends to water solubility. Addition of boron phosphate to a sPEEK solution gave microporous composite membranes in which submicron particles of BPO_4 were embedded in the pore walls [145], presumably resulting from phase inversion with water associated with the inorganic phase. Note that dispersion of a powder into a solution, followed by phase inversion, leads to entrapment of the inorganic phase in the pore walls, unlike the microporous sPEEK membranes incorporating in situ prepared ZrP described above, where the inorganic material is located as a coating on the pore surfaces or within the wall. Further, no mention was made of an improved dimensional stability, although the conductivity of composites sPEEK–BPO_4 was in every case significantly increased over that of sPEEK. Boron phosphate stabilised with silica has been incorporated into polyimide by incorporation into polyamic acid followed by thermal treatment [146].

4
Preparation of Composite Membranes Using Polymer Solutions or Dispersions as Reaction Medium for In Situ Formation of the Inorganic Component

An important limitation of the use of a pre-formed membrane as template for inorganic particle growth, as described in Sect. 2, is the content of inorganic material that can be formed. The approach by which metal ions are introduced into the ionomer membrane by ion exchange is in effect limited by the IEC of the constituent polymer. Since excessive increase in the IEC is detrimental to the mechanical stability of the membrane in contact with water, simply boosting the degree of sulfonation displaces the problem. Using sPEEK of IEC 1.3 meq g^{-1}, up to 30 wt. % of ZrP may be formed by ion exchange/precipitation [29]. When the inorganic component is formed within a polymer solution there is in principle no limit to the composition, within the bounds of solubility considerations of the inorganic precursors in the solvent system used. While this approach has been largely developed for the preparation of hybrid inorganic–organic membranes for other applications, it has as yet been relatively little used in the fuel cell membrane field. Some of the first work in this area using Nafion® was carried out by Zoppi and Nunes, when Nafion®–silica membranes containing 6–54 wt. % silica were prepared by addition of TEOS (and, in some cases 1,1,3,3-tetramethyl-1,3-diethoxydisiloxane, TMDES, to introduce flexible segments) to propanol/water Nafion® dispersions. The Nafion® microstructure forms at the same time as the condensation/polymerisation of the silica and, although diverse morphologies were observed, the particle size was generally large and in the micron range, with polymer-rich regions in the membranes devoid of silica. Even at only moderate levels of inorganic content, these Nafion®–SiO_2 membranes were reported to be brittle, while incorporation of 15–20 wt. % TMDES improved flexibility and elasticity [147]. In dry argon, the conductivity of Nafion®–SiO_2 membranes prepared according to this approach and dried under vacuum at 30 °C for 1 week was $< 10^{-5}$ S cm^{-1} [148], a set of conditions outside of the PEMFC application window for membranes based on current PFSA. A similar approach has been developed in recent work, with important differences in that DMAc was used as reaction medium (recovery of "dry" Nafion® from propanol/water and redispersion in DMAc) and the cast membranes were thermally treated at up to 120 °C [149]. The water uptake at 20 °C of these membranes increased with silica content (5–15 wt. %), but the proton conductivity followed an opposite trend that was pursued throughout the measurement range to 80 °C. Despite this, DMFC performance at 60 °C was higher with the MEA based on Nafion®–SiO_2 (5 wt. %) than reference recast Nafion® due to reduced methanol crossover.

The size of Nafion aggregates in suspension relative to dissolved molecular inorganic precursors does not favour formation of nanoscale integration of components, in particular when no specific interaction operates between them that could lead to an increase of the interface region, and act to avoid segregation of the inorganic and organic phases on removal of solvents from a cast film. Sulfonated polyaromatic polymers, on the other hand, form solutions in a variety of solvents having a range of dielectric constants, and this provides a process parameter of which use can be made in generating hybrid membranes comprising different inorganic component morphologies. This route was further opened up by Rozière, Jones and co-workers by adjunction of molecular inorganic precursors having polarity opposite to that of the polymer to favour self-assembly by ionic interaction. An acid-catalysed sol–gel type reaction takes place between TEOS and aminophenyltrimethoxysilane (APTMOS) in a solution of sPEEK (e.g. in NMP) to give silicon oxide in which the surface is functionalised by aminophenyl groups. Our work showed that for silica contents in the range up to ca. 25–30 wt. %, small cluster-type morphology is observed, the size of which depends on the relative amounts of silicon source (TEOS) and cross-linking agent (APTMOS). For example, elongated particles of length 50 to 100 nm are formed in hybrid membranes containing 20 wt. % silicon oxide prepared with TEOS/APTMOS in weight ratio 9:1, while on increasing the proportion of APTMOS (TEOS/APTMOS weight ratio 8:2) growth is limited to much smaller spherical silica particles around 10–20 nm in diameter.

This ionic cross-linking $-NH_3^+-{^-}O_3S-$ no longer suffices in ensuring high dispersion above inclusion of ca. 30 wt. % silica, in the absence of other favourable factors. In the present work, high dielectric constant solvents were used to solvate ion pairs via electrostatic interactions without hindering ionic

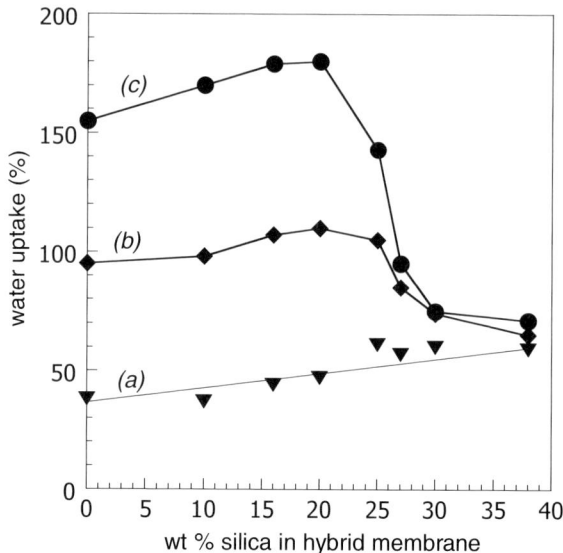

Fig. 13 The dependence of water uptake at 25, 80 and 120 °C on silica content of sPEEK–silica membranes

interaction at the interface [26]. In this way, with the use of DMSO as solvent (dielectric constant 30), transparent sPEEK–silicon oxide membranes could be obtained containing up to 50 wt. % SiO_2. The importance of the ionic cross-linking in controlling membrane swelling in water is demonstrated by the results shown in Fig. 13, in which the water uptake at 25, 80 and 120 °C is plotted as a function of the weight percent of silica in the nanocomposite membrane. Water uptake for unmodified sPEEK membranes, although moderate at 25 °C, is high (90–150 wt. %) at between 80 and 120 °C. In sPEEK–silica membranes, water uptake at these temperatures is sharply attenuated when the silica content is above ca. 30 wt. %, and approaches at 120 °C a value close to that of unmodified sPEEK membrane at 25 °C. This decrease in water uptake, most marked at high temperature, is important in the context of the recognised contribution of membrane swelling/contraction cycles to mechanical fatigue and membrane failure in an operating fuel cell. The explanation for these observations is found at a microstructural level, since all evidence points to the presence of interpenetrating inorganic and polymer networks. From TEM, homogeneously distributed inorganic and organic regions were identified, each having dimension ca. 4 nm. After removal of the sPEEK constituent from the nanocomposite membrane (thermal treatment or dissolution), the inorganic component could be recovered as a self-supported film (Fig. 14) that nitrogen adsorption–desorption measurements revealed to be mesoporous, with pore dimension 4–5 nm. The inorganic nanoporous structure is a replica of the organisation that the sPEEK polymer adopted

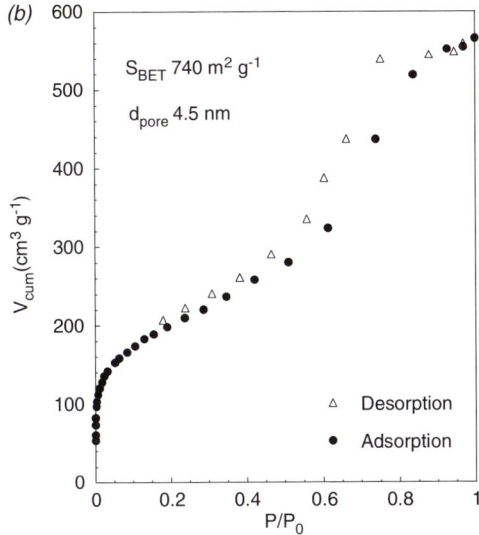

Fig. 14 a Self-supported silica film recovered from hybrid sPEEK–silica (50 wt. %) after thermal removal of sPEEK component. Film thickness: 40 μm. **b** Nitrogen adsorption–desorption isotherms of silica film

in the nanocomposite membrane, and the elevated silica surface area (BET surface area $740\,m^2\,g^{-1}$) provides good evidence for a highly extensive interface region between inorganic and organic polymer components. In this arrangement, sPEEK is essentially "confined" within a continuous reinforcing silica network formed in situ, which limits water swelling and dimensional change. However, even at much lower silica content the ionic interaction still influences the mechanical properties: the maximum strength of a sPEEK membrane of 39 MPa drops by a factor of 5 when 10 wt. % silica is formed in situ from TEOS only, but including a proportion of 9:1 or 9:2 TEOS/APTMOS increases the strength to 32 and 37 MPa, respectively. An MEA based on sPEEK–silica (10 wt. %) has provided one of the first examples of long-term fuel cell operation of a sulfonated polyaromatic membrane when 1000 h of operation were achieved under conditions of full humidification at 90 °C, a cell voltage of 600 mV being produced under constant current density of $0.5\,A\,cm^2$ [150].

The proton conductivity in sPEEK–silica membranes depends on both their silica content and the proportion of APTMOS ionic cross-linking agent since, as described above for hybrid membranes incorporating aminophenyl-functionalised colloidal silica, the membrane effective IEC is affected by the presence of basic aminophenyl groups. Figure 15 displays proton conductivity over the temperature range to 100 °C of membranes containing up to 50 wt. % silica. Using a new fluorine-containing PBI, covalent bonding between functionalised silica and benzimidazole nitrogen has given a hybrid membrane with high modulus and low methanol crossover, although insufficient proton conductivity [151].

The in situ construction of the inorganic component within a cast polymer solution is not limited to metal oxides and in practice a range of other inorganic materials can be formed depending on the choice of precursor(s) incorporated in the polymer solution, and the nature of post-treatment following solvent removal. Rozière and Jones and co-workers have developed nanocomposite membranes in which zirconium phosphate is formed from zirconyl propionate introduced into a DMAc solution of sPEEK, by immersion of the cast film, after solvent removal, into phosphoric acid. This approach provides a robust synthetic route that can be generalised to other ionomers, and allows the amount of ZrP to be readily varied, even up to ca. 40–50 wt. %.

In common with the sPEEK–silica systems described above, the morphology of zirconium phosphate formed from a multicomponent solution con-

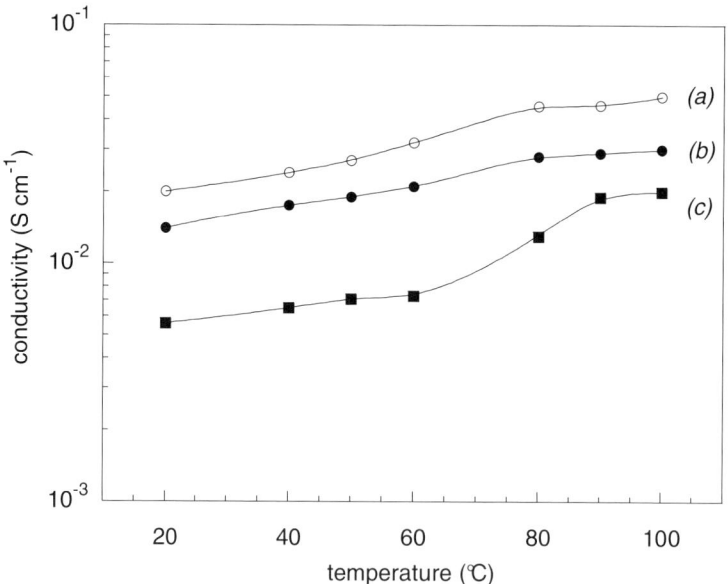

Fig. 15 Temperature dependence of proton conductivity at 100% RH of sPEEK (a) and sPEEK–silica membranes containing 20 (b) and 45 wt. % (c) silica

taining zirconyl propionate and sPEEK can be induced to evolve from discrete particulate to an extended network as the zirconium phosphate content increases. The self-supported films of ZrP that are recovered when sPEEK is removed from the composite membrane again provide a means of interpreting, in terms of interpenetrating sPEEK–ZrP networks, the reduced water uptake and dimensional change of sPEEK–ZrP membranes. Mesoporous, high surface area zirconium phosphate is being sought for applications in separation and catalysis [152], and this route, in addition to its utility for fuel cell ionomer membrane preparation, also represents an original approach to such a material (BET surface area 390 $m^2 g^{-1}$, pore diameter 5 nm), moreover in a shaped and self-supported form.

sPEEK–ZrP membranes prepared in this way and containing 20 wt.% ZrP have been operated in DMFC single cells with air and oxygen feeds at temperatures up to 150 °C, using non-commercial electrodes of metal loading 1.2 mg cm^{-2} (anode, PtRu; cathode, Pt) [153]. The power density increases with temperature up to 150 °C using an oxygen feed, with particularly strong performance improvement between 100 and 110 °C. The maximum power density, reached at a potential close to 0.4 V at 150 °C, increases from 70 mW cm^{-2} (with air oxidant, 50 mW cm^{-2}) at 90 °C to 180 mW cm^{-2} at 150 °C (Fig. 16). With air as oxidant, the maximum power density (130 mW cm^{-2}) was obtained at 130 °C. These observations are of particular note given the low catalyst loading, and the results compare favourably with those that were obtained in the same fuel cell hardware using Nafion®-117 and E-TEK electrodes (2 mg metal cm^{-2}) where, although the maximum power density was higher (250 mW cm^{-2} with oxygen), the maximum power density decreased above 130 °C, with irreversible deterioration of the Nafion®-117 membrane.

Particularly promising in the context of high-temperature fuel cell operation is the exceptional long-term stability of a DMFC stack comprising 30 of the above MEAs. This DMFC stack (0.7 kW power with air feed) was operated with daily load (open circuit to 0.4 A cm^{-2}) and temperature cycling (room temperature to 130 °C), with 8 h operation at 130 °C and shutdown overnight. The stack was run in this way for two periods of a month, with an interim shutdown for 1 month between runs. An initial oscillatory behaviour, attributed to electrode stabilisation, progressed to stable power output after around 20 days of temperature/load cycling, and recovery of stack voltage was excellent even after the shutdown period of 1 month. These results are remarkable both in the stability of stack performance and the durability of the components under severe operating conditions (130 °C, air and methanol mixture pressures, stop/start operation), and are all the more encouraging in that no optimisation was made of the interface between membrane and electrodes. These results may be compared with those obtained with Nafion®–ZrP prepared by ion exchange/precipitation in cast Nafion membranes (Sect. 2.2), where power densities of ca. 90 mW cm^{-2} at

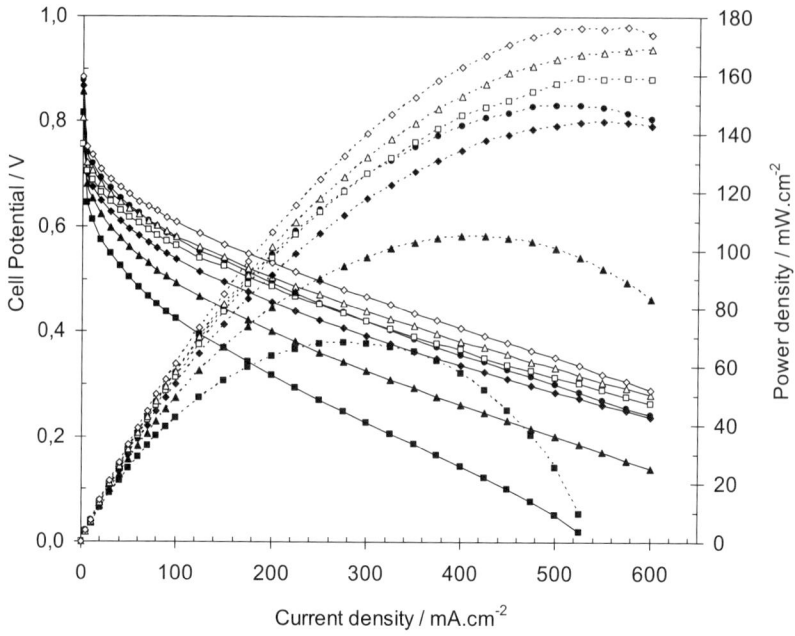

Fig. 16 Single fuel cell characterisation at 90–150 °C of membrane electrode assemblies prepared using sPEEK–ZrP (20 wt.% ZrP prepared in situ in sPEEK solution). Non-commercial electrodes, 1.2 mg cm^{-2} PtRu anode, 1.2 mg cm^{-2} Pt at cathode. 1 M methanol feed, 2.5 bar; oxygen, 3 bar. Measurements at ■ 90 °C, ▲ 100 °C, ♦ 110 °C, ●120 °C, □ 130 °C, △ 140 °C, ◇ 150 °C. *Unbroken line* is current density vs. cell voltage; *dashed line* is power density vs. cell voltage

130 °C (Pt/Ru 1.85 mg cm^{-2} at the anode, 0.37 mg cm^{-2} Pt at the cathode, methanol 1.5 M) and 260 mW cm^{-2} at 140–150 °C have been reported for Nafion-117–ZrP with air feed, but no information was provided on MEA durability [72, 76].

In a related approach, Alberti and Casciola described the addition of a solution in propanol of zirconyl propionate, phosphoric acid and sulfophenylphosphonic acid to PFSA or sPEK in suspension/solution [86]. The removal of solvents from cast films leads to membranes giving signals characteristic of O_3POH and $O_3PC_6H_4SO_3H$ groups, at positions expected for ZrSPP, presumably of very low particle size since no characteristic X-ray diffraction lines were observed. The proton conductivity of sPEK (IEC 0.9 meq g^{-1}) increased from 2×10^{-4} to 7×10^{-4} and 5×10^{-3} S cm^{-1} for membranes containing 20 and 33 wt.% ZrSPP, respectively, at 100 °C and 70% RH. Even the better value is at the lower limit for fuel cell use, but use of a sPEK of slightly higher IEC would overcome this difficulty.

5
Conclusions and Outlook

There is increasing evidence that the main advantage conferred by the presence of an inorganic component, in terms of both fuel cell performance at higher temperature and lower RH, and MEA durability and lifetime, lies in improvement in mechanical properties. This effect is a result both of specific interactions between the inorganic and organic components of which hydrogen bonding is the most ubiquitous, and of an internal scaffolding effect played by the presence of inorganic particles and networks, which exercises a stabilising influence on the ionic cluster volume, which is then less susceptible to changes in degree of hydration. Charge-driven assembly of inorganic and organic components of different polarity (sulfonic acid functionalised polymer, basic functionalised inorganic material, or vice versa), or of different relative acidity, allows highly dispersed and homogeneous nanocomposites to be obtained. Hydrogen bonding and ionic cross-linking are most effective and most prevalent when the organic/inorganic interface region is extensive, a condition satisfied when the inorganic phase is present as nanometric particles or as an extended system forming a co-continuous network with the organic polymer. Different strategies for nanocomposite fuel cell membrane preparation may be distinguished, broadly divided into those making use of a cast membrane as a template for inorganic particle growth, and those in which an inorganic material is either added to, or formed within, a polymer solution or dispersion. While the former makes use of existing hydrophilic regions as preferential sites for particle growth within an existing membrane architecture and microstructure, the latter opens up the opportunity of modifying or orienting the resulting microstructure, by mutual adaptation of the inorganic structure and polymer chain organisation. In this sense, the preparation route is critically important in defining the properties of the final membrane, as are thermal post-treatments when use has been made of a polymer dispersion. In general, proton conductivity is less sensitive to membrane preparation route than are mechanical properties and membrane lifetime, and conductivity measurements taken alone inadequately predict MEA fuel cell performance and durability. Many recent reports have succeeded in their endeavour to identify under which conditions inorganic–organic membranes provide properties superior to those shown by the polymer-only counterpart, and there is every reason to be optimistic that MEAs based on nanocomposite membranes have a role to play in the highly strategic operation conditions of low RH at 110–130 °C. Current hurdles persist: membrane electrical resistance and long-term durability under fuel cell operation and, in this context, in-depth studies of ageing and degradation under realistic operation conditions are still needed to enable further materials improvements and alignment with current targets.

References

1. Rozière J, Jones DJ (2003) Ann Rev Mater Res 33:503
2. Wieser C (2004) Fuel Cells 4:245
3. Mauritz KA, Moore RB (2004) Chem Rev 104:4535
4. Doyle M, Rajendran G (2003) In: Vielstich W, Lamm A, Gasteiger H (eds) Handbook of fuel cells: fundamentals, technology and applications, vol 3. Wiley, Chichester, p 351
5. Gubler L, Gürsel SA, Scherer GG (2005) Fuel Cells 3:317
6. Jones DJ, Rozière J (2001) J Membr Sci 185:41
7. Nolte R, Ledjeff K, Bauer M, Mülhaupt R (1993) J Membr Sci 83:211
8. Wang F, Hickner M, Kim YS, Zawodzinski TA, McGrath JE (2002) J Membr Sci 197:231
9. Bauer B, Jones DJ, Rozière J, Tchicaya L, Alberti G, Casciola M, Massinelli L, Peraio A, Besse S, Rammuni E (2000) J New Mater Electrochem Syst 3:93
10. Zaidi SMJ, Mikhailenko SD, Robertson GP, Guiver MD, Kaliaguine S (2000) J Membr Sci 173:17
11. Glipa X, Haddad ME, Jones DJ, Rozière J (1997) Solid State Ionics 97:323
12. Ariza MJ, Jones DJ, Rozière J (2002) Desalination 147:183
13. Besse S, Capron P, Diat O, Gebel G, Jousse F, Marsacq D, Pinéri M, Marestin C, Mercier R (2002) J New Mater Electrochem Syst 5:109
14. Fang J, Guo X, Harada S, Watari T, Tanaka K, Kita H, Okamoto KI (2002) Macromolecules 35:9022
15. Wainright JS, Wang J-T, Savinell RF, Litt M, Moaddel H, Rogers C (1994) Proc Electrochem Soc 94:255
16. Wainright JS, Litt MH, Savinell RF (2003) In: Vielstich W, Lamm A, Gasteiger H (eds) Handbook of fuel cells: fundamentals, technology and applications, vol 3. Wiley, Chichester, p 436
17. Glipa X, Mula B, Jones DJ, Rozière J (1998) In: Sequeira CAC, Moffat J (eds) Chemistry, energy and the environment. Royal Society of Chemistry, Cambridge, UK, p 249
18. Glipa X, Bonnet B, Mula B, Jones DJ, Rozière J (1999) J Mater Chem 9:3045
19. Xiao L, Zhang H, Jana T, Scanlon E, Chen R, Choe E-W, Ramanathan LS, Yu S, Benicewicz BC (2005) Fuel Cells 5:287
20. Kerres J (2005) Fuel Cells 5:230
21. Jones DJ, Grasset F, Marrony M, Rozière J, Glipa X (2005) Fuel Cell Bull, p 12
22. Yang Y, Holdcroft S (2005) Fuel Cells 5:171
23. Jones DJ, El Haddad M, Mula B, Rozière J (1996) Environ Res Forum 1/2:115
24. Kjaer J, Yde-Andersen S, Knudsen NA, Skou E (1991) Solid State Ionics 46:169
25. Tchicaya-Bouckary L, Jones DJ, Rozière J (2002) Fuel Cells 2:40
26. Jones DJ, Rozière J (2003) In: Vielstich W, Lamm A, Gasteiger H (eds) Handbook of fuel cells: fundamentals, technology and applications, vol 3. Wiley, Chichester, p 447
27. Kreuer KD (2001) J Membr Sci 185:29
28. Savadogo O (2004) J Power Sources 127:135
29. Bonnet B, Jones DJ, Rozière J, Tchicaya L, Alberti G, Casciola M, Massinelli L, Bauer B, Peraio A, Rammuni E (2000) J New Mater Electrochem Syst 3:87
30. Rozière J, Jones DJ (2001) In: Büchi FN, Scherer GG, Wokaun A (eds) Proceedings of the 1st European PEFC Forum, Lucerne, p 145
31. Alberti G, Casciola M (2003) Ann Rev Mater Res 33:129
32. Jones DJ, Rozière J (1997) In: Dyer A, Hudson MJ, Williams PA (eds) Progress in ion exchange: advances and applications. Royal Society of Chemistry, Cambridge, UK, p 16

33. Jones DJ (2000) In: Wilson I (ed) Encyclopedia of separation science. Academic Press, New York, p 3501
34. Gierke TD, Munn GE, Wilson FC (1981) J Polym Sci B 19:1687
35. Rubetat L, Gebel G, Diat O (2004) Macromolecules 37:7772
36. Rubetat L, Rollet A-L, Gebel G, Diat O (2002) Macromolecules 35:4050
37. Litt MH (1997) Polym Prepr 38:80
38. Haubold H-G, Vad T, Jungbluth H, Hiller P (2001) Electrochim Acta 46:1559
39. Gebel G (2000) Polymer 41:5829
40. Moore RB, Martin CR (1988) Macromolecules 21:1334
41. Moore RB, Martin CR (1989) Macromolecules 22:3594
42. Sone Y, Ekdunge P, Simmonson D (1996) J Electrochem Soc 143:1254
43. Kreuer KD, Paddison SJ, Spohr E, Schuster M (2004) Chem Rev 104:4637
44. Paddison SJ (2003) Ann Rev Mater Res 33:289
45. Gummaraju RV, Moore RB, Mauritz KA (1996) J Polym Sci B 34:2383
46. Apichatachutapan W, Moore RB, Mauritz KA (1996) J Appl Polym Sci 62:417
47. Mauritz KA (1998) Mater Sci Eng C 6:121
48. Shao PL, Mauritz KA, Moore RB (1996) J Polym Sci B 34:873
49. Deng Q, Wilkie CA, Moore RB, Mauritz KA (1998) Polymer 39:5961
50. Mauritz KA, Stefanithis ID, Davis SV, Scheetz RW, Pope RK, Wilkes GL, Huang H-H (1995) J Appl Polym Sci 55:181
51. Deng Q, Cable KM, Moore RB, Mauritz KA (1996) J Polym Sci B 34:1917
52. Grot W, Rajendran G (1996) WO 9629752
53. Jung DH, Cho SY, Peck DH, Shin DR, Kim JS (2002) J Power Sources 106:173
54. Baradie B, Dodelet JP, Guay D (2000) J Electroanal Chem 489:101
55. Adjemian KT, Srinivasan S, Benziger J, Bocarsly AB (2002) J Power Sources 109:356
56. Adjemian KT, Lee SJ, Srinivasan S, Benziger J, Bocarsly AB (2002) J Electrochem Soc 149:A256
57. Bonnet B, Jones DJ, Rozière J, Tchicaya L, Alberti G, Casciola M, Massinelli L, Bauer B, Peraio A, Rammuni E (1999) In: Savadogo O (ed) 3rd international symposium on new materials for electrochemical systems, Montreal, p 244
58. Alberti G, Casciola M (1992) In: Colomban P (ed) Proton conductors (solids, membranes and gels). Cambridge University Press, Cambridge, UK, p 238
59. Jones DJ, Leloup J-M, Ding Y, Rozière J (1993) Solid State Ionics 61:117
60. Glipa X, Leloup J-M, Jones DJ, Rozière J (1997) Solid State Ionics 97:227
61. Alberti G, Casciola M (1997) Solid State Ionics 97:177
62. Clearfield A, Oskarsson A, Oskarsson C (1972) Ion Exch Membr 1:91
63. Alberti G, Torraca E (1968) J Inorg Nucl Chem 30:317
64. Benhamza H, Barboux P, Bouhaouss A, Josien FA, Livage J (1991) J Mater Chem 1:681
65. Costantino U, Gasparoni A (1970) J Chromatogr 51:289
66. Clayden NJ (1987) J Chem Soc Dalton Trans, p 1877
67. Rozière J (2001) In: Savadogo O (ed) New materials for electrochemical systems IV. Ecole Polytechnique de Montreal, Montreal, p 355
68. Tchicaya L (2000) PhD thesis, University Montpellier II
69. Clearfield A, Vaughan PA (1956) Acta Crystallogr 9:555
70. Toth LM, Lin JS, Felker LK (1991) J Phys Chem 95:3106
71. Jones DJ, Rozière J (1991) In: Hasnain S (ed) X-ray absorption fine structure. Ellis Horwood, London, p 405
72. Yang C, Srinivasan S, Arico AS, Creti P, Baglio V, Antonucci V (2001) Electrochem Solid State Lett 4:A31

73. Costamagna P, Yang C, Bocarsly AB, Srinivasan S (2002) Electrochim Acta 47:1023
74. Yang C, Srinivasan S, Bocarsly AB, Tulyani S, Benziger J (2004) J Membr Sci 237:145
75. Bauer F, Willert-Porada M (2004) J Membr Sci 233:141
76. Bauer F, Willert-Porada M (2005) J Power Sources 145:101
77. Trouffy-Boutry D, De Geyer A, Guetaz L, Diat O, Gebel G (2007) Macromolecules 40:8259
78. Alberti G, Casciola M, Capitani D, Donnadio A, Narducci R, Pica M, Sganappa M (2007) Electrochim Acta 52:8125
79. Song M-K, Hwang J-S, Kim Y-T, Rhee H-W, Kim J (2003) Mol Cryst Liq Cryst 407:421
80. Song M-K, Kim Y-T, Hwang J-S, Ha HY, Rhee H-W (2004) Electrochem Solid State Lett 7:A127
81. Grasset F (2005) PhD thesis, University Montpellier II
82. Song S-W, Torkelson JM (1995) J Membr Sci 98:209
83. Jones DJ, Grasset F, Marrony M, Rozière J (2005) In: Bossel U (ed) Fuel cell forum. European Fuel Cell Forum, Lucerne
84. Kolde J (1995) Proceedings of the 1st international symposium on proton conducting membrane fuel cells 1. Electrochem Soc Proc 23:93
85. Si Y, Kunz HR, Fenton JM (2004) J Electrochem Soc 151:A623
86. Alberti GCM, Pica M, Tarpanelli T, Sganappa M (2005) Fuel Cells 5:366
87. Huang L-N, Chen L-C, Yu TL, Lin H-L (2006) J Power Sources 161:1096
88. Chen S-L, Krishnan L, Srinivasan S, Benziger J, Bocarsly AB (2004) J Membr Sci 243:327
89. Chen S-L, Xu K-Q, Dong P (2005) Chem Mater 17:5880
90. Peled E, Duvdevani T, Melman (1998) Electrochem Solid State Lett 1:210
91. Peled E, Livshits V, Rakhman M, Aharon A, Duvdevani T, Philosoph M, Feiglin T (2004) Electrochem Solid State Lett 7:A507
92. Reichman S, Duvdevani T, Aharon A, Philosoph M, Goldnitsky D, Peled E (2006) J Power Sources 153:228
93. Casciola M, Alberti G, Ciarletta A, Cruccolini A, Piaggio P, Pica M (2005) Solid State Ionics 176:2985
94. Amarilla JM, Rojas RM, Rojo JM, Cubillo MJ, Linares MJ, Acosta JL (2000) Solid State Ionics 127:133
95. Curtin DE, Lousenberg RD, Henry TJ, Tangeman PC, Tisack ME (2004) J Power Sources 131:41
96. Watanabe M, Uchida H, Emori M, Stonehart P (1996) J Electrochem Soc 143:3837
97. Watanabe M, Uchida H, Emori M (1998) J Phys Chem B 102:3129
98. Antonucci PL, Arico AS, Creti P, Ramunni E, Antonucci V (1999) Solid State Ionics 125:431
99. Baglio V, Arico AS, Di Blasi A, Antonucci V, Antonucci PL, Licoccia S, Traversa E, Serraino Fiory F (2005) Electrochim Acta 50:1241
100. Dimitrova P, Friedrich KA, Vogt B, Stimming U (2002) Solid State Ionics 150:115
101. Arico AS, Baglio V, Di Blasi A, Modica E, Antonucci PL, Antonucci V (2004) J Power Sources 128:113
102. Adjemian KT, Dominey R, Krishnan L, Ota H, Majsztrik P, Zhang T, Mann J, Kirby B, Gatto L, Velo-Simpson M, Leahy J, Srinivasan S, Benziger JB, Bocarsly AB (2006) Chem Mater 18:2238
103. Arico AS, Baglio V, Antonucci V, Nicotera I, Oliviero C, Coppola L, Antonucci PL (2006) J Membr Sci 270:221
104. Arico AS, Baglio V, Di Blasi A, Creti P, Antonucci PL, Antonucci V (2003) Solid State Ionics 161:251

105. Baglio V, Di Blasi A, Arico AS, Antonucci V, Antonucci PL, Nannetti F, Tricoli V (2005) Electrochim Acta 50:5181
106. Elliot JA, Hanna S, Elliot AMS, Cooley GE (2001) Polymer 42:2251
107. Slade RCT, Pressman H, Skou E (1990) Solid State Ionics 38:207
108. Mioc U, Davidovic M, Tjapkin N, Colomban P, Novak A (1991) Solid State Ionics 46:103
109. Meng F, Horan J, Malers JL, Dec SF, Herring AM, Turner JA (2005) ACS Div Fuel Chem Symp Prepr 50:435
110. Sweikart MA, Herring AM, Turner JA, Williamson DL, McCloskey BD, Boonrueng SR, Sanchez M (2005) J Electrochem Soc 152:A98
111. Tazi B, Savadogo O (2000) Electrochim Acta 45:4329
112. Tian H, Savadogo O (2005) Fuel Cells 5:375
113. Malhotra S, Datta R (1997) J Electrochem Soc 144:23
114. Ramani V, Kunz HR, Fenton JM (2004) J Membr Sci 232:31
115. Ramani V, Kunz HR, Fenton JM (2005) Electrochim Acta 50:1181
116. Kim YS, Wang F, Hickner M, Zawodzinski TA, McGrath JE (2003) J Membr Sci 212:263
117. Zhang Y, Zhang H, Bi C, Zhu X (2007) Electrochim Acta 53:4096
118. Asensio JA, Borros S, Gomez-Romero P (2003) Electrochem Commun 5:967
119. Staiti P, Minutoli M, Hocevar S (2000) J Power Sources 90:231
120. Ponce ML, Prado LASdA, Silva V, Nunes SP (2004) Desalination 162:383
121. Ponce ML, Prado L, Ruffmann B, Richau K, Mohr R, Nunes S (2003) J Membr Sci 217:5
122. Honma I, Nomura S, Nakajima H (2001) J Membr Sci 2001:83
123. Kim J-D, Honma I (2004) Electrochim Acta 49:3429
124. Kim J-D, Honma I (2004) Electrochim Acta 49:3179
125. Yamada M, Honma I (2004) J Phys Chem B 108:5522
126. Kim J-D, Honma I (2005) Solid State Ionics 176:979
127. Kim J-D, Honma I (2005) Solid State Ionics 176:547
128. Vernon DR, Meng F, Dec SF, Williamson DL, Turner JA, Herring AM (2005) J Power Sources 139:141
129. Kim J-D, Mori T, Honma I (2006) J Electrochem Soc 153:A508
130. Song M-K, Park S-B, Kim Y-T, Kim K-H, Min S-K, Rhee H-W (2004) Electrochim Acta 50:639
131. Gaowen Z, Zhentao Z (2005) J Membr Sci 261:107
132. Karthikeyan CS, Pereira-Nunes S, Prado LASA, Ponce ML, Silva H, Ruffmann B, Schulte K (2005) J Membr Sci 254:139
133. Schuster MFH, Meyer WH, Schuster M, Kreuer KD (2004) Chem Mater 16:329
134. Rhee CH, Kim HK, Chang H, Lee JS (2005) Chem Mater 17:1691
135. Lee W, Kim H, Kim TK, Chang H (2007) J Membr Sci 292:29
136. Alberti G, Casciola M, Costantino U (1985) J Colloid Interface Sci 107:256
137. Silva VS, Weisshaar S, Reissner R, Ruffmann B, Vetter S, Mendes A, Madeira LM, Nunes S (2005) J Power Sources 145:485
138. Alberti G, Casciola M, Pica M, Jones DJ, Rozière J, Bauer B (2002) Italien Patent IT PG 2002 A 0013
139. Kuan H-C, Wu C-S, Chen C-H, Yu Z-Z, Dasari A, Mai Y-W (2006) Electrochem Solid State Lett 9:176
140. Casciola M, Alberti G, Donnadio A, Pica M, Marmottini F, Bottino A, Piaggio P (2005) J Mater Chem 15:4262

141. Alberti G, Costantino U, Casciola M, Ferroni S, Massinelli L, Staiti P (2001) Solid State Ionics 145:249
142. Alberti G, Casciola M, D'Alessandro E, Pica M (2004) J Mater Chem 14:1910
143. Kim Y-T, Song M-K, Kim K-H, Park S-B, Min S-K, Rhee H-W (2004) Electrochim Acta 2004:645
144. Baradie B, Poinsignon C, Sanchez JY, Piffard Y, Vitter G, Bestaoui N, Foscallo D, Denoyelle A, Delabouglise D, Vaujany M (1998) J Power Sources 74:8
145. Mikhailenko SD, Zaidi SMJ, Kaliaguine S (2001) Catal Today 67:225
146. Morin A, Mosdale R, Gautier L, Jones DJ, Rozière J (2004) In: Mosdale R, Stevens P (eds) France–Deutschland fuel cell conference, Forbach, p 323
147. Zoppi RA, Yoshida IVP, Nunes SP (1997) Polymer 39:1309
148. Zoppi RA, Nunes SP (1998) J Electroanal Chem 445:39
149. Jiang R, Kunz HR, Fenton JM (2006) J Membr Sci 272:116
150. Jones DJ, Rozière J, Tchicaya L, Bonnet B (2001) In: Watanabe M (ed) Proc 2nd international fuel cell workshop, Kofu, p 147
151. Chuang S-W, Hsu SL-C, Liu Y-H (2007) J Membr Sci 305:353
152. Jiménez-Jiménez J, Maireles-Torres P, Olivera-Pastor P, Rodriguez-Castellon E, Jiménez-Lopez A, Jones DJ, Rozière J (1998) Adv Mater 10:808
153. Jones DJ, Rozière J, Marrony M, Lamy C, Coutanceau C, Léger JM, Hutchinson H, Dupont M (2005) Fuel Cell Bull, p 12

Subject Index

Acid-bearing polymers 55, 59, 61
–, membranes 63
Additives 180
Alkaline fuel cells (AFC) 4, 57, 144
Aminophenyltrimethoxysilane (APTMOS) 253
Atomistic simulations 15

BAPFDS 96
Base polymer 169
BDSA/NTDA/ODA 95
9,9-Bis(4-aminophenyl)fluorine-2,7-disulfonic acid (BAPFDS) 96
2,2-Bis(trifluoromethyl)-4,5-difluoro-1,3-dioxole 140
1,2-Bis(vinyl phenyl)ethane (BVPE) 66, 169
BPSH 73–77, 87–91, 97, 104, 248

Catalyst layers (CLs) 17
Catalyst-coated membrane (CCM) 142
Charge-driven self-assembly 222
Chemical degradation 142
Chemical microstructure 55
Chemical stability 206
Chlor-alkali electrolysis 132
Cluster network model 41, 131, 223
Composite membranes 242
Conductivity 201
Copolymers, microstructures 62
Cross leak degradation 127
Crosslinking 180

Degree of grafting 163
Differential scanning calorimetry (DSC) 192
Direct methanol fuel cells (DMFC) 4, 58, 130, 146, 206, 236

–, performance 209
Disodium 3,3′-disulfonate-4,4′-dichlorodiphenyl sulfone 75
Divinylbenzene (DVB) 66
DMFCs 58, 146, 236
Dose rate 171

Electron transfer (ET) 17
Empirical valence bond (EVB) models 37
Equivalent weights (EW) 10
ETFE-g-PSSA 64, 66, 87–93, 102
Ethylene–tetrafluoroethylene copolymer (ETFE) 133
Excess mobility 26

FEP, radiolysis 165
FEP-g-polyacrylic acid 173, 187, 191
FEP-g-PSSA 64, 66
Fuel cell application 196
Fuel cell membranes, matrices for aqueous proton transfer 16
Fuel cell testing 204
Fuel cells, alkaline (AFC) 4, 57, 144
–, direct methanol (DMFC) 4, 58, 130, 146, 206, 236
–, inorganic–organic membranes 219
–, molten carbonate (MCFC) 4, 57
–, performance 203, 207
–, phosphoric acid (PAFC) 4, 57
–, polymer electrolyte (PEFCs) 1, 4, 7, 157, 196
–, solid oxide (SOFC) 4, 57

Gas diffusion electrodes (GDEs) 2, 5, 60
Gas diffusion layers (GDLs) 103
Graft mapping 185
Graft polymerization 162
Grafted films, membranes 185

Grafting, degree of 163
Grafting medium 177
Grafting parameters 167
Grafting temperature 176
Grotthus mechanism 11, 199

Heteropolyacids (HPAs) 246
Hexafluoropropylene 164, 173
High-temperature membranes 127, 138
Hybrid inorganic–organic membranes, preformed membranes 225
Hybrid ionomeric membrane 219

Inorganic particles and networks 219
Ion-containing polymers 55
Ion-exchange capacity (IEC) 11, 58, 147, 184, 197, 231
–, PFSA membranes 129
Ionomer membranes 15
–, morphology/microstructure 222
–, templates, inorganic particle growth 222
Ionomers, perfluorinated 18
Irradiation dose 171

Keggin ions 246

Macroscopic network models 41
MEAs 1, 8, 58, 130, 203, 220
–, fabrication 204
–, fuel cell performance 111
Mechanical integrity 207
Melt flow index (MFI) 170
Membrane development 134
Membrane electrode assemblies (MEA) 1, 8, 58, 130, 203, 220
Membrane operation, cell 43
Membrane properties 8, 196
Membranes, catalyst-coated (CCM) 142
–, short side chains 134
–, structural complexity 19
Metal(IV) phosphates 227, 250
N,N-Methylene-bis-acrylamide 181
Microstructure 59
Molten carbonate fuel cells (MCFC) 4, 57
Monomer concentration 174
Morphology 55

Nanocomposite membrane 219
Naphthalenic polyimides, sulfonated 79

Network models and theory 15
NTDA 96

ODA 96

P(VDF-co-HFP)-b-SPS 93
PBI, acid-doped 220
–, sulfonated 16
PEEK, sulfonated 16, 66, 227, 249
PEEKK, sulfonated 16, 74
PEFCs, bulk properties 1
–, conductivity 11
–, durability 12
–, interfacial properties 1
–, perfluorinated ionic polymers 127
–, perfluorinated membranes 128, 132
PEK, sulfonated 16, 250
PEM fuel cell membrane 219
PEMs, block copolymer 78
–, graft copolymer 85
–, morphological studies 69
–, randomly sulfonated 69
–, state-of-the-art 18
Perfluorinated ionic polymers 127
Perfluorinated membranes (PEFCs) 132
Perfluorinated sulfonic acid (PFSA) 13, 127, 131
Perfluorosulfonic acid ionomers (PFSIs) 63
Perfluorosulfonyl vinyl ether (PSVE) 133
PFSA 13, 127
–, structure 131
PFSA dispersions 147
PFSA membranes 127, 130, 220, 243
PFSA-type membranes, DMFC 146
Phosphoantimonic acid 251
Phosphomolybdic acid 248
Phosphoric acid fuel cells (PAFC) 4, 57
Phosphotungstic acids (PTA) 248
Poly(arylene ether sulfone), disulfonated 75
Poly(ethylene-alt-tetrafluoroethylene) (ETFE) 166
Poly(tetrafluoroethylene) (PTFE), irradiation, degradation 164
Poly(tetrafluoroethylene-co-hexafluoropropylene) (FEP) 165
Poly(trifluorostyrene sulfonic acid) (PTFSSA) membranes 76

Subject Index

Poly(vinylidene fluoride) (PVDF), irradiation 166
Polyarylenes 58
Polybenzimidazoles (PBI) 11, 220, 248
Polyimides 58, 220
–, sulfonated naphthalenic 79
Polymer electrolyte fuel cells (PEFCs) 1, 4, 7, 157, 196
Polymer electrolyte membrane fuel cells (PEMFC) 57
Polyphosphazenes 58
Polystyrene, radiation-grafted 58
Polystyrene sulfonic acid (PSSA) 187
Polysulfones, sulfonated 66, 68, 220, 241
Polytetrafluoroethylene (PTFE) 66, 129
–, reinforcement 129
Pore morphology and proton conductance 15
Preirradiation grafting 163
Proton-conducting polymer membrane 1
Proton conductivity 55, 87, 201
Proton exchange membrane 1, 8, 157
Proton transport/transfer 15, 17, 100
–, models, mesoscopic scale 30
–, polymer–water interface 31
–, water/aqueous networks 26
PS-g-macPSSA 95
PS-r-PSSA 95
PTFSSA 64, 76–78, 87–93, 102, 115
–, Pb^{2+}-stained 76, 78
PVDF 242
PVDF-g-polystyrene 185
PVDF-g-PSSA 64, 66, 81, 94

Quantum effects 29

Radiation effects 161, 164
Radiation grafted membranes 160, 293
Radiation grafting 157, 162
Radiation-grafted polystyrene 58
Reactant permeability 206
Reinforcement technology 127, 144
Representative elementary volume elements (REVs) 44

Silicotungstic acid (STA) 246, 247
Solid electrolyte 1
Solid oxide fuel cells (SOFC) 4, 57
Solid polymer electrolyte (SPE) 7
sPEEK 16, 66, 227, 249
sPEEKK 16, 74
sPEEK–ZrP 235, 258
sPEK 16, 250
SPS 64, 80, 93
SPSU-b-PVDF 94, 96
SSEBS 64, 80, 87–94
SSIBS 64, 80, 93
Sulfonated naphthalenic polyimides 79
Sulfonated poly(arylene ether sulfone) 73–77, 87–91, 97, 104, 248
Sulfonated poly(p-phenylene)s 66
Sulfonated polybenzimidazoles 66
Sulfonated polyimides (SPIs) 66
Sulfonated polyphosphazenes 66
Sulfonated polystyrene (SPS) 64, 93
Sulfonated polysulfones 66, 68
Sulfonation 184
Sulfonimide membranes 138
Surface morphology, atomic force microscopy 191
Swelling clays 249

Teflon-like backbones 18
TEOS 252
Terpolymers 140
Tetrafluoroethylene 132
– / hexafluoropropylene 164
Thermogravimetric analysis (TGA) 192
TMDES 252
Translocation of hydronium 28
Triallylcyanurate (TAC) 66, 169, 181
Triflic acid monohydrate 33
Trifluoromethane sulfonic acid monohydrate (TAM) 33
Trifluorostyrene (TFS) 168, 173
Trifluorostyrene–trifluorostyrene copolymer, sulfonated (PTFSSA) 64
Tungstomolybdic acid 246
Tungstophosphoric acid 246

Ultrasonic attenuation spectroscopy (UAS) 148

Vapour induced phase separation (VIPS) 238

Water management 12, 103
Water sorption 11

Water states, water management 204
Water transport 106
Water uptake 199

X-ray microprobe analysis (XMA), graft distribution 185

X-ray photoelectron spectroscopy (XPS) 188

Zirconium phosphate sulfophenylphosphonates 251
Zirconium species 230, 250, 256

Printing: Krips bv, Meppel, The Netherlands
Binding: Stürtz, Würzburg, Germany